Localization of Putative Steroid Receptors

Volume I
Experimental Systems

Editors

Louis P. Pertschuk, D.O.
Associate Professor
Department of Pathology
State University of New York
Downstate Medical Center
and
Attending Pathologist
Department of Pathology
Kings County Hospital Center
Brooklyn, New York

Sin Hang Lee, M.D., F.R.C.P.(C)
Attending Pathologist
Department of Pathology
Hospital of St. Raphael
and
Associate Clinical Professor
Department of Pathology
Yale University School of Medicine
New Haven, Connecticut

CRC Press
Taylor & Francis Group
Boca Raton London New York

CRC Press is an imprint of the
Taylor & Francis Group, an **informa** business

First published 1985 by CRC Press
Taylor & Francis Group
6000 Broken Sound Parkway NW, Suite 300
Boca Raton, FL 33487-2742

Reissued 2018 by CRC Press

Library of Congress Cataloging in Publication Data
Main entry under title:

Localization of putative steroid receptors.

Bibliography: p.
Includes index.
Contents: v. 1. Experimental systems — v. 2.
Clinically oriented studies.
1. Cancer—Endocrine aspects. 2. Steroid hormones—
Receptors. I. Pertschuk, Louis P., 1925-
II. Lee, Sin Hang, 1932- . [DNLM: 1. Receptors,
Steroid—analysis. 2. Histocytochemistry—methods.
3. Breast Neoplasms—analysis. WK 150 L811]
RC268.2.L63 1985 616.99'449071 84-20049
ISBN 0-8493-6048-X (v. 1)
ISBN 0-8493-6049-8 (v. 2)

A Library of Congress record exists under LC control number: 84020049

Publisher's Note
The publisher has gone to great lengths to ensure the quality of this reprint but points out that some imperfections in the original copies may be apparent.

Disclaimer
The publisher has made every effort to trace copyright holders and welcomes correspondence from those they have been unable to contact.

ISBN 13: 978-1-315-89504-8 (hbk)
ISBN 13: 978-1-351-07414-8 (ebk)

Visit the Taylor & Francis Web site at http://www.taylorandfrancis.com and the
CRC Press Web site at http://www.crcpress.com

PREFACE

For the past two decades, the study of steroid hormone receptors has been almost entirely pursued by investigators whose primary expertise has been in the field of steroid biochemistry. To these workers, who have contributed so much of the information upon which our present knowledge is based, the medical and scientific communities are greatly indebted.

However, for a full understanding of the mechanism of steroid action at the cellular level, it has become necessary to know not only the precise cells in which steroids exert their effects, but also the subcellular compartments in which the specific steroid binding proteins are concentrated. Since biochemical assays are performed on homogenized tissue extracts, they possess certain inherent limitations. In addition, as sophisticated equipment is required for their performance, there are large areas of the world in which they cannot be performed. Therefore many patients (in particular with breast cancer) are unable to reap the benefits which might accrue if information about tumor receptors was available.

The past few years have witnessed the development of morphologic assays designed to assess steroid hormone binding. The prime purpose of these techniques (since they are performed on intact tissue sections) has been to obtain information not readily gleaned from conventional assays. Of more importance, most are designed in such a manner that they could be executed in any hospital pathology laboratory. Many of the morphologic assays covered in this work have been developed by pathologists, often with considerable service duties related to patient care, and in many cases without research funds. It is not surprising that these assays have become subjects of much criticism. Nonetheless, as evidenced throughout this work, a number of enthusiastic researchers have obtained quite impressive results and countered much of the criticism. More recently, some research laboratories specializing in steroid biochemistry have also begun to direct an increasing share of effort and resources towards the development of morphological assays, including the use of monoclonal antibodies which can trace receptor antigens *in situ*. As a consequence, there no longer exists a sharp boundary between the fields of biochemistry and pathology.

These volumes represent the "state-of-the-art" in morphological methods for detection of steroid receptors. No final solution is presented or intended. Indeed, at present a definite answer to the question "where is the receptor located?" cannot be given. Furthermore, there are a number of discrepancies, some serious, which prevail.

In Volume I we have attempted to collect current available methods and experimental approaches which might be useful in solving present enigmas. Some physician authors, not being content to wait until complete answers are available, have applied many of the new methodologies in clinical research projects designed to improve medical practice. These attempts we have placed in Volume II. Of necessity, there is occasional overlap with newer techniques and experimental data in the latter volume.

It is our hope that the background and recent advances presented here will stimulate further experimentation and new innovations, and spur clinical trials by all investigators with an interest in steroid hormones, cell biology, and clinical oncology.

Louis P. Pertschuk
Sin Hang Lee

THE EDITORS

Louis P. Pertschuk, D. O., is Associate Professor of Pathology at State University of New York, Downstate Medical Center, and Attending Pathologist at the Kings County Hospital Center, Brooklyn, New York.

Dr. Pertschuk received an A.B. in Biology from New York University in 1946 and graduated from the Philadelphia College of Osteopathic Medicine in 1950. Following a year of internship he entered private general practice until 1970 when he began a residency in pathology at the State University-Kings County Hospital Center. He was a Fellow in Pathology at the Memorial Sloan-Kettering Cancer Center in 1973 and 1974 and was certified in Anatomic Pathology by the American Board of Pathology in 1974.

Dr. Pertschuk is a Fellow of the College of American Pathologists and the American Society of Clinical Pathology. He is a member of the American Association of Pathologists, the International Academy of Pathology, the New York Academy of Sciences, and the American Association for the Advancement of Science. He has published 67 research papers, 43 scientific abstracts, and contributed 19 chapters to various books. His current research interests include the development and study of different methodologies for detection of the sites of action of steroid hormones, especially in human neoplasia.

Sin Hang Lee, M.D., is Attending Pathologist at the Hospital of St. Raphael, and an Associate Clinical Professor of Pathology at Yale University in New Haven, Connecticut.

Born in Hong Kong, Dr. Lee is a 1956 graduate of Wuhan Medical College, the People's Republic of China. After graduation, he worked in Sichuan Medical College, China, as a microbiologist before he joined the Department of Pathology at the University of Hong Kong in 1961 as a demonstrator. He finished his residency in pathology in the U.S. at the New York Hospital-Cornell Medical Center, and was appointed Instructor in Pathology at Cornell Medical College in 1966. In the same year, he was certified by the American Board of Pathology, and was made a Fellow of the Royal College of Physicians of Canada. Dr. Lee spent the year of 1967 to 1968 as a Pathology Fellow at Memorial Hospital for Cancer and Allied Diseases. Thereafter, he served as Assistant Professor of Pathology at McGill University (1968 to 1971) and subsequently Associate Professor of Pathology at Yale University (1971 to 1973). In addition to the current hospital and academic appointments in New Haven, Dr. Lee also holds a guest professorship at Wuhan Medical College, the People's Republic of China.

Dr. Lee's professional and research interests range from surgical pathology to electron microscopic localization of enzymes and mycoplasma antigens. As a surgical pathologist, he saw an urgent need for a practical histochemical approach to the identification of breast cancer cells rich in estrogen receptors, and has devoted much of his free time during the past few years to develop the hydrophilic fluorescent estradiol technique, which has become the most popular histochemical assay of estrogen receptors to date.

CONTRIBUTORS

J.P.A. Baak, M.D., Ph.D.
Pathological Institute
Free University Hospital
Amsterdam, The Netherlands

Alberto Bagni, M.D.
Resident in Pathology
Department of Pathological Anatomy and Histopathology
University of Ferrara
Ferrara, Italy

Etienne-Emile Baulieu, M.D., Ph.D.
Professor
Department of Biochemistry
University of Paris South
Bicetre, France

Christopher C. Benz, M.D.
Assistant Professor of Medicine
Cancer Research Institute
University of California at San Francisco
San Francisco, California

Maximilian Binder, Ph.D.
Assistant Professor
Institute for Tumorbiology and Cancer Research
University of Vienna
Vienna, Austria

Manfred Boehm, M.D.
Research Associate
Institute for Tumorbiology and Cancer Research
University of Vienna
Vienna, Austria

Richard H. Buell, M.D.
Lady Davis Institue for Medical Research
Sir Mortimer B. Davis-Jewish General Hospital
Montreal, Quebec, Canada

Giovanni Bussolati, M.D.
Professor
Department of Pathology
University of Turin Medical School
Turin, Italy

Anne C. Carter, M.D.
Visiting Professor
Department of Medicine
Division of Endocrinology
State University of New York
Downstate Medical Center
Brooklyn, New York

Peizhen Chen, M.D.
Associate Professor
Department of Surgery
Sichuan Medical College
Chengdu, Sichuan, China

Klaus Czerwenka, M.D.
Head, Department of Gynecology and Obstetrics
Histology and Cytology Unit
University of Vienna
Vienna, Austria

Karen S. Byer Eisenberg, R.N., M.P.S.
Assistant Nursing Director
Department of Pathology
State University of New York
Downstate Medical Center
Brooklyn, New York

Vincenzo Eusebi, M.D.
Associate Professor
Department of Pathology
University of Bologna Medical School
Bologna, Italy

Guidalberto Fabris, M.D.
Associate Professor
Department of Pathological Anatomy and Histopathology
University of Ferrara
Ferrara, Italy

Joseph G. Feldman, M.D.
Associate Professor
Department of Preventive Medicine
State University of New York
Downstate Medical Center
Brooklyn, New York

Bernard Fisher, M.D.
Professor of Surgery
Department of Surgery
University of Pittsburgh School of Medicine
Pittsburgh, Pennsylvania

Jean-Marie Gasc, Ph.D.
Maitre-Assistant
Institute of Embryology
Nogent-sur-Marne, France

N. Gunduz, Ph.D.
Assistant Professor of Surgical Research
Department of Surgery
University of Pittsburgh School of Medicine
Pittsburgh, Pennsylvania

Wedad Hanna, M.D., F.R.C.P.(C)
Assistant Professor
Department of Pathology
University of Toronto
Staff Pathologist
Women's College Hospital
Toronto, Ontario, Canada

David L. Ingram, A.A.P.A.
Research Assistant, Pathology Associate
Department of Pathology
Duke University Medical Center
Durham, North Carolina

Raimund Jakesz, M.D.
Assistant Professor
Department of Surgery
University of Vienna
Vienna, Austria

Charmayne Jesik, Ph.D.
Research Associate
Department of Urology
Northwestern University Medical School
Chicago, Illinois

Roland Kolb, M.D.
Associate Professor
Department of Surgery
University of Vienna
Vienna, Austria

Pietro Lampertico, M.D.
Head, Anatomical Pathology Laboratory
Unitã Socio-Sanitaria Locale
Busto Arsizio, Italy

Chung Lee, Ph.D.
Associate Professor
Department of Urology
Northwestern University Medical School
Chicago, Illinois

J. Lindeman, M.D., Ph.D.
Head, Department of Pathology
S.S.D.Z.
Delft, The Netherlands

Richard J. Macchia, M.D.
Associate Professor of Urology
Department of Urology
State University of New York
Downstate Medical Center
Brooklyn, New York

Marayart Mangkornkanok
Assistant Professor
Department of Pathology
Northwestern University Medical School
Chicago, Illinois

Elisabetta Marchetti, M.D.
Assistant Professor
Department of Pathological Anatomy and Histopathology
University of Ferrara
Ferrara, Italy

Andrea Marzola, M.D.
Associate Professor
Department of Pathological Anatomy and Histopathology
University of Ferrara
Ferrara, Italy

Mauricio Maturana
Mathematician and Statician
Institute of Public Health
University of Chile Medical School
Santiago, Chile

Kenneth McCarty, Jr., M.D., Ph.D.
Associate Professor of Pathology
Assistant Professor of Medicine
Duke University Medical Center
Durham, North Carolina

Kenneth McCarty, Sr., Ph.D.
Professor of Biochemistry
Duke University Medical Center
Durham, North Carolina

Zhijia Mei, M.D.
Instructor
Department of Oncology
Sichuan Medical College
Chengdu, Sichuan, China

C.J.L.M. Meijer, M.D., Ph.D.
Professor
Pathological Institute
Free University Hospital
Amsterdam, The Netherlands

Miguel A. Mena
Biologist
Laboratory of Experimental Endocrinology
Department of Experimental Morphology
University of Chile Medical School
Santiago, Chile

Xiangin Meng, M.D.
Associate Professor
Department of Pathology
Sichuan Medical College
Chengdu, Sichuan, China

Italo Nenci, M.D.
Professor of Pathology
Director, Department of Pathological Anatomy and Histopathology
University of Ferrara
Ferrara, Italy

Kathleen I. Pritchard, M.D., F.R.C.P.(C)
Assistant Professor, Department of Medicine
University of Toronto
Staff Physician, Women's College Hospital
Toronto, Ontario, Canada

Patrizia Querzoli, M.D.
Resident in Pathology
Department of Pathological Anatomy and Histopathology
University of Ferrara
Ferrara, Italy

Georg Reiner, M.D.
Department of Surgery
University of Vienna
Vienna, Austria

Angel Rodriguez
Biologist
Laboratory of Experimental Endocrinology
Department of Experimental Morphology
University of Chile Medical School
Santiago, Chile

Elizabeth A. Saffer, B.A.
Senior Research Assistant
Department of Surgery
University of Pittsburgh School of Medicine
Pittsburgh, Pennsylvania

Julia Sensibar, B.Sc.
Research Assistant
Department of Urology
Northwestern University Medical School
Chicago, Illinois

Juergen Spona, Ph.D.
Head, Endocrine Research Unit
Department of Gynecology and Obstetrics
University of Vienna
Vienna, Austria

Franca Stagni, D.Sc.
Biologist
Anatomical Pathology Laboratory
Unità Socio-Sanitaria Locale
Busto Arsizio, Italy

Andrei N. Tchernitchin, M.D.
Head, Laboratory of Experimental Endocrinology
Department of Experimental Morphology
University of Chile Medical School
Santiago, Chile

Gilles Tremblay, M.D.
Senior Pathologist
Royal Victoria Hospital
Professor of Pathology
McGill University
Montreal, Quebec, Canada

J. van Marle, Ph.D.
Senior Scientist
Department of Pharmacology
University of Amsterdam
Amsterdam, The Netherlands

Rosemary A. Walker, M.D., M.R.C. Pathology
Senior Lecturer
Department of Pathology
University of Leicester
Leicester, England

Israel Wiznitzer, M.D.
Postdoctoral Research Associate
Yale University School of Medicine
New Haven, Connecticut

Xianyun Yao, M.D.
Associate Professor
Department of Pathology
Sichuan Medical College
Chengdu, Sichuan, China

TABLE OF CONTENTS

Volume I: Experimental Systems

Chapter 1
An Overview 1
S. H. Lee and L. P. Pertschuk

Chapter 2
Radioautographic Localization of Estrogen Receptors in the Rat Uterus: a Tool for the Study of Classical and Nontraditional Mechanisms of Hormone Action 5
A. N. Tchernitchin, M. A. Mena, A. Rodriguez, and M. Maturana

Chapter 3
Fluoresceinated Estrone Binding by Normal and Neoplastic Cells of Human and Animal Origin 39
N. Gunduz, E. Saffer, and B. Fisher

Chapter 4
Histochemical Study of Estrogen Receptors in the Rat Uterus With a Hydrophilic Fluorescent Estradiol Conjugate 59
S. H. Lee

Chapter 5
Estrogen Receptors and Hormone Responsiveness in Serially Transplanted Mammary Tumors in Rats 85
C. Lee, C. Jesik, M. Mangkornkanok, J. Sensibar, and S. H. Lee

Chapter 6
Flow Cytometric Analysis of Fluorescent Estrogen Binding in Cancer Cell Suspensions 95
C. Benz, I. Wiznitzer, and S. H. Lee

Chapter 7
Sex Steroid Action Mechanism by Cytochemistry in Normal and Neoplastic Target Tissues 111
E. Marchetti, A. Marzola, A. Bagni, P. Querzoli, G. Fabris, and I. Nenci

Chapter 8
Immunohistochemical Studies with Antibodies to the Chicken Oviduct Progesterone Receptor 125
J.-M. Gasc and E.-E. Baulieu

Chapter 9
Methods for the Quantification of Histochemical Steroid Binding Assays 143
L. P. Pertschuk

Chapter 10
A Two-State Immunocytochemical Method for Putative Estrogen Receptor Analysis .. 161
V. Eusebi and G. Bussolati

Chapter 11
Estradiol-BSA Conjugates for Estrogen Receptor Localization: the Vienna Experience 167
M. Binder, K. Czerwenka, M. Boehm, J. Spona, R. Jakesz, R. Kolb, and G. Reiner

Index 183

TABLE OF CONTENTS

Volume II: Clinically Oriented Studies

Chapter 1
The Application of Thaw-Mount Autoradiography for the Localization of Putative Estrogen Target Cells in Human Mammary Lesions 1
R. H. Buell and G. Tremblay

Chapter 2
Morphologic Detection of Multiple Steroid Binding Sites in Breast Cancer by Histochemistry and by Immunocytochemistry with Monoclonal Antireceptor Antibody: Relationship to Endocrine Response 15
L. P. Pertschuk, K. B. Eisenberg, A. C. Carter, and J. G. Feldman

Chapter 3
A Fluorescent Histochemical Study of Steroid Receptors in Human Breast Cancer 37
S. H. Lee

Chapter 4
Histochemical Sex Steroid Hormone Receptor Assay and Treatment of Breast Cancer in Chinese Patients 51
P. Chen, X. Yao, Z. Mei, and X. Meng

Chapter 5
Cancer of the Breast: Clinicopathological Correlations in Primary, Recurrent, and Metastatic Tumors with Histochemically Determined Hormone Receptors 63
P. Lampertico and F. Stagni

Chapter 6
How to Validate Histochemical Techniques as Predictors of Hormonal Response 85
W. Hanna and K. I. Pritchard

Chapter 7
Localization of Steroid Binding in Prostatic Carcinoma by Histochemistry: Therapeutic Implications 93
L. P. Pertschuk, R. J. Macchia, and K. B. Eisenberg

Chapter 8
Binding of Labeled Estrogen-Albumin Conjugates in Breast Cancer 113
J. van Marle, J.P.A. Baak, C.J.L.M. Meijer, and J. Lindeman

Chapter 9
Peroxidase Labeling of Estrogen Binding Sites in Breast Cancer 133
R. A. Walker

Chapter 10
Critical Issues in the Evaluation of Histochemical and Biochemical Methods for Steroid Receptor Analysis 149
K. S. McCarty, Jr., D. L. Ingram, and K. S. McCarty, Sr.

Index 165

Chapter 1

AN OVERVIEW

Sin Hang Lee and Louis P. Pertschuk

TABLE OF CONTENTS

I. An Overview 2

References 4

I. AN OVERVIEW

Since the first successful synthesis of (^{3}H) estradiol by Jensen and Jacobson,[1] receptors of sex steroid hormones — in particular, estrogen — have become a subject of major interest of biochemists and oncologists.[2-9] By quantitative determination of the high-affinity estrogen binding proteins, i.e., the receptors in tissue homogenates, many steroid target tissues have been identified and hormone responsive cancers studied. As all organs and tumors are composed of different cell types which may possess different levels of receptor protein and respond to hormonal stimulation with different degrees of sensitivity, it is self-evident that in order to study the mechanism of steroid hormone action in depth, an identification of the true target cells, namely, those rich in steroid receptors, must be made. This can be most conveniently accomplished by the examination of intact tissue sections by histologic means.

Recently, it has been shown that there are several types of high-affinity estrogen binding proteins or estrogen receptors.[10-13] These molecules not only have different molecular conformations, but may be isolated in different cellular compartments and require different experimental conditions for their detection. This review represents the state-of-the-art of these histochemical methods and aims to present both rationale and merits of the procedures that have been used for the demonstration of steroid hormone receptors and steroid binding activity *in situ*. As with the development of any new methodology, advances often build on error as well as on correct assessments. Some of the data presented here will eventually prove to be correct. Other interpretations may have to be modified or reassessed as new information becomes available.

The classical morphologic approach to localize steroid binding cells is to inject radiolabeled hormones into animals and trace the target cells in endocrine responsive tissue in order to visualize the concentrated radioactive hormone. Alternatively, incubation of tissue fragments in vitro is carried out in a solution of radiolabeled ligand. A variety of steroid target cells have been located in this manner. The chapters contributed by Tchernitchin et al. and by Buell and Tremblay illustrate the application of these techniques in experimental situations and their potential usefulness in a clinical laboratory.

Because of the simplicity of the procedures, the most widely used histochemical techniques for localizing cells rich in receptor are the fluorescent steroid compounds with or without a hydrophilic molecular carrier. Using this type of reagent, S. H. Lee has studied cytoplasmic estrogen binding activity in the luminal epithelial cells of the rat uterus and presented evidence that the binding visualized is due to a substance which behaves as the classical estrogen receptor. Several groups from different countries report on their experience with the application of these methods to the study of cancerous tissue in both breast and prostate. Their results indicate that histochemical assays may be useful as either a supplementary or alternative method to biochemistry for the identification of hormone responsive neoplasms. In other chapters, the potential for modification of these procedures by introducing enzymatic markers is discussed. This would allow for permanent tissue sections.

Quantitation of steroid binding activity of individual cancer cells in tissue sections by the human eye lacks precision. In one chapter, a more analytical approach using computer-assisted microfluorometry is discussed with interesting results. An even more precise way of quantitation is the measurement of bound ligand in isolated cells by flow cytometry. Benz and co-workers have laid down the basic technical procedure for this purpose. This type of approach might be very useful for measuring receptor in cancer cells from malignant effusions.

Utilizing an experimentally induced, hormone-dependent rat mammary carcinoma, C. Lee has studied estrogen binding heterogeneity with a combined biochemical and histochemical approach. The data presented may have important implications in the study of hormone-responsive human breast cancer as well as in treatment. In attempting to understand the mechanism of action of intact steroid hormone target cells it should be of great value to follow the events both before and after the steroid molecules enter into the cells. Several chapters deal with the use of fluorescent steroids for this purpose.

The most recent development in this rapidly expanding area is the use of antibodies directed against steroid receptor proteins. In one chapter work on the localization of progesterone receptor is reported, while in another the employment of monoclonal antiestrophilin antibodies is discussed and compared with fluorescent histochemistry. Results may be important for the prognostication of endocrine response in breast cancer as well as in furthering an understanding of steroid hormone action.

It would be ideal if the morphologic approaches to a study of cellular function *in situ* would neatly coincide with the parameters laid down by the steroid biochemists. If this were so morphology would lend an important new dimension to biochemistry. Unfortunately, this is not the case. Failure to employ the exact conditions defined for biochemical receptor assays which rely on the use of radiolabeled isotopes and scintillation counting, with rigid criteria designed to discriminate specific binding proteins from tissue or blood contaminants, has caused some authorities to reject the entire histochemical approach out of hand. The material presented in these volumes demonstrates that the frontier of our knowledge of steroid receptors is still advancing and continues to be reshaped. The discrepancies that exist between the biochemical and histochemical approaches do not seem to be irreconcilable and the two systems may eventually be shown to be complementary. With our presently incomplete knowledge, it is inevitable that various contributors should have different conceptions as to the significance of their findings even though they may use similar techniques and obtain similar results. The title of this book admits this uncertainty and reflects the opinion of its editors that an open mind must be maintained.

Current dogma, based on conventional biochemical assays, states that steroids enter all cells freely in vivo by apparent passive diffusion and are only retained in target cells possessing specific cytoplasmic receptor proteins. Only after receptor and steroid combine does the complex move into the nucleus, bind to chromatin, and induce gene activation. However, the demonstration of a potential plasma membrane binding site, as discussed by Marchetti et al., and the failure of monoclonal antireceptor antibody to detect cytoplasmic antigen[14-16] indicate that a revision in our concept of steroidal mechanisms of action may be necessary. This is not surprising since it is obvious that biochemical assays performed on homogenates of tissue may introduce artifactual distortions even more so than any histological method. Consequently, it is uncertain whether biochemical dogma should be accepted as the standard for evaluation of the morphological methods, although the issues raised by the biochemists as expressed in the chapter by McCarty et al. require very serious consideration.

We have allowed the interchangeable use of the term "receptor" and "binding site" in these volumes because of the uncertainties noted above. On some occasions the term "receptor" is used to refer to the specific, high-affinity binding proteins, while on others, perhaps to binding proteins of lower affinity, still with high specificity. In other circumstances this term may also be applied to denote an antigenic site(s) on receptor molecules. We ask for the reader's indulgence and tolerance in understanding our reasons for permitting this interchange.

As of this writing, no single histochemical technique can be regarded as perfect or as fully validated. Readers must use discretion and judgment in selecting those methodologies they might wish to pursue.

REFERENCES

1. **Jensen, E. V. and Jacobson, H. I.,** Basic guides to the mechanism of estrogen action, *Recent Prog. Horm. Res.*, 18, 387, 1962.
2. **Gorski, J., Toft, D., Shyamala, G., Smith, D., and Notides, A.,** Hormone receptors: studies on the interaction of estrogen with the uterus, *Recent Prog. Horm. Res.*, 24, 45, 1968.
3. **Jensen, E. V. and DeSombre, E. R.,** Mechanism of action of the female sex hormones, *Annu. Rev. Biochem.*, 41, 203, 1972.
4. **Jensen, E. V., Mohla, S., Gorell, T. A., and DeSombre, E. R.,** The role of estrophilin in estrogen action, *Vitam. Horm.*, 32, 89, 1974.
5. **Gorski, J. and Gannon, F.,** Current models of steroid hormone action: a critique, *Annu. Rev. Physiol.*, 38, 425, 1976.
6. **Jensen, E. V., Block, G. E., Smith, S., Kyser, K., and DeSombre, E. R.,** Estrogen receptors and breast cancer response to adrenalectomy, *Natl. Cancer Inst. Monogr.*, 34, 55, 1971.
7. **McGuire, W. L., Vollmer, E. P., and Carbone, P. P., Eds.,** *Estrogen Receptors in Human Breast Cancer,* Raven Press, New York, 1975.
8. **Heuson, J. C., Longeval, E., Mattheiem, W. H., Deboel, M. C., Sylvester, R. J., and Leclercq, G.,** Significance of quantitative assessment of estrogen receptors for endocrine therapy in advanced breast cancer, *Cancer*, 39, 1971, 1977.
9. **McGuire, W. L., Horwitz, K. B., Pearson, O. H., and Segaloff, A.,** Current status of estrogen and progesterone receptors in breast cancer, *Cancer*, 39, 2934, 1977.
10. **Clark, J. H., Markaverich, B., Upchurch, S., Eriksson, H., Hardin, J. W., and Peck, E. J., Jr.,** Heterogeneity of estrogen binding sites: relationship to estrogen receptors and estrogen responses, *Recent Prog. Horm. Res.*, 36, 89, 1980.
11. **Eriksson, H. A., Hardin, J. W., Markaverich, B., Upchurch, S., and Clark, J. H.,** Estrogen binding in the rat uterus: heterogeneity of sites and relation to uterotrophic response, *J. Steroid Biochem.*, 12, 121, 1980.
12. **Thomas, T., Leung, B. S., Yu, W. E. Y., and Kiang, D. T.,** Diverse mechanisms of estrogen receptor activation, *Fed. Proc. Fed. Am. Soc. Exp. Biol.*, 42, 1877, 1983.
13. **Wiehle, R. D. and Wittliff, J. L.,** Multiple forms of estrogen receptors during differentiation of the mammary gland of the rat, *Fed. Proc. Fed. Am. Soc. Exp. Biol.*, 42, 1877, 1983.
14. **Greene, G. L., Nolan, C., Engler, J. P., and Jensen, E. V.,** Monoclonal antibodies to human estrogen receptor, *Proc. Natl. Acad. Sci. U.S.A.*, 77, 5115, 1980.
15. **Greene, G. L. and Jensen, E. V.,** Monoclonal antibodies as probes for estrogen receptor detection and characterization, *J. Steroid Biochem.*, 16, 353, 1982.
16. **King, W. J. and Greene, G. L.,** Monoclonal antibodies to estrogen receptor localize receptor in the nucleus of target cells, *Nature (London)*, 307, 745, 1984.

Chapter 2

RADIOAUTOGRAPHIC LOCALIZATION OF ESTROGEN RECEPTORS IN THE RAT UTERUS: A TOOL FOR THE STUDY OF CLASSICAL AND NONTRADITIONAL MECHANISMS OF HORMONE ACTION

Andrei N. Tchernitchin, Miguel A. Mena, Angel Rodríguez, and Mauricio Maturana

TABLE OF CONTENTS

I. Introduction 6

II. Radioautographic Methods as a Tool for the Study of Estrogen Receptors 8
- A. Aim, Possibilities, and Limitations of Radioautography 8
  - 1. The Dry Radioautographic Technique for Diffusible Compounds .. 9
  - 2. The Wet Radioautographic Technique 9
  - 3. Radiohistochemical Methods for the Characterization of Receptors 9
- B. Experimental Principles of Radioautography 9
- C. The Radioautographic Techniques 10
  - 1. The Dry Radioautographic Technique for Diffusible Compounds 10
  - 2. The Wet Radioautographic Technique 10
  - 3. Radiohistochemical Procedures for the Characterization of Receptors 11
- D. Receptor Criteria as Applied to Radioautography 11

III. Radioautographic Studies on Estrogen Binding by Uterine Cells 13
- A. Early Studies 13
- B. Radioautographic Localization of Uterine Cytosol-Nuclear Estrogen Receptors 13
- C. Radioautographic Localization of Uterine Eosinophil Estrogen Receptors 14
- D. Radiohistochemical Studies for the Characterization of Estrogen Receptors 16
  - 1. Extraction of Estrogens Bound to the Eosinophil Estrogen Receptors 16
  - 2. Evaluation of Receptor Affinity by Radiohistochemical Procedures 16
  - 3. Histochemical Receptor Structure Studies 18

IV. Correlation with Biological Response: Evidence for the Role of Different Kinds of Estrogen Receptors in Hormone Action 18

V. Conditions Dissociating Responses Mediated by the Different Mechanisms of Estrogen Action 20

VI. Proposed Mechanisms Explaining Eosinophil-Mediated Responses to Estrogen in the Uterus 23
- A. Recognition of the Uterus by Eosinophils in the Presence of Estrogen ... 23
- B. Mechanisms for the Eosinophil Mediated Responses in the Uterus 23

VII. A Model Explaining the General Biological Mechanism of Cell-Cell or Cell-Target Organ Recognition Via Specific Receptors 27

VIII. Concluding Remarks 27

References 29

I. INTRODUCTION

Estrogen action in the rat uterus has frequently been taken as a model for the investigation of the mechanisms of steroid hormone action. It is well known that estrogens, as well as other steroid hormones, induce a selective stimulation of target organs. This specific stimulation has suggested the existence in target organs of specific molecules, called hormone receptors, which are able to recognize the hormone, receive the hormonal signal, and translate it to a series of events that end in the specific effects of hormone stimulation in target organs. The preferential accumulation and retention of highly labeled steroid hormones by target tissues, that was first described for hexestrol in goats and sheep[1] and for estradiol in rats,[2] provided the first experimental evidence suggesting the presence of these receptors in target organs only. Further studies from a number of laboratories led to characterization of hormone receptors, and to the development of the first hypothesis for steroid hormone action that was assumed, for some time, to be the only mechanism of action possible for steroid hormones (see References 3 and 4 for a review). According to this widely accepted hypothesis, the interaction of estrogens with the cytosol-nuclear receptor system involves a two-step mechanism.[3] The estrogenic steroid enters the uterine cell and combines with the binding unit of an extranuclear receptor protein, called the cytosol receptor. The estrogen-cytosol receptor complex undergoes temperature-dependent conversion to a new form, which is translocated to the nucleus.[3] A weak migration of estrogen receptors to the nucleus has also been shown to occur in the absence of estrogen;[5-7] estradiol was proposed to enhance the nucleotropy of the receptor.[5] After entering the nucleus, the estrogen-nuclear receptor complex interacts with a specific acceptor site, apparently in the chromatin.[3] This interaction induces genomic activation (or derepression), resulting in an increased transcription of specific messenger RNA and subsequent synthesis of specific proteins and biochemical, morphological, and functional differentiation of target cells.[3]

Most biochemical studies that led to the above hypothesis considered the uterus as a single cell-type organ with only one kind of receptor for estrogens, mediating all the responses to hormone stimulation through the same mechanism of hormone action.[3] The biochemical techniques that have been used for the study of estrogen receptors in uterine homogenates cannot discriminate among the various uterine cell types, therefore, they cannot be applied to investigate whether all cell types possess the same kind of estrogen receptor or if all cell types are targets for hormone action. In this context, radioautographic techniques provided valuable tools for the investigation of hormone receptors separately in individual target cells.[8-23] For estrogens, the first radioautographic studies demonstrated predominant binding by the nuclei of various uterine cell types after labeled hormone injection to adult ovariectomized or intact immature rats.[8,10-13] This binding, which was assumed to reflect the existence of the cytosol-nuclear estrogen receptor system, was not in disagreement with the dogma of a single type of estrogen receptor in the different uterine cell types.

The first evidence for the existence of different kinds of estrogen receptors mediating a separate group of responses to estrogen through an independent mechanism of hormone action was obtained in our laboratory by experiments using the radioautographic technique, with the finding of high-affinity specific estrogen binding by uterine eosinophil leukocytes in the rat,[9,14-16,24-28] Syrian hamster,[29] and human.[30,31] Our reports of specific and saturable estrogen binding by rat uterine eosinophils after in vivo administration of ^{3}H-estradiol[15,16,27,28] or after uterine cryostat section incubation with very low concentrations (0.2 to 0.6 n*M*) of the labeled steroid[9,14,25] were confirmed by others,[32-35] suggesting the existence of specific hormone receptors.[9,14,24] Subsequent biochemical studies have confirmed the existence of high-affinity and low-capacity estrogen receptors in rat and human eosinophil leukocytes.[36,37] As will be discussed in detail below, these receptors are assumed to mediate some early nongenomic responses to estrogen stimulation.[14,15,38-44]

Further reports from our laboratory and elsewhere described several other pieces of evidence suggesting difference in the action of estrogens and of other steroid hormones. For estrogens, the following findings are not in agreement with the classical hypothesis of a single mechanism of estrogen action mediating all hormone effects, and can only be explained if the existence of multiple mechanisms of estrogen action is accepted: firstly, in addition to the well-known cytosol-nuclear estrogen receptor system, estrogen receptors were found in the membrane and in the cytoplasm of blood and uterine eosinophil leukocytes,[9,15,24,32-37] in the surface of uterine blood vessels,[45] as well as in vascular endothelial cell cultures,[46] in the cytoplasmic membranes of uterine cells,[47-50] and in the microsomal fraction of target cells;[51] the existence of type II cytoplasmic and nuclear estrogen binding sites[52] and of specific antiestrogen receptors[53-56] was also observed. Secondly, in addition to the effects known to be induced through genomic activation (or derepression) by the hormone, to the increase in cyclic GMP levels,[57-60] to the induction of type II nuclear receptors,[52] and to the controversial lysosome migration to the nucleus,[61] the following nongenomic responses to estrogen stimulation, which are not counteracted by actinomycin D,[43,62-65] were described in the uterus: cyclic AMP-mediated[62,63,66-71] and prostaglandin-mediated[64,65,72-75] parameters of estrogen stimulation, a very early migration of eosinophil leukocytes to the uterus after estrogen administration,[40] which is determined by estrogen levels in the blood but not in the uterus,[76] and a group of estrogenic responses that is selectively suppressed by conditions blocking estrogen-induced migration of eosinophils to the uterus.[77-80] The latter can be elicited in the absence of estrogen stimulation by conditions inducing a migration of eosinophils to the organ, such as treatment with actinomycin D for 26 hr,[43] or with indomethacin,[81] or a intrauterine instillation of saline physiological solution.[76] Thirdly, in disagreement with the classical hypothesis of a single mechanism of hormone action for the steroid, the effects of estrogens can be separated into at least three independent groups of responses which can be dissociated under various experimental conditions; each group can be selectively stimulated, inhibited, or completely blocked without interfering with the other groups of responses to estrogen (see below).

Based on the above evidence, we have proposed that the separate groups of estrogenic responses in the rat uterus are mediated by at least three (or perhaps more) independent mechanisms of hormone action:[42-44] the first group of responses (estrogen-induced increases in uterine RNA and protein synthesis, the increase in the content of some specific uterine enzymes, and biochemical, morphological, and functional differentiation of target cells) would be mediated by the cytosol-nuclear estrogen receptor system, through hormone-receptor complex interaction with genome;[3] these responses, that are selectively suppressed by actinomycin D,[43,64,65,82-84] can be considered as the genomic responses to estrogen stimulation.[3,43,82] The second group of responses (uterine eosinophilia, edema, increase in vascular permeability, release of histamine, and, possibly, antiimmune protection of the blastocyst against its rejection) would be mediated by the uterine eosinophil leukocytes;[14,15,38-44] eosinophils were proposed to migrate from the blood to the uterus under estrogen stimulation by a nongenomic mechanism,[43] possibly mediated by estrogen receptors in uterine eosinophils.[40,42-45] A third group of responses (including increases in uterine glycogen content and in a specific cervico-vaginal antigen) was proposed to be mediated by cyclic AMP;[62,63,66-71] the role of cyclic AMP in estrogen action remains, however, controversial.[85,86]

In addition to the existence of the "nontraditional" mechanisms of estrogen action mediating separate groups of responses to hormone stimulation, recently published data[22,87-98] and evidence that will be discussed below do not support the traditionally accepted concept that all the classical cytosol-nuclear estrogen receptors are similar. As already shown for glucocorticoid receptors,[99-102] data rather suggest that different kinds of cytosol-nuclear estrogen receptors may exist in the various target cell types.

Considering the existence of various different estrogen receptor systems mediating separate groups of responses through independent mechanisms, the known difference in cytosol-nuclear estrogen receptor level regulation in the various target cell types,[27,28,103-108] and the possibility of differences between cytosol-nuclear estrogen receptors from the various target cell types, it follows that all receptor systems and the different responses to hormone stimulation must be investigated in every individual target cell type or target organ region if a complete understanding of all the mechanisms involved in hormone action is intended. Several not completely successful attempts to study individual cell types or tissue layers have been performed: cell types were separated mechanically[106,109,110] or enzymatically,[111-113] but these techniques did not provide pure cell preparations; tissue culture procedures assured pure cell preparations,[90] but their receptor levels and their responses to hormone stimulation were different from those in the in vivo situation[114,115] since the cells were not in their physiological environment. Radioautography proved to be the most valuable technique for studies of receptors in individual cell types; further, it may provide important new possibilities for the elucidation of several endocrine phenomena in target organs. The results, however, should be correlated to the responses mediated by these receptors in the same cell types or target organ regions, for which, in many cases, appropriate techniques are still not available.

This chapter describes the radioautographic techniques that have been used in the study of estrogen receptors. It describes the localization of cytosol-nuclear estrogen receptors in the rat uterus and analyzes the differences in receptor level regulation in the various target cell types. It describes the radioautographic studies that allowed the finding and subsequent detailed characterization of estrogen receptors in uterine eosinophil leukocytes. It further discusses supportive biochemical and morphological evidence that suggested a role for eosinophil leukocytes in the mediation of a separate group of responses to estrogen through an independent mechanism. It suggests the importance of the consideration of all cell types and of all receptor systems involved in estrogen action for the complete understanding of the processes involved. Finally, it foresees new possible applications of radioautography in future studies on hormonal action.

II. RADIOAUTOGRAPHIC METHODS AS A TOOL FOR THE STUDY OF ESTROGEN RECEPTORS

A. Aim, Possibilities, and Limitations of Radioautography

The biochemical methods used for the study of steroid hormone receptors cannot evaluate estrogen binding characteristics separately in the different target organ cell types. Therefore, they cannot be applied to investigate if all cell types are targets of hormone action or possess the same kind of estrogen receptor, or whether all cells within the same cell type possess hormone receptors. As discussed above, cell separation procedures or tissue culture of pure cell preparations are not the most convenient procedures for the study of hormone receptors under physiological conditions. Furthermore, the biochemical procedures used for the study of receptors may cause artifactual translocation of the receptors or may show a kind of binding which does not occur in the living animal under physiological conditions.

Radioautography proved to have solved most of the above problems. For this purpose, the following three variants of the radioautographic technique have been developed and applied successfully for the study of the different receptor systems.

1. The Dry Radioautographic Technique for Diffusible Compounds

This technique is the most convenient procedure for the study of the cytosol-nuclear steroid receptors, although all other types of binding and receptors can be visualized by this method.[8,10-13,19] It can be applied to the analysis of tissue samples from animals that have been injected with physiological amounts of tritium-labeled hormones; in this case, it is able to localize the label in the exact place it was located in vivo at the time of tissue sample collection, without the danger of artifactural translocation. It cannot, however, discriminate between the hormone that is specifically bound to its receptors, the steroid that is bound to a lower-affinity binding protein, the nonspecifically bound hormone, and the free steroid. Considering the existence of different kinds of estrogen receptors (i.e., different kinds of cytosol-nuclear receptors, cytosol and nuclear type II binding proteins, etc.), it may not allow the identification of the different kinds of receptors within the same cell type, unless additional studies are performed (see below). Besides, radioautography does not inform as to the chemical nature of the localized radioactivity, i.e., whether the label corresponds to the unmodified compound or to its metabolite(s).

A simultaneous administration of excess of unlabeled hormone can suppress by competition the binding of the labeled hormone by low capacity (saturable) receptors, a procedure that is frequently used in receptor studies. To eliminate most free or loosely bound steroid without considerable loss of specifically bound hormone to receptors, tissue samples may be collected at longer times after the in vivo hormone administration; in this case the lower-affinity but high-capacity binding proteins carried by the blood flow extract comparatively more unbound or loosely bound steroid than receptor bound hormone. For in vitro experiments, this extraction by the blood flow can be replaced by submitting the tissue to either a superfusion with plasma proteins[26,31] or a postincubation with several changes of human plasma.[116]

2. The Wet Radioautographic Technique

This is a procedure adapted for the study of the eosinophil estrogen receptors.[9,14] Glass slide-mounted cryostat tissue sections can be incubated with the labeled steroid under different experimental conditions, and then dipped in melted radioautographic emulsion. The advantage of this procedure is that it allows the simultaneous study of thousands of target organ sections from the same animal, consequently, many experimental conditions may be compared within the same animal. The limitation of this technical approch is that it only allows the study of hormone binding to nondiffusible binding proteins, whose binding properties are not destroyed by freezing, but does not allow the study of the cytosol-nuclear receptors that are extracted from the tissue sections during the procedure.[15]

3. Radiohistochemical Methods for the Characterization of Receptors

These methods are histochemical reactions performed with the target tissue.[25,117-119] The effects of these reactions on hormone receptors can be visualized as changes in labeled hormone binding and/or retention, studied radioautographically. In addition, any other procedure that, beyond simple morphologic localization of hormone binding in tissues, uses radioautography for the characterization of receptor affinity, capacity, and other properties, can also be considered as a radiohistochemical method. Considering the possibility of difference between estrogen receptors among the various target cell types, this kind of analysis may be applied in future studies for the evaluation of the affinity spectra of individual cell-type estrogen receptors for different estrogenic and antiestrogenic compounds (see below).

B. Experimental Principles of Radioautography

Radioautography is a very sensitive technique for the study of radioactive substances in tissues that can detect up to a few hundred labeled molecules per cell. The information transmitted by the radiation is stored in the radioautographic emulsion and accumulates during the exposure time. The average number of disintegrations required to obtain the latent image for a silver grain depends on the isotope, the section thickness, the type of radioautographic emulsion, the exposure time, the developing process, and many other factors. For instance, when 3-μm-thick methacrylate sections on stripping film were used, 16 disintegrations were estimated to result in a silver grain.[120] Different kinds of radioautographic emulsions, developers, times, and temperatures of the developing process[121] or modifications of the technique[12] changes its sensitivity and yields a different number and size of radioautographic grains.

For the successful application of radioautography, it was recommended that:[19] (1) the range of radiation of the isotopic label and the type of emulsion used must be commensurate with the expected structural resolution; (2) the compound under study should be labeled at high specific activity with the radioisotope in chemically stable positions; (3) the labeled compound should be radiochemically pure at the time of administration; and (4) the chemical nature of the radioactivity localized in the radioautogram should be characterized by radiochemical analysis of tissue extracts or by radioautographic studies with competitors which preferably undergo different metabolism. For the localization of diffusible substances not chemically bound to stable constituents, in addition to the above recommendations it is fundamental to avoid translocation of the radioactive compound and/or its metabolites (mainly due to the use of fluids or solvents during tissue and radioautogram preparation, the lack of label and cell constituent immobilization by histological fixatives, or to post-mortem changes).[19]

C. The Radioautographic Techniques

1. The Dry Radioautographic Technique for Diffusible Compounds

This technique is an adaptation for radioautography[8,10-13,19] of the Altmann and Gersh freeze-drying method,[122] that has been devised to reduce diffusion, extraction, or translocation of the components of the cellular structures. After the administration of the labeled compound to an animal in vivo or after an in vitro tissue sample incubation with the radioactive compound, the specimens are placed on top of copper mounts containing small amounts of finely minced liver (used to adhere the specimens to the mounts),[8,10,19] quickly frozen in liquid propane,[13] propane-butane,[116] or freon,[20,21,123,124] cooled by liquid nitrogen, and stored in liquid nitrogen. Cryostat tissue sections (4 to 6 μm thick) are freeze-dried at about −30 to −40°C in a Cryopump® (Thermovac Industries, Copiague, L.I., N.Y.) containing Molecular Sieve (Linde A5), by preevacuating the pump for 10 to 30 min and then cooling the cryosorbtion chamber of this pump with liquid nitrogen.[8,10,19] This procedure increases the performance of the originally designed Jansen's freeze-drying apparatus.[125] Alternatively, the sections can be directly freeze-dried at −30 to −40°C with a vacuum produced by an oil diffusion pump backed by a mechanical vacuum pump.[102,124] The vacuum, after approximately 6 hr at −30 to −40°C and one additional hour at room temperature, should be broken by dry gas and the freeze-dried tissue sections must be kept in a desiccator containing a desiccant such as anhydrous calcium sulfate (Drierite). Subsequently, in a low-humidity dark room, the freeze-dried sections are placed on small Teflon® squares and pressed on to glass slides which had previously been coated with radioautographic emulsion and allowed to dry.[8,10,19] The radioautograms are then kept in light-safe boxes at −30°C during the exposure time (usually between 1 and 18 months) and afterwards developed for 30 to 45 sec in Kodak® D-19 developer at 20°C (the size of radioautographic grains increases by increasing developing time), fixed in Kodak® acid hypo fixer, stained, and submitted to further histological procedure.

2. The Wet Radioautographic Technique

This technique is an adaptation for radioautography of a steroid enzyme histochemical method.[126] It has been devised for the study of hormone binding to nondiffusible binding proteins whose binding properties are not destroyed by freezing, particularly, the eosinophil estrogen receptors.[9,14] Cryostat sections of a tissue that has been frozen as for dry radioautography are attached to glass slides by momentary thawing, and, subsequently, stored at −30°C. Before the incubation the slides are allowed to reach room temperature (the sections become dry) and 0.2 to 0.5 mℓ of the incubation medium containing the labeled steroid is dropped on the sections. After 10 min of incubation at room temperature, the sections are washed in water to extract unbound or loosely bound label, dried at room temperature, and dipped in Kodak® NTB-3 radioautographic emulsion melted at 38°C. The emulsion is allowed to dry for about 12 hr at room temperature and then the slides are transferred to light-proof boxes containing desiccant and stored at −30°C. At the end of the exposure time, the radioautograms are developed in Kodak® D-19 developer, fixed in Kodak® hypo fixer, washed in water, stained, and submitted to further histological procedure.[9,14]

3. Radiohistochemical Procedures for the Characterization of Receptors

These are methods that, beyond simple localization of hormone binding in tissues, use radioautography as a tool for the histochemical investigations of receptors. For the study of the eosinophil estrogen receptors, the effects on estrogen binding of ethanol extraction of receptor-bound estrogens,[24] of incubation with different concentrations of unlabeled steroids,[14] of iodination in alkaline media, of the protective effect of receptor-bound estrogen against receptor inactivation by iodination,[25] of SH group blockage,[117] of a treatment with urea to partially unfold the receptor molecule to expose residues to blockers,[118] and of tryptophan residue blockage,[119] studied by the wet radioautographic technique, were reported. For the study of the cytosol-nuclear receptors (dry radioautography is required), similar procedures may be applied to small tissue samples that have not been frozen, provided the reagents penetrate inside alive, intact cells.

A quantitative evaluation of radioautograms of sections that have been incubated with different concentrations of labeled steroids can be used for the evaluation of the affinity of the hormone for its receptors. For this purpose, here, we propose that data may be expressed in plots similar to that proposed by Scatchard,[127] where bound steroid concentration is expressed as radioautographic grains per target cell (adjusted to a standard exposure time) and free hormone concentration is expressed as its initial concentration in the incubation medium. This is possible, based on the assumption that free hormone concentration almost equals total steroid in the incubation medium, since the volume of the incubation medium largely exceeds that of the tissue sections, and the total amount of specific and nonspecific binding components is very low. Figure 3 shows the application of this method. Data from tissues incubated with different concentrations of unlabeled competitors added to a constant concentration of the labeled hormone may be used as well (Figure 4A). Similar studies may be done in vivo for the comparison of the affinities of receptors from different cell types; in this case, data may be plotted as bound steroid (as radioautographic grains) divided by the dose (in nmol/kg body weight) as a function of bound steroid (Figure 4B). The free labeled hormone concentration within the tissue is not known, but it can be assumed that it is similar for the different cell types within the same organ, therefore, data from different cell types are comparable. This approach may reveal differences between affinities of the receptors from the different target cell types; further, the correlations between affinity data from individual cell types with dose-response data from the same cell types may be investigated, which is fundamental for the ascription of individual hormonal responses to individual receptor types.

D. Receptor Criteria as Applied to Radioautography

In hormone localization studies using radioautography, the question arises whether or not label concentration in any place reveals the existence of specific hormone receptors. For the correct interpretation of radioautographic data and the identification of steroid receptors, it is important to keep in mind the following criteria that are commonly used to define hormone receptors.

High-affinity binding — This is a very important requirement for hormone action, considering the low steroid levels in the blood and tissue fluids. The hormone may induce receptor-mediated responses only if its affinity for the receptors is in the range of hormone levels in the blood. This does not preclude, however, the possibility of hormone-receptor interactions of weaker affinity if hormone or receptor levels are elevated. Steroids are also bound to low-affinity binding proteins in blood and tissues. In radioautographic studies, high-affinity binding may be demonstrated when the tissue is incubated with very low steroid concentrations or very low doses of the labeled hormone are administered in vivo.

Limited binding capacity (saturability) — The number of hormone receptors per cell is very low, therefore, their capacity to bind the steroids is saturable at relatively low hormone concentrations. Since hormone binding to its receptor is a prerequisite for hormone action, the biological response to steroid hormones is also a saturable phenomenon. The levels of saturability, as it will be shown below, may change under physiological conditions and may be regulated by a number of factors, determining the magnitude of the biological response. It is the saturability that separates the high-affinity, specific binding from the low-affinity, nonspecific binding. In radioautographic studies, the limited binding capacity may be investigated in competition experiments, where excess of unlabeled steroid blocks hormone binding by its receptors.

Steroid hormone specificity — Each steroid hormone receptor is specific for its high-affinity binding to its own class of hormones. The concept of specificity implies high-affinity binding, since a steroid, when at very high concentrations, may also bind to receptors of different classes of hormones. At lower but still nonphysiologic concentrations, steroid receptors from some target organs may not be absolutely specific for their own class of hormones;[128] this phenomenon may also occur at physiologic hormone concentrations,[129] suggesting differences among the receptors from the different cell types. Accordingly, for an accurate characterization of any receptor, its spectrum of affinity should be described for the different compounds of its own class of hormones and also for compounds that do not belong to it. For glucocorticoids, a number of different receptors were reported differing in their spectra of affinity for different steroids;[99-102] the possibility of heterogeneity may also exist for estrogen receptors in the different target cell types. Therefore, the spectra of affinity of estrogen receptors need to be investigated for each individual cell type. In radioautographic studies, steroid receptor specificity may be demonstrated in competition experiments as a decrease in hormone binding under the effect of different concentrations (or doses) of the competitor(s).

Tissue specificity — Biochemical studies show that not all tissues are targets for hormonal action and, therefore, possess receptors. Radioautography may provide additional information, since not all cell types within a target organ are targets for hormone action, and not every cell of each target cell type must necessarily have hormone receptors at the moment of the investigation. Further, tissue specificity may be a quantitative rather than a qualitative difference between target and nontarget tissues, since the so-called nontarget tissues may contain low levels of receptors.

Correlation with biological response — It is assumed that hormone binding to a receptor results in a biological response. Since there are different kinds of hormone receptors and of responses to hormonal stimulation for each kind of hormone, it is very important to investigate the correlations between hormone binding to receptors and the specific tissue responses to a hormone. In many systems, a defect of a response to hormone stimulation may be due to an absence of receptors (e.g., in the pseudohermaphroditic male mouse).[130] One has to keep in mind, however, that post-receptor defects[131] or other deficiencies[77,78] may preclude the elicitation of a biological response. The correlation of hormone binding with biological response is the most difficult and most important receptor criteria to establish.

III. RADIOAUTOGRAPHIC STUDIES ON ESTROGEN BINDING BY UTERINE CELLS

A. Early Studies

The early attempts to use radioautography for the localization of estrogen receptors have encountered difficulties, because of the diffusibility of steroids and their high solubility in the organic solvents commonly used in histology, and the reported results on ^{3}H-estrogen binding by uterine tissues were conflicting: radioactivity was found in the lumen of glandular tubes in contact with apical poles of cells,[132] in the cytoplasm of the luminal epithelium,[133] in the nuclei of endometrial and glandular cells,[134] or at the apex and base of luminal cells.[135]

To avoid the action of organic solvents that would displace or extract bound estrogen by organic solvents used in standard tissue preparations, two different approaches were used: (1) the incubation of glass slide-mounted cryostat sections of adult rat uterus with ^{3}H-estradiol and subsequent process by the wet radioautographic technique,[9] and (2) the in vivo administration of the labeled hormone to adult ovariectomized or immature intact rats and subsequent process of the tissues by the dry radioautographic technique for diffusible compounds.[8,10-13,19] The tissue processed by the wet radioautographic technique displayed label in the cytoplasm and in the cell surface of uterine eosinophil leukocytes, with no labeling of the nuclei of other uterine cell types.[9] The binding by eosinophils had a high affinity and specificity for estrogens but not for steroids without estrogenic action, and a limited binding capacity.[14,24] The tissues obtained from adult ovariectomized or intact immature rats and processed by the dry radioautographic technique displayed label concentration by the nuclei of luminal and glandular epithelial, stromal and muscular uterine cells,[8,10-13,19] with no binding by the uterine eosinophils.[11,13] This discrepancy can be explained since there are not eosinophils in the uteri of adult ovariectomized or intact immature rats and these cells migrate to the organ after estrogen administration.[24,40,77,136-138] Further studies of estrogen binding in intact adult animals by dry radioautography demonstrated binding by both the nuclei of the different uterine cell types and the eosinophils.[15,16,27,28] To explain the discrepancy about nuclear binding, uterine tissue sections from ^{3}H-estrogen-injected adult rats, processed by the dry radioautographic technique, were extracted with water immediately after mounting the dry sections on radioautographic emulsion. This procedure extracted label from the nuclei of uterine cells but not from eosinophils.[15] It was suggested that label is lost from the nuclei of sections processed by the wet radioautographic technique, because cytosol-nuclear hormone-receptor complexes are extracted from the tissue during the procedure.[15] The above studies suggested the existence of two different kinds of estrogen receptors.

B. Radioautographic Localization of Uterine Cytosol-Nuclear Estrogen Receptors

Estrogen binding by the nuclei of uterine cells has classically been assumed to reflect the existence of cytosol-nuclear receptors. Dry radioautographic studies of estrogen binding in vivo by the uterus of the intact immature or the adult ovariectomized rat have shown label localization in the nuclei of luminal and glandular epithelial, stromal, and smooth muscular uterine cells.[8,10-13,19] In the adult nonovariectomized rat, glandular epithelial, stromal, and muscular cells display nuclear concentration of radioactivity (Figure 1).[16,17,27,28] Luminal epithelial cells show nuclear concentration of radioactivity in rats in proestrus, estrus, or in estrogen-pretreated animals, but not in diestrous, progesterone-pretreated, or estrogen + progesterone-pretreated rats (Figure 1).[17,27,107] It was suggested that progesterone interferes with either cytosol receptor synthesis or replenishment in the uterine luminal epithelial cells

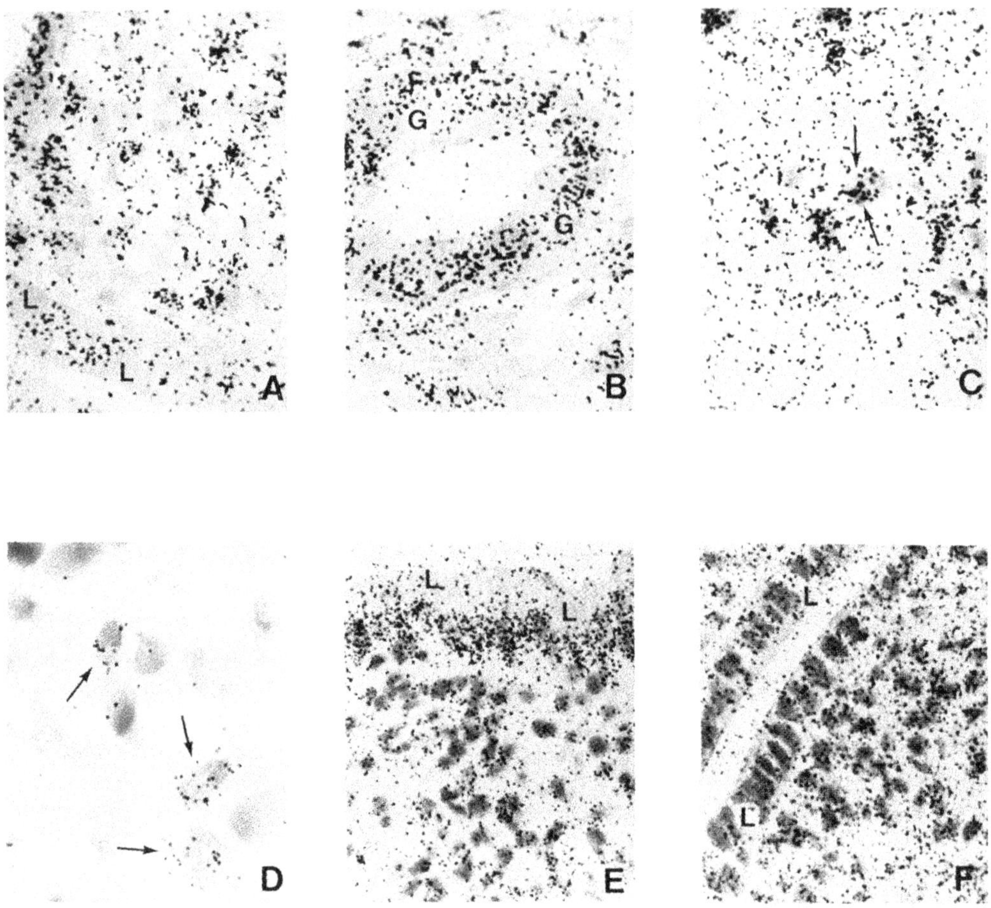

FIGURE 1. Dry radioautograms of in vivo binding of ^{3}H-estradiol by uterine cells of diestrous (A, B, C, and D), estrogen pretreated (E), and estrogen + progesterone pretreated (F) adult rats. Arrows, uterine eosinophils; L, luminal epithelium; and G, glandular epithelium. D corresponds to a freeze-dried section that was extracted with water before exposure. (A, B, C, and D, magnification × 1000; E and F, magnification × 420.) (A, B, C, and D reprinted with permission from the *Journal of Steroid Biochemistry*, 4, 41, Tchernitchin, A. and Chandross, R., In vivo uptake of estradiol 17β by the uterus of the mature rat, Copyright 1973, Pergamon Press, Ltd., New York; E and F reproduced with permission from *Experientia*, 32, 1069, Tchernitchin, A., Effect of progesterone on the in vivo binding of estrogens by uterine cells, Copyright 1976, Springer-Verlag, New York.)

but not in the other uterine cell types.[27] This interference correlates with the selective blockage of estrogen action by progesterone in this cell type only.[27,107,139] The dynamics of estrogen binding by the nuclei of the different uterine cell types at different times after intravenous administration of the steroid has been reported,[28] and a similar action of progesterone on luminal epithelium was found.[28] In vitro dry radioautographic studies of estrogen binding by uterine tissue after incubation with the labeled steroids revealed, for the rat[26] and human,[31] similar results to that obtained after in vivo radiolabeled hormone administration.

C. Radioautographic Localization of Uterine Eosinophil Estrogen Receptors

Studies on ^{3}H-estrogen binding by uterine defrosted cryostat sections in vitro using the wet radioautographic technique demonstrated hormone binding by the eosinophil leukocytes

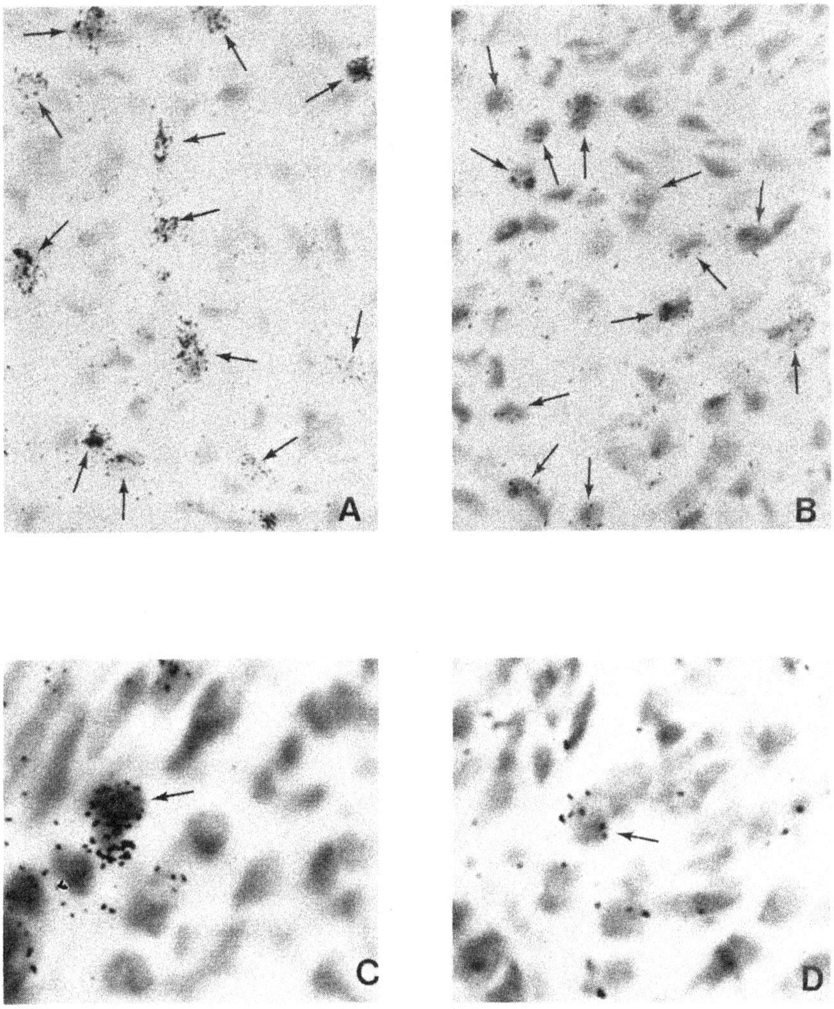

FIGURE 2. Wet radioautograms of ^{3}H-estradiol binding by uterine eosinophils in the absence (A and C) or presence (B and D) of excess of unlabeled estradiol. Arrows, uterine eosinophils. (A and B, magnification × 600; C and D, magnification × 1100.)

of the adult rat (Figure 2),[9,14,24,25] Syrian hamster,[29] and human.[30] These findings were confirmed by dry radioautography after in vivo administration of ^{3}H-estrogen to rats,[15,16,27,28] or the in vitro incubation of rat[26] or human[31] uterine tissue samples. Reports from other laboratories, using different experimental approaches, have confirmed estrogen binding by uterine eosinophil leukocytes.[32-35]

The following data strongly suggest that estrogen binding by uterine eosinophils is due to the presence of receptors in these cells: (1) the very low ^{3}H-estrogen concentrations in the incubation medium (0.2 to 0.6 n*M*) needed to show hormone binding by the eosinophils[14,24,25] suggest the existence of high-affinity binding; (2) the binding of three different labeled estrogens (estradiol-17β, estrone, and estriol),[140] but not of tritiated progesterone, testosterone, corticosterone, or aldosterone,[24,151] by eosinophils and the failure of excess of unlabeled progesterone, testosterone, or corticosterone added to the incubation medium to decrease, by competition, the binding of tritiated estradiol,[14] suggests specificity of binding for estrogens only; and (3) the dose-dependent decrease in ^{3}H-estrogen binding in eosinophils by nonradioactive estradiol-17β, estrone, estriol,[14] or equilin[140] added to the incubation medium (Figure 2) suggests saturability due to low-capacity binding by the receptors. Further, estrogen binding by eosinophils decreases in animals with increased blood estrogen levels, suggesting competition with physiologic levels of endogenously produced estrogens.[24] The existence of specific estrogen receptors in rat uterine and human blood eosinophil leukocytes has been confirmed biochemically, and a constant of dissociation of 0.56 n*M* was calculated for estradiol.[36,37] A number of agents interfere with estrogen binding by eosinophils: thyroid hormones[80,141] and insulin[79] decrease ^{3}H-estradiol binding, while theophylline increases it;[142] these agents selectively modify estrogenic responses mediated by eosinophils. The correlations of estrogen binding by eosinophils and uterine eosinophilia with the biological responses will be analyzed below.

D. Radiohistochemical Studies for the Characterization of Estrogen Receptors

1. Extraction of Estrogens Bound to the Eosinophil Estrogen Receptors

Ethanol extracts bound estrogens from the eosinophil receptors, without impairing their subsequent ability to bind the steroid.[24] In rats with low estrogen levels in the blood, such extraction does not modify subsequent estrogen binding; in animals with high hormone levels in the blood, in which estrogen binding is very low due to competition, ethanol extraction increases subsequent binding to nearly the same values as in hypoestrogenic animals.[24]

2. Evaluation of Receptor Affinity by Radiohistochemical Procedures

The first studies for the comparison of the affinities of different estrogens for their receptors in eosinophils were performed by competition between ^{3}H-estradiol and different concentrations of unlabeled estrogens, using the wet radioautographic technique (Figure 2).[14] These studies suggested that the eosinophil receptors have a higher affinity for estriol than for estradiol,[14] in contraposition to the known higher affinity of the cytosol-nuclear receptors for estradiol.[143] Subsequent dry radioautographic studies of uteri incubated by superfusion with labeled estradiol or estriol and, subsequently, exchanged (by an additional superfusion) with unlabeled estradiol or estriol, confirmed the above affinity characteristics for the eosinophil and for the cytosol-nuclear receptors.[26]

For a more accurate characterization of the eosinophil estrogen receptors, uterine cryostat sections were incubated with different concentrations of ^{3}H-estradiol (with or without excess of unlabeled steroid) and processed by the wet radioautographic technique.[143a] Figure 3A shows the resulting saturation analysis and Figure 3B shows a plot of bound steroid (as radioautographic grains) divided by total hormone concentration (which nearly equals free) as a function of bound. Data suggest the existence of two specific binding components, the higher-affinity component has a dissociation constant of about 0.4 n*M*. This value is similar to the previously reported K_d = 0.56 n*M* in biochemical studies that considered the existence of only one binding component in eosinophils.[36,37]

Bound steroid in eosinophils can also be expressed as a decrease in ^{3}H-estradiol binding in the presence of different concentrations of nonradioactive steroids. Figure 4A shows a plot constructed with previously published data[14] with a decrease in ^{3}H-estradiol binding by eosinophils in the presence of different concentrations of nonradioactive estradiol and estriol; it also suggests the existence of two binding components for eosinophils, and, in agreement with previous reports,[14,26] shows that the higher-affinity component has higher affinity for estriol than for estradiol.

For the evaluation of the affinity of estrogen receptors in the nuclei of different target cell types, we have constructed a plot with radioautographic data from a study of estrogen binding in vivo by the nuclei of various target cell types, reported by Holderegger et al.[22] Figure 4 shows the total bound steroid divided by the dose of estrogen (in nmol/kg body

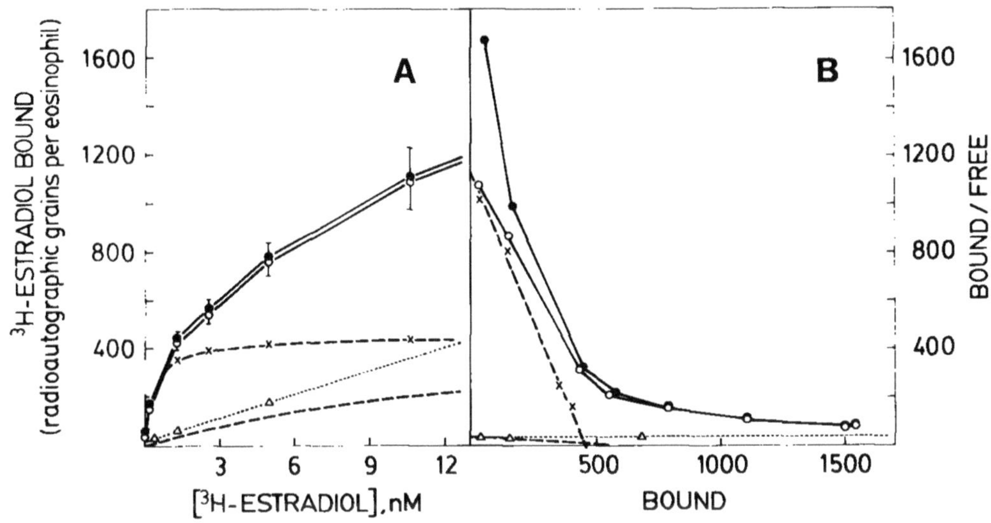

FIGURE 3. Saturation (A) and Scatchard (B) plots for the analysis of in vitro binding of ^{3}H-estradiol by rat uterine eosinophils in cryostat sections, using radioautographic data. Total binding per eosinophil (○) is expressed as *radioautographic grains per eosinophil* (per year of exposure) minus background. The total number of radioautographic grains per eosinophil including the background (●) is shown for reference. The free hormone concentrations are assumed to nearly equal the total steroid concentrations in the incubation media (see text). The two specific binding components of different affinities (broken lines) and the calculated values for specifically bound steroid to the higher affinity component of K_d = 0.4 n*M* (x) are also shown. The regression for nonspecific binding data, evaluated as label in eosinophils in the presence of excess of nonradioactive estradiol (△), is shown as the dotted line. (Modified from Reference 143a.)

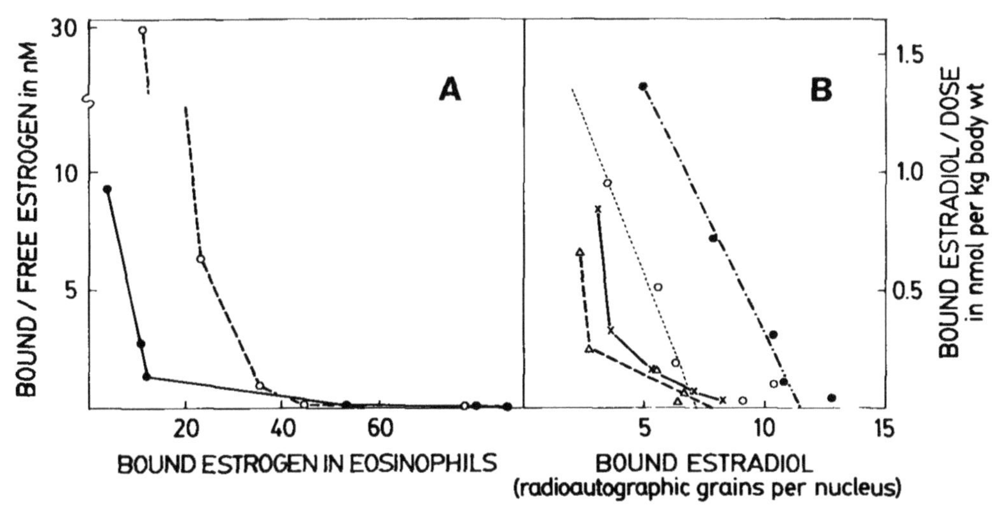

FIGURE 4. Application of the radiohistochemical method of receptor affinity evaluation for the analysis of previously reported radioautographic data. (A) Scatchard analysis of binding of nonradioactive estradiol (●) and estriol (○) by rat uterine eosinophils, using data that have been previously reported[14] as % decrease in ^{3}H-estradiol binding by uterine eosinophils after preincubation with different concentrations of the nonradioactive steroids. (B) Receptor affinity evaluation using radioautographic data from a study of in vivo estrogen binding by the nuclei of vaginal stromal (△), uterine stromal (x), uterine glandular epithelial (○) and pituitary lactotrope (●) cells reported by Holderegger et al.[22] Background and nonspecific binding data are not available. The total bound steroid (as radioautographic grains per nucleus) divided by the dose of estrogen administered in vivo (as nmol/kg body weight) is plotted as a function of the total bound steroid (instead of bound/free vs. bound as in the classical Scatchard plot) for each cell type.

weight) plotted as a function of total bound steroid in four different cell-type nuclei. The binding by the uterine glandular and pituitary lactotrope cell nuclei is evidently different from that of uterine stromal and vaginal stromal cells. In the two epithelial cell types, only one high-affinity component can be detected, while in the two stromal cell types, data suggest the existence of two specific binding components. It can be suggested that estradiol has higher affinity for the higher-affinity component of the stromal cells (presumably, cytosol-nuclear receptors) than for that of the epithelial cells. Although this needs to be confirmed with a wider range of doses, it is in agreement with data reported elsewhere,[22,87-98] suggesting difference between the cytosol-nuclear estrogen receptors from the different target cell types.

3. Histochemical Receptor Structure Studies

The study of estrogen receptor configuration and structure, especially at the site related with hormone recognition and binding, offers an approach to the mechanism of hormone action. For the cytosol-nuclear estrogen receptors, the interaction of estradiol with receptor sites in the rat uterus depends critically on the presence of sulfhydryl groups, since sulfhydryl group reagents (iodoacetamide, *N*-ethylmaleimide, or *p*-hydroxymercuribenzoate) abolish estrogen binding and disrupt existing estrogen-receptor complexes.[144] In contradistinction to above findings, a treatment of uterine cryostat sections with iodoacetate or with *N*-ethylmaleimide does not impair subsequent binding by the eosinophil estrogen receptors.[117] This differential action provides evidence for a different nature of both estrogen binding proteins, in addition to other differences, such as those related to their affinities for estriol.

Iodination of calf uterine cytosol-nuclear receptors in an alkaline media destroys their ability to bind estradiol; iodination is ineffective when the hormone is interacting with the binding site, thus, protecting it from a direct effect.[145] Similar results were obtained for the eosinophil estrogen receptors in a study performed with uterine cryostat sections.[25] It was found that iodination destroys the ability of the eosinophil receptors to bind estradiol, and that the steroid bound to the receptors protects them against the inactivation by iodination.[25] It can be concluded that iodination interacts with either a thyrosyl, a histidyl, or a tryptophyl residue (s) that is (are) involved in hormone-receptor interaction at or near the site of steroid binding in both the cytosol-nuclear and the eosinophil estrogen receptors.

Further studies investigated the action of tryptophan group blockers on estrogen binding by eosinophils using wet radioautography. It was shown that *N*-bromosuccinimide slightly impaires estrogen binding by unloaded eosinophil estrogen receptors.[119] This impairment is greatly enhanced in the presence of urea,[119] by probably unfolding or splitting the active site or adjacent area of the receptor,[118] thus, increasing the number of tryptophan residues exposed to the action of the blocker.[119]

Urea,[118] or guanidinium chloride,[146] denaturants that probably act by unfolding proteins, impair estrogen binding by eosinophils; this impairment is enhanced if estrogen receptors are unoccupied, suggesting a protective action of bound estrogen against the action of these denaturants. The mechanism of protection is still not well understood.

IV. CORRELATION WITH BIOLOGICAL RESPONSE: EVIDENCE FOR THE ROLE OF DIFFERENT KINDS OF ESTROGEN RECEPTORS IN HORMONE ACTION

There is always a relationship between estrogen binding by any receptor and the responses mediated by that receptor. Estrogens exhibiting a high affinity for one of the receptor systems only are strong hormones for the induction of estrogenic responses mediated by that receptor system only. Any condition selectively modifying estrogen binding by any receptor system selectively modifies the responses mediated by that receptor. The following differences among the estrogen receptor systems, that correlate to differences among the biological

Table 1
AFFINITY/POTENCY RELATIONSHIP OF ESTRADIOL AND ESTRIOL

Affinity/Potency Relationship

	Estradiol	Estriol
Affinity for		
Cytosol-nuclear receptors	+ + + +	+
Eosinophil receptors	+ + + +	+ + + + +
Potency of		
Genomic responses	+ + + +	+
Eosinophil-mediated responses[a]	+ + + +	+ + + + +

[a] Eosinophil-mediated responses: uterine eosinophilia, edema, increase in vascular permeability, histamine release.

responses that are mediated by these receptors, suggest the existence of multiple independent mechanisms of hormone action.

The cytosol-nuclear receptors and the genomic response to estrogen — The affinity spectrum of the cytosol-nuclear estrogen receptors for different estrogenic agents correlates well with the potency spectrum of the same compounds to induce the genomic responses to estrogen. Table 1 shows that estradiol-17β displays both a high affinity for the cytosol-nuclear receptors and a strong potency to induce the genomic responses. Estriol has a lower affinity for the cytosol-nuclear receptors.[26,143,147] and is a weaker estrogen than estradiol for the early genomic responses[138,148] including the very early increase in induced protein (IP) synthesis;[147] since estriol-receptor complexes dissociate faster than estradiol, estriol is even a much weaker estrogen for the prolonged genomic action.[66,149,150] Diethylstilbestrol binds more strongly than estradiol to cytosol-nuclear receptors; accordingly, this compound is stronger for the genomic responses.[151] The affinity/potency correlations for several other estrogenic compounds are discussed elsewhere.[14,148]

The eosinophil estrogen receptors and the eosinophil-mediated nongenomic responses — Table 1 shows that estradiol-17β displays a strong affinity for the eosinophil receptors and is a strong estrogen for the eosinophil-mediated responses. In contradistinction to reported data for the cytosol-nuclear mediated genomic response, estriol displays a higher affinity than estradiol for the eosinophil receptors,[14,26,140] and is the most potent estrogen for hormone-induced water imbibition,[138,149,150] increase in vascular permeability,[148,152] and release of histamine[153] in the uterus. There are not affinity data available for diethylstilbestrol, but it is slightly weaker than estradiol for fluid imbibition.[151] The dissociation between the potencies of other estrogens for the eosinophil-mediated and the genomic responses is analyzed elsewhere.[14,148]

Insulin[79] and thyroid hormones[80,141] decrease estrogen binding by eosinophils; accordingly, these hormones selectively block eosinophil-mediated responses to estrogen in the uterus. Theophylline increases estrogen binding by eosinophils,[142,154] explaining the potentiation of estrogen-induced edema under conditions (i.e., in the immature rat), wherein the eosinopenic effect of theophylline is not a limiting factor.[154]

Miscellaneous receptors — The affinity spectrum for membrane estrogen receptors is different from that of cytosol-nuclear or eosinophil receptors. While estradiol displays a high affinity for these receptors and estriol has a lower affinity, diethylstilbestrol does not bind (or binds very weakly) to this receptor system.[49,50,155] For type II receptors, scarce information is available: nonradioactive estradiol and diethylstilbestrol inhibits binding by type II cytosol receptors; estradiol but not estriol (unless very high doses are used) causes the appearance of type II nuclear receptors; diethylstilbestrol, estradiol, and estriol compete with labeled estradiol for binding to type II nuclear receptors.[93,156,157] It is still not clear what responses to estrogen stimulation are mediated by membrane or by type II receptors.

Miscellaneous responses not ascribed to known receptors — The receptors mediating estrogen-induced increases in uterine cAMP, cGMP, glycogen levels, or prostaglandin-mediated responses have not yet been identified. An approach to this identification may be the study of the potency spectra of different estrogens for these responses and their comparison to affinity spectra of known receptors. Estradiol is as potent as estriol for the increase in uterine glycogen content,[138] uterine hyperemia,[158] and the increase in uterine cGMP content;[58,159] and diethylstilbestrol is slightly less efficient than estradiol for the increases in glucose oxidation[151] and cGMP levels.[159] Further studies are necessary to ascribe these effects to known estrogen receptors.

V. CONDITIONS DISSOCIATING RESPONSES MEDIATED BY THE DIFFERENT MECHANISMS OF ESTROGEN ACTION

The dissociation of responses to estrogen stimulation by a number of agents or experimental conditions into several groups of independent parameters, which can be selectively stimulated, inhibited, or completely blocked without interfering with the other groups of responses, is in agreement with our hypothesis of multiple and independent mechanisms of estrogen action. In addition to affinity/potency dissociations already analyzed, the following agents or conditions dissociate the responses to estrogen: blockage of genomic activation by actinomycin D, interference with cAMP level increase by propranolol, prostaglandin biosynthesis inhibition by indomethacin, blockage of histamine receptors, progesterone, theophylline, local or systemic estrogen levels, and interference with estrogen-induced migration of eosinophil leukocytes to the uterus (colchicine, eosinopenic conditions). The action of these agents or conditions is summarized in Figure 5.

Actinomycin D, a genomic activation blockader at the level of transcription, completely blocks genomic responses to estrogen,[43,64,65,82-84] without interfering with estrogen binding by cytosol-nuclear receptors.[160,161] Indeed, it blocks 60% of cytosol receptor replenishment[162-164] and prevents nuclear processing of estrogen receptor.[165] Actinomycin D does not interfere with the estrogen-induced migration of eosinophil leukocytes to the uterus,[43] or the eosinophil-mediated water imbibition of the uterus,[43] nor does it block either the increase in glycogen content,[43,62] or the hyperemic[64,65] responses in the uterus. It moderately decreases the increase in cervico-vaginal antigen[63] and blocks the increase in cGMP.[166,167]

Estrogen induces an increase in uterine adenylate cyclase activity[168] and in cAMP content.[169] Exogenous cAMP increases uterine glycogen content.[62,68] Propranolol suppresses the estrogen-induced increases in uterine cAMP levels[170,171] and in cervico-vaginal antigen,[70] but it does not interfere with the hormone-induced increase in uterine glycogen content[85] and does not either modify estrogen-induced genomic responses,[71,171] uterine eosinophilia,[71] eosinophil-mediated responses,[71] uterine hyperemia,[72] or the increase in cGMP.[58,172] It may be suggested that cyclic AMP is involved in some but not all estrogenic responses, probably as a separate mechanism of estrogen action.

Indomethacin, a prostaglandin biosynthesis inhibitor, completely blocks estrogen-induced uterine hyperemia.[73,74] It does not inhibit estrogen-induced uterine eosinophilia,[74] water imbibition,[74,173] the increase in vascular permeability,[174] or the increase in uterine cGMP,[175] suggesting that these responses are not mediated by prostaglandins.

Progesterone does not compete with estrogen for the cytosol-nuclear receptors,[139,176] but interferes with their *de novo* synthesis or replenishment in some cell types only[16,27,28,107] selectively blocking the genomic responses in the same cell types.[27,107,139] Progesterone does not compete either with estrogen for the eosinophil estrogen receptors,[14] does not block the

AGENT OR CONDITION	CYTOSOL-NUCLEAR RECEPTORS	ESTROGEN BINDING BY CYTOSOL-NUCLEAR RECEPTORS	UTERINE GENOMIC RESPONSES TO ESTROGEN (MEDIATED BY CYTOSOL-NUCLEAR RECEPTORS)	ESTROGEN BINDING BY EOSINOPHILS	BLOOD EOSINOPHIL LEVELS	ESTROGEN-INDUCED UTERINE EOSINOPHILIA	UTERINE EDEMA, INCREASE IN VASCULAR PERMEABILITY, HISTAMINE RELEASE, ETC. (RESPONSES MEDIATED BY EOSINOPHILS)	INCREASE IN cAMP	GLYCOGEN CONTENT & RELATED ENZYMES INCREASE	CERVICO-VAGINAL ANTIGEN	UTERINE HYPEREMIA (MEDIATED BY PROSTAGLANDINS?)	INCREASE IN cGMP	TYPE II NUCLEAR RECEPTOR
ACTINOMYCIN D	= OR ↓	=	↓↓↓	?	?	=	=	?	=	↓	=★☆	↓↓	?
PROPRANOLOL	?	?	=	?	?	=	=	↓↓↓	=	↓↓↓	=☆	=	?
INDOMETHACIN	?	?	?	?	?	=	=	?	?	?	↓↓	=	?
PROGESTERONE	↓↓↓ IN SOME CELL-TYPES ONLY	= (DOES NOT COMPETE WITH ESTRADIOL)	↓↓↓ IN SOME CELL-TYPES ONLY; ↓ SOME RESPONSES	= (DOES NOT COMPETE WITH ESTRADIOL)	?	DOES NOT BLOCK MIGRATION OF EOSINOPHILS BUT DEGRANULATES THEM	↑ INTENSITY, BUT ↓ PERSISTENCE OF EDEMA	↓↓	↓↓	?	↓	↓↓↓	↓↓
COLCHICINE	DOES NOT BLOCK RECEPTOR MIGRATION TO THE NUCLEUS	=	?	?	?	↓↓↓ (DOES NOT INTERFERE WITH RECOGNITION OF THE UTERUS BY EOSINOPHILS; BLOCKS THEIR MIGRATION TO EXTRAVASCULAR SPACE)	↓↓↓	?	?	?	?	?	?
GLUCOCORTICOIDS	=	=	=	=	↓↓↓	↓↓↓	↓↓↓	↓↓	= OR ↓↓	?	=	= ?	↓↓↓
THYROID HORMONES	=	=	=	↓↓	↓	↓↓	↓↓	?	?	?	?	?	?
INSULIN	=	=	↑	↓	↓↓	↓↓	↓↓	?	?	?	?	?	?
THEOPHYLLINE	?	?	= OR ↑	↑↑	↓↓	↓↓	↑ OR ↓ ACCORDING TO PREDOMINANCE OF EFFECT ON BLOOD EOSINOPHILIA OR ESTROGEN BINDING BY EOSINOPHILS	↑↑	↑↑	?	?	?	?
AGE OF RATS: 10 DAYS	=	=	=	?	↓↓↓	↓↓↓	↓↓↓	?	?	?	?	?	?
INTRALUMINAL ESTROGEN INJECTION INTO ONE UTERINE HORN	=	FOLLOWS LOCAL ESTROGEN LEVELS	DEPENDS ON LOCAL ESTROGEN LEVELS IN UTERINE TISSUE		=	DEPENDS ON ESTROGEN LEVELS IN THE BLOOD	EDEMA DEPENDS ON ESTROGEN LEVELS IN THE BLOOD	?	?	?	?	?	?
H_1 ANTAGONISTS	?	?	?	?	?	=	=	?	?	↓	= OR ↓	?	?
H_2 ANTAGONISTS	?	?	?	?	?	?	=	?	?	=	=	?	?

FIGURE 5. Agents or conditions dissociating the responses to estrogen in the rat uterus. The action of these agents or conditions is shown on the different groups of responses to estrogen, on the estrogen receptors, on estrogen binding, and on the levels of eosinophil leukocytes in the blood, as: =, no interaction or no modification; ↓, moderate decrease; ↓↓, important decrease; ↓↓↓, drastic decrease or complete blockage; ↑, moderate increase; ↑↑, important increase; ?, unknown; ★, data for the rabbit; ☆, data for sheep. For details and references, see text.

migration of eosinophils to the uterus,[177] but degranulates them,[177-180] shortening their life in the uterus, thus, increasing the intensity but decreasing persistence of edema.[177] Progesterone inhibits the late increase in uterine cAMP content[181] and the increase in glycogen,[182] and decreases hyperemia.[183,184] Progesterone completely blocks the increase in cGMP,[59,60] except when it is administered as pretreatment for 4 days.[172] Progesterone does not compete with membrane estrogen receptors,[49,155] but blocks half of the stimulation of type II nuclear receptors.[157]

Histamine has been proposed as involved in estrogen action.[185] H_1 histamine receptor antagonists do not interfere with estrogen-induced uterine eosinophilia,[186] edema,[186,187] or hyperemia,[72,187] although slight decreases were reported for the hyperemic response[73] and the increase in the cervico-vaginal antigen in vitro.[63] H_2 histamine antagonists do not either interfere with the estrogen-induced uterine edema,[187] hypermia,[73,187] or the cervico-vaginal antigen increase in vitro.[63]

Estrogen binding by cytosol-nuclear receptors depends on local estrogen levels in target organs.[3] Accordingly, the uterine genomic responses to estrogen depend on local hormone levels in the uterus.[76,188] On the contrary, estrogen-induced uterine eosinophilia and edema are dependent on estrogen levels in the blood but not in the uterus.[76]

Estrogens induce a massive migration of eosinophils to the uterus, which can be detected as soon as 5 min after the intravenous administration of the hormone to immature animals.[40] Once eosinophils enter uterine extravascular space, they are assumed to mediate several parameters of estrogen stimulation independently from genomic activation, cyclic AMP, or prostaglandins. According to this hypothesis, estrogen-induced uterine edema and other eosinophil-mediated responses should be dependent on the number of eosinophils migrating to the uterus under the effect of hormone stimulation. It has been shown that estrogen-induced uterine eosinophilia and edema are dose-dependent responses[138] and that below the maximal uterine edematous response level, there is always a correlation between uterine eosinophilia and edema,[43,71,76-80,138,154,177] independently from the extent of the genomic response to estrogen. Conditions blocking eosinophil penetration into uterine extravascular space, such as agents blocking cell mobility, young age blood eosinopenia, and eosinopenic hormones (glucocorticoids, thyroid hormones, and insulin), also block the edematous response and other eosinophil-mediated parameters of estrogen stimulation.

Colchicine, an agent blocking cell mobility by a disassembly of the microtubular system, does not interfere with the recognition of uterine blood vessels by eosinophils in the presence of estrogens, but blocks their migration through the endothelial lining to uterine extravascular space,[189] and also blocks uterine edematous responses to estrogen.[189,190] Colchicine does not interfere with estrogen binding by cytosol receptors nor receptor complex translocation to the nucleus.[191]

Young age blood eosinopenia is a physiologic condition under which there exist only few eosinophils in the blood.[136] Under these conditions, estrogen can bind the cytosol receptors and translocate them to the nucleus.[192,193] Estrogen induces the genomic response to the hormone but not uterine eosinophilia or edema.[78]

Glucocorticoids do not modify estrogen binding by eosinophils,[14] but induce a dramatic blood eosinopenia due to a massive migration of eosinophils to lymphoid organs.[194] This limits the availability of eosinophils for migration to the uterus under estrogen stimulation and blocks uterine eosinophilia,[77] edema,[77,152,195,196] the increase in vascular permeability,[152,174] and implantation.[197] Glucocorticoids do not modify estrogen binding by cytosol-nuclear receptors,[157,176,198] estrogen-receptor complex translocation to the nucleus,[157,198] or receptor replenishment.[157] They do not interfere with the genomic responses to estrogen[77,196] or hyperemia,[195] but decrease the cAMP increase[199] and completely block the appearance of type II nuclear receptors.[157,198] Glucocorticoids per se induce an important increase in glycogen content in the uterus, estradiol + cortisol produces a slightly greater increase in uterine glycogen level, reaching the same level as in animals treated with estradiol alone.[77] The glucocorticoid hypersecretion that occurs under treatment with low doses of actinomycin D[200] blocks estrogen-induced water imbibition (actinomycin D per se does not block this response in adrenalectomized animals[43]), but does not counteract the increase in cyclic GMP levels.[166]

Thyroid hormones[80] and insulin[79] were reported to induce a moderate blood eosinopenia, in addition to the decrease in estrogen binding by eosinophils.[79,80,141] Either one of these hormones greatly decrease estrogen-induced uterine eosinophilia[79,80,201] and water imbibition.[79,80] Thyroid hormones or insulin do not modify estrogen binding and retention by the cytosol-nuclear receptors and receptor translocation to the nucleus.[201-204] Thyroid hormones do not interfere with the genomic responses to estrogen;[203,205] insulin slightly potentiates them.[206,207]

Theophylline is a cAMP-phosphodiesterase inhibitor that increases estrogen-induced increase in cAMP,[208] potentiates the action of estradiol on several uterine glycolytic enzymes,[209] and potentiates the increase in glycolytic enzyme content and glucose metabolism induced by cAMP.[68] It has been suggested that cyclic AMP modulates the action of the cytosol-nuclear receptor-mediated mechanism of estrogen action,[210] but theophylline either does not modify or only slightly potentiates uterine genomic responses to estrogen.[154] Theophylline increases estrogen binding by eosinophils,[141,154] but induces blood eosinopenia.[154] The resulting action of theophylline on estrogen-induced uterine water imbibition may be its potentiation[154] or inhibition[211] according to predominance of the effect on blood eosinophil levels or estrogen binding by eosinophils.

VI. PROPOSED MECHANISMS EXPLAINING EOSINOPHIL-MEDIATED RESPONSES TO ESTROGEN IN THE UTERUS

A. Recognition of the Uterus by Eosinophils in the Presence of Estrogen

In addition to the existence of estrogen receptors in the cytoplasm and in (or near) the plasma membrane of the eosinophil leukocytes, saturable estrogen binding was found in the wall of small uterine blood vessels but not in the blood vessels of nontarget organs for estrogens.[45] Figure 6 summarizes the proposed mechanism explaining the specificity of recognition of uterine blood vessel wall by eosinophil leukocytes in the presence of estrogen, based in the existence of these surface receptors.[44] The finding that estrogen-induced uterine eosinophilia and edema are determined by estrogen levels in the blood but not in the uterus[76] supports this hypothesis.

B. Mechanisms for the Eosinophil-Mediated Responses in the Uterus

The proposed mechanisms explaining estrogenic responses mediated by eosinophils are summarized in Figure 7. Following the attachment of eosinophils to uterine vascular endothelium, they penetrate into uterine extravascular space, redistribute within the uterus, undergo well-defined ultrastructural changes, and subsequently degranulate,[38] releasing their content, which includes their specific granules known as peroxidasosomes,[212] and a number of enzymes and agents that are involved in estrogen action in the uterus. There are different ways to interfere with the above process: the migration of eosinophils to the uterus can be blocked by eosinopenic conditions, their penetration into uterine extravascular space can be blocked with colchicine,[189] their redistribution in the uterus can be modified by progesterone,[177] thyroid hormones,[80] or colchicine,[189] and their degranulation is increased by progesterone.[177-180] The above agents selectively interfere with the eosinophil-mediated estrogenic responses.

The description and characterization of the agents that are released from the eosinophils are reviewed in detail elsewhere.[38,42-44,136,212-219] Of these agents, it has been suggested that

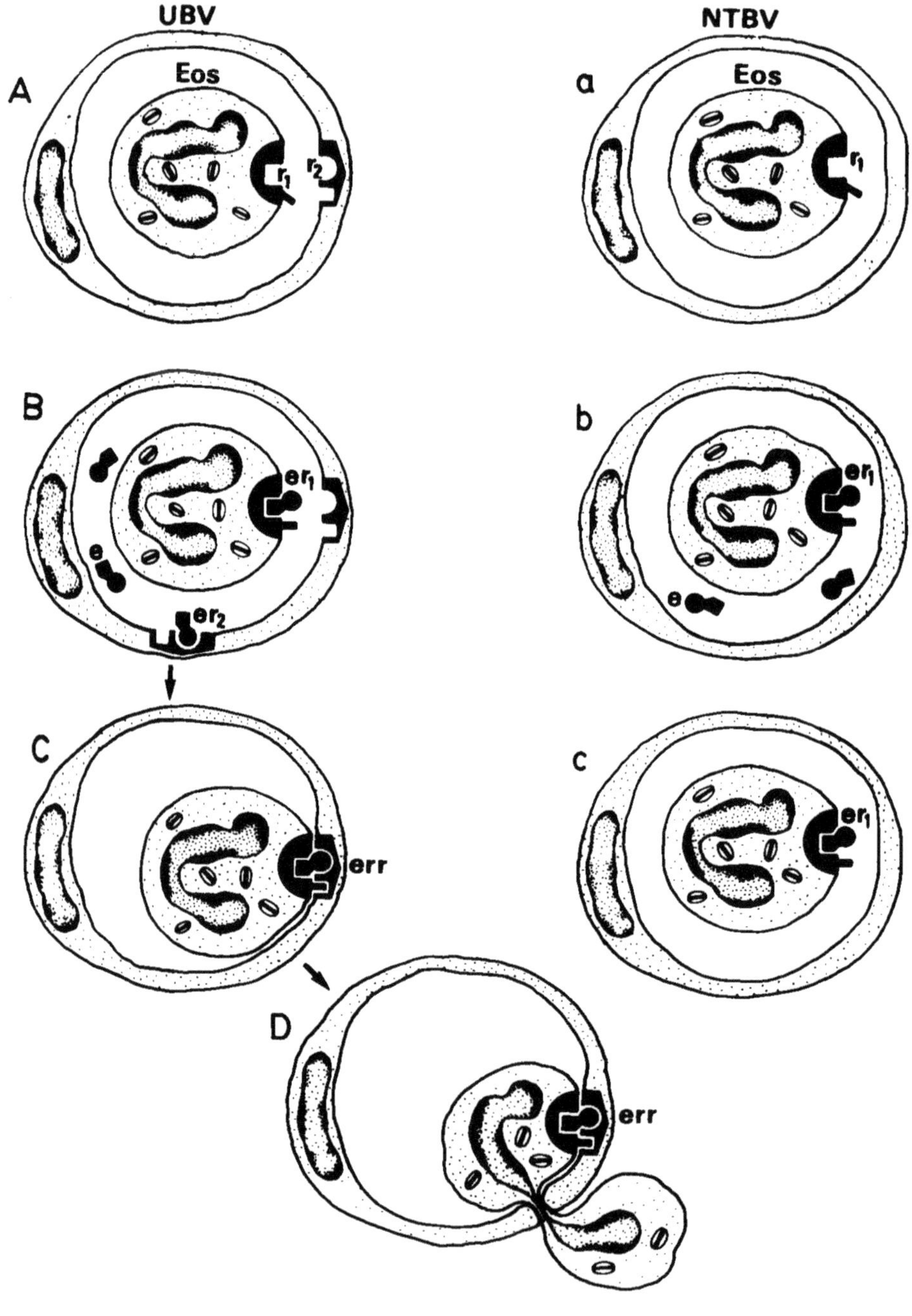

FIGURE 6. Hypothesis explaining estrogen-induced recognition of uterine blood vessels by eosinophil leukocytes. Eos, eosinophil leukocytes; UBV, uterine blood vessel; NTBV, nontarget organ blood vessel; e, estrogen molecule; r_1, receptor for estrogens in the surface of the eosinophil; r_2, receptor for estrogens and/or for eosinophil estrogen-receptor complexes in the uterine blood vessel; er, estrogen-receptor complex in the surface of the eosinophil (er_1) or in the uterine blood vessel (er_2); err, uterine blood vessel and eosinophil estrogen-bireceptor complex, binding the eosinophil to the blood vessel. Eosinophil leukocytes in a uterine blood vessel (A) and in a nontarget organ blood vessel (a). In the presence of estrogen, hormone-receptor complexes are formed with estrogen receptors in the surface of eosinophils (B and b) and in the surface of uterine blood vessels (B). Eosinophil estrogen-receptor complexes recognize unloaded receptors from uterine blood vessels and form estrogen-bireceptor complexes (C). This causes eosinophil attachment to UBV (C) but not to NTBV (c), the initial step for eosinophil migration through endothelial lining in target organs only (D).

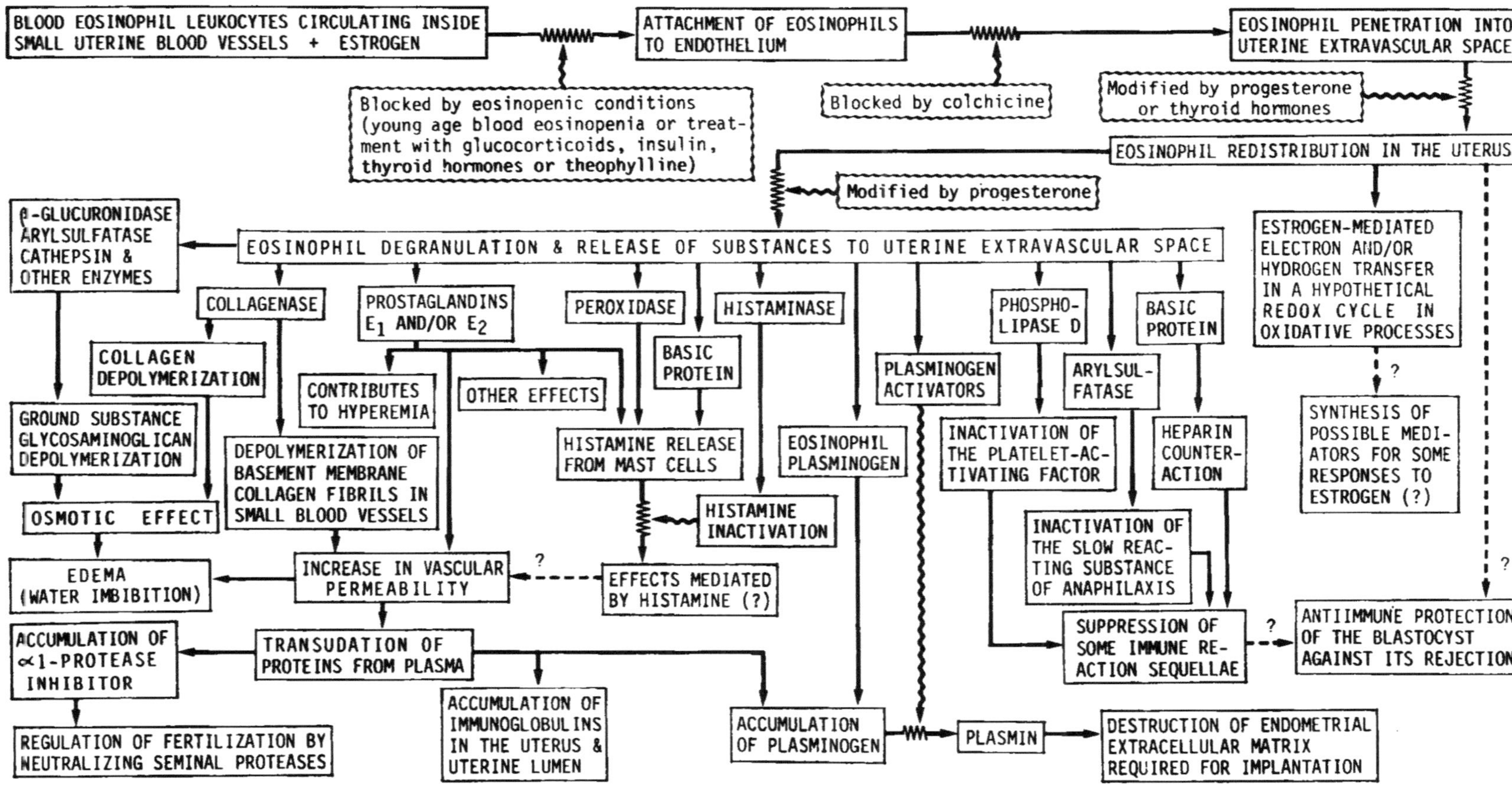

FIGURE 7. Proposed mechanisms explaining estrogenic responses mediated by eosinophils. The sequence of known events explaining the effects that were proposed to be mediated by eosinophils is shown. For details and references, see text.

β-glucuronidase, arylsulfatase, cathepsin, and other enzymes released from eosinophils to uterine stroma may be responsible for the depolymerization of uterine ground substance glycosaminoglycans,[38,41-44] observed after estrogen stimulation.[220] Further, it was reported that eosinophil collagenase degrades type-specific collagen,[221] which is concordant with the biochemical finding of type-specific collagen degradation[222] and the ultrastructural observation that collagen fibers disaggregate and disappear[223] in the uterus after estrogen stimulation. The hypothesis of the mediation of estrogen-induced uterine collagen depolymerization by collagenase released from eosinophils is supported by the observation (shown in Figure 12 of our previous report[38]) that the disappearance of collagen occurs first in the immediate vicinity of uterine eosinophils. Uterine glycosaminoglycan and collagen depolymerization would osmotically increase uterine extracellular water content (edema, also called water imbibition). Released eosinophil collagenase may also cause a depolymerization and disaggregation of basement membrane collagen fibrils in small uterine blood vessels, noted following estrogen treatment.[223] Perhaps this contributes to the increase in vascular permeability induced by estrogen, in addition to the release of prostaglandin E_1 and/or E_2 from eosinophils.[224]

The increase in vascular permeability, in addition to its contribution to edema, causes transudation of proteins from plasma that accumulate in uterine tissues (and may be transferred into uterine lumen).[152,174] Among them are the immunoglobulins IgA, IgM, and IgG (of which IgG but not IgA or IgM moves to the lumen),[152] α1-protease inhibitor,[225] and plasminogen.[174] The α1-protease inhibitor accumulation is not blocked by puromycin;[225] the plasminogen uptake by the uterus is not counteracted by either puromycin or by indomethacin but is completely blocked by prednisolone.[174] The immunoglobulin accumulation is blocked by dexamethasone.[152] In addition, estriol is a stronger estrogen than estradiol for the immunoglobulin accumulation in the uterus.[152] The interference with glucocorticoids but not with protein synthesis or prostaglandin biosynthesis inhibitors and the higher potency of estriol is in agreement with the hypothesis of the mediation of these effects by eosinophils. It was suggested that α1-protease inhibitor regulates fertilization by neutralizing seminal proteases,[226] and that plasma plasminogen, in addition to the eosinophil plasminogen,[225] when transformed to plasmin under the effect of plasminogen activators (eosinophils were also shown to possess a plasminogen activator[227]), is required for the destruction of the extracellular matrix in the endometrium required for implantation.[228,229]

The eosinophil basic protein (similarly to what was shown for peritoneal mast cells[230]), the eosinophil peroxidase,[231] and/or prostaglandins E_1 and/or E_2[224] may cause the release of histamine from uterine mast cells. The role of histamine is still not clear although it was proposed that it may be involved in decidualization and ovum implantation,[197] in the stimulation of superoxide anion production by eosinophils,[232] and may contribute to the increase in vascular permeability. The activity of the histamine released from mast cells may, in turn, be limited by degradation by the eosinophil histaminase.

Eosinophils are known to release a few agents that suppress some immune reaction sequelae: the eosinophil phospholipase D inactivates the platelet-activating factor;[233] the eosinophil arylsulfatase inactivates the slow-reacting substance of anaphylaxis,[234,235] although the slow-reacting substance bioactivity of leukotrienes C_4 and D_4 is also decreased by eosinophil peroxidase;[236] and the eosinophil basic protein counteracts the heparin of mast cells.[237] The action of the above agents released from eosinophils and the known cortisol-induced migration of eosinophils to lymphoid tissues,[194] where they were proposed to have a role in the antiimmune action of glucocorticoids,[42,179,194] suggest the possibility that eosinophils in the uterus may protect blastocyst from its rejection as a homograft.[42] Indeed, intrauterine skin homograft rejection can be prevented by estrogen administration.[238]

Finally, eosinophil peroxidase was proposed to participate in the activity of estrogen as an intermediate hydrogen and electron carrier[38] in which hydrogen peroxide may act as a terminal hydrogen acceptor in a hypothetical redox cycle.[239] This possibility awaits further investigations, as it may result in the synthesis of possible mediators for some responses to estrogen.

VII. A MODEL EXPLAINING THE GENERAL BIOLOGICAL MECHANISM OF CELL-CELL OR CELL-TARGET ORGAN RECOGNITION VIA SPECIFIC RECEPTORS

Evidence is discussed in this chapter for the involvement of estrogen receptors in estrogen-induced migration of eosinophils into uterine stroma, where they mediate a separate group of responses to hormone stimulation through an independent mechanism. Similarly, glucocorticoids induce a massive migration of eosinophil leukocytes from the blood to the thymus, spleen, and lymph nodes,[194] where these cells undergo changes similar to those described for eosinophils in the uterus under estrogen stimulation.[240] Accordingly, it was suggested that eosinophil leukocytes in lymphoid organs might be involved in some effects of glucocorticoids, including immunoregulation.[42,179,194] It is not yet known whether glucocorticoid-induced migration of eosinophils to lymphoid organs is mediated by the same mechanism proposed for estrogen-induced eosinophil mobilization or by the release of eosinophilotactic substances under the effect of glucocorticoid stimulation. The existence of glucocorticoid receptors in eosinophils[241] favors the first possibility.

Eosinophils also migrate to tissues where histamine (in the horse)[242] or an eosinophilotactic substance produced by T lymphocytes[243,244] is present. Neutrophil leukocytes are known to migrate to the vagina under the effect of progesterone. Mesenteric lymph node lymphoblasts or lymphocytes migrate to the mammary gland during the latter stages of pregnancy and during lactation.[245-248] Other blood cells are known to migrate to target tissues under the effect of specific stimuli. Embryonic cells also migrate to specific locations at precise stages of development under the effect of known or unknown substances. Based on the above grounds summarized in Figure 8, we proposed that the simultaneous presence of hormone receptors in the migrating cells and in small blood vessels (or other cell types) of the target tissue explain the general biological phenomenon of tissue specificity for cell migration to target tissues for different migrating cells under various specific stimuli.[44] Therefore, the migrating cells could accomplish their function in target tissues where their presence is required.

VIII. CONCLUDING REMARKS

For a long time, most studies on the mechanisms of estrogen action have considered the uterus as a single cell-type organ, with only one receptor system and one mechanism of action for the hormone possible. The uterus, however, consists of many different cell types, and each one behaves differently. The discovery of differences in the action of estrogen, implying multiple and independent mechanisms of action for the hormone, suggests that all mechanisms should be considered in any biochemical, physiological, or clinical study related with steroid hormone action. Scientists should not close their eyes to them if an understanding of the complete process is intended.

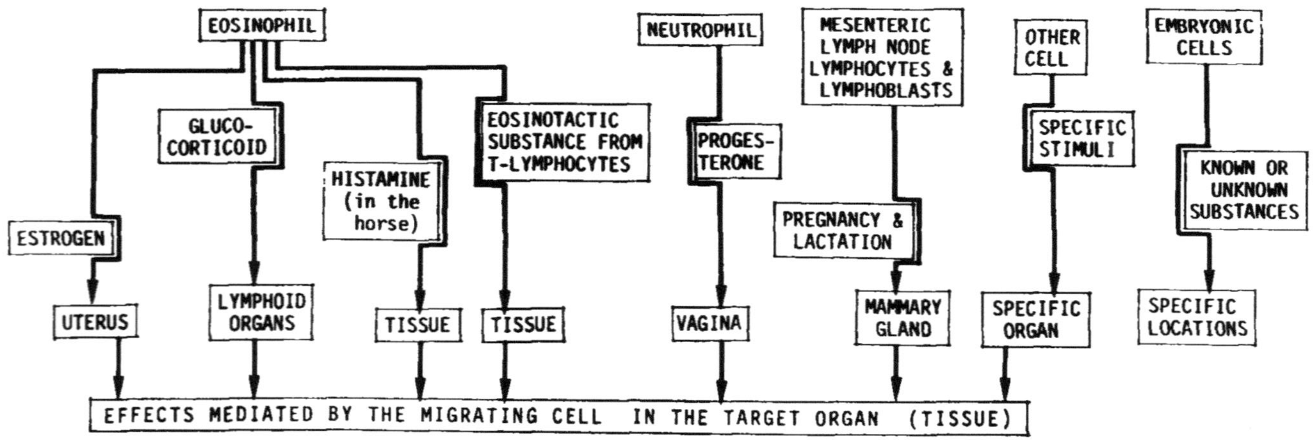

FIGURE 8. General model for various "hormone-migrating cell-target organ" systems. Some examples are shown for hormone (or tactic substance)-target cell interactions that determine target cell-specific migration to target organs (or tissues), so that target cells could accomplish their function at these locations. For details and references, see text.

REFERENCES

1. **Glascock, R. F. and Hoekstra, W. G.,** Selective accumulation of tritium-labeled hexestrol by the reproductive organs of immature goats and sheep, *Biochem. J.*, 72, 673, 1959.
2. **Jensen, E. V. and Jacobson, H. I.,** Basic guides to the mechanism of estrogen action, *Recent Prog. Horm. Res.*, 18, 387, 1962.
3. **Jensen, E. V. and DeSombre, E. R.,** Mechanism of action of the female sex hormones, *Annu. Rev. Biochem.*, 41, 203, 1972.
4. **Baulieu, E. E., Atger, M., Best-Belpomme, M., Corvol, P., Courvalin, J. C., Mester, J., Milgrom, E., Robel, P., Rochefort, H., and De Catalogne, D.,** Steroid hormone receptors, *Vitam. Horm.*, 33, 649, 1975.
5. **Jungblut, P. W., Hughes, A., Gaues, J., Kallweit, E., Maschler, I., Parl, F., Sierralta, W., Szendro, P. I., and Wagner, R. K.,** Mechanisms involved in the regulation of steroid receptor levels, *J. Steroid Biochem.*, 11, 273, 1979.
6. **Lévy, C., Mortel, R., Eychenne, B., Robel, P., and Baulieu, E. E.,** Unoccupied nuclear oestrogen-receptor sites in normal human endometrium, *Biochem. J.*, 185, 733, 1980.
7. **Martin, P. M. and Sheridan, P. J.,** Towards a new model for the mechanism of action of steroids, *J. Steroid Biochem.*, 16, 215, 1982.
8. **Stumpf, W. E. and Roth, L. J.,** High resolution autoradiography with dry mounted, freeze-dried frozen sections. Comparative study of six methods using two diffusible compounds [^{3}H]-estradiol and [^{3}H]-mesobilirubinogen, *J. Histochem. Cytochem.*, 14, 274, 1966.
9. **Tchernitchin, A.,** Autoradiographic study of (6,7-^{3}H) oestradiol-17β incorporation into rat uterus, *Steroids*, 10, 661, 1967.
10. **Stumpf, W. E.,** Subcellular distribution of ^{3}H-estradiol in the rat uterus by quantitative autoradiography. A comparison between ^{3}H-estradiol and ^{3}H-norethynodrel, *Endocrinology*, 83, 777, 1968.
11. **Stumpf, W. E.,** Too much noise in the autoradiogram?, *Science*, 163, 958, 1969.
12. **Stumpf, W. E.,** Autoradiographic techniques for the localization of hormones and drugs at the cellular and subcellular level, *Acta Endocrinol. Suppl.*, 153, 205, 1971.
13. **Stumpf, W. E., Baerwaldt, C., and Sar, M.,** Autoradiographic cellular and subcellular localization of sexual steroids, in *Basic Actions of Sex Steroids on Target Organs*, Hubinont, P. O., Leroy, F., and Galand, P., Eds., S. Karger, Basel, 1971, 3.
14. **Tchernitchin, A.,** Radioautographic study of the effect of estradiol-17β, estrone, estriol, progesterone, testosterone and corticosterone on the *in vitro* uptake of 2,4,6,7-^{3}H estradiol-17β by uterine eosinophils of the rat, *Steroids*, 19, 575, 1972.
15. **Tchernitchin, A. and Chandross, R.,** *In vivo* uptake of estradiol-17β by the uterus of the mature rat, *J. Steroid Biochem.*, 4, 41, 1973.
16. **Tchernitchin, A.,** Radioautography of estradiol-17β *in vivo* under different hormonal conditions, *Eur. J. Obstet. Gynecol. Reprod. Biol.*, 4 (Suppl.), S99, 1974.
17. **Uriel, J., Bouillon, D., Aussel, C., and Loisillier, F.,** Localisation autoradiographique des cellules fixatrices de catécholamines chez le rat et chez l'homme, *C. R. Acad. Sci. Paris*, 279D, 575, 1974.
18. **Strum, J. M., Feldman, D., Taggart, B., Marver, D., and Edelman, I. S.,** Autoradiographic localization of corticosterone receptors (type III) to the collecting tubule of the rat kidney, *Endocrinology*, 97, 505, 1975.
19. **Stumpf, W. E.,** Techniques for the autoradiography of diffusible compounds, in *Methods in Cell Biology*, Vol. 13, Prescott, D. M., Ed., Academic Press, New York, 1976, 171.
20. **Tchernitchin, A., Wenk, E. J., Southren, A. L., and Vittek, J.,** Radioautography of dexamethasone in rabbit gingiva and buccal mucosa, *J. Dent. Res.*, 59, 2100, 1980.
21. **Tchernitchin, A. N. and Tchernitchin, N.,** Glucocorticoid localization by radioautography in the rabbit eye following systemic administration of ^{3}H-dexamethasone, *Experientia*, 37, 1120, 1981.
22. **Holderegger, C., Keefer, D. A., Babler, W., and Langman, J.,** Quantitative autoradiographic analysis of in situ estrogen concentration by individual cell types of mouse target organs: dose-uptake study, *Biol. Reprod.*, 25, 719, 1981.
23. **James, C. R. H. and Cotlier, E.,** Fate of insulin in the retina: an autoradiographic study, *Br. J. Ophthalmol.*, 67, 80, 1983.
24. **Tchernitchin, A.,** Radioautographic analysis of 6,7-^{3}H-estradiol-17β uptake in rat uterus following extraction of endogenous estrogens, *Steroids*, 15, 799, 1970.
25. **Tchernitchin, A., Tchernitchin, X., and Bongiovanni, A. M.,** Iodination of estrogen receptors in rat uterine eosinophils, *J. Steroid Biochem.*, 4, 401, 1973.
26. **Tchernitchin, A.,** Effect of superfusion with human male serum, bovine serum albumin or non radioactive estrogens on the retention of tritiated estradiol-17β and estriol by the rat uterus, *J. Steroid Biochem.*, 5, 481, 1974.

27. **Tchernitchin, A.,** Effect of progesterone on the in vivo binding of estrogens by uterine cells, *Experientia*, 32, 1069, 1976.
28. **Tchernitchin, A., Tchernitchin, X., Collao, C., and Rodríguez, A.,** Dynamics of estrogen binding by uterine cells in vivo, *Experientia*, 34, 134, 1978.
29. **Mena, M. A., Rodríguez, A., Unda, C., and Tchernitchin, A.,** The eosinophil-estrogen receptor system in the Syrian hamster uterus, *IRCS Med. Sci.*, 5, 170, 1977.
30. **Tchernitchin, A., Hasbún, J., Peña, G., and Vega, S.,** Radioautographic analysis of 6,7-^{3}H estradiol 17β uptake in human endometrium, *Proc. Soc. Exp. Biol. Med.*, 137, 108, 1971.
31. **Tchernitchin, A., Tseng, L., Stumpf, W. E., and Gurpide, E.,** Radioautographic study of human endometrium superfused with estradiol-17β, estrone, estriol and progesterone, *J. Steroid Biochem.*, 4, 451, 1973.
32. **Brokelmann, J.,** Peroxidase-associated binding of estradiol by rat uterus, *J. Histochem. Cytochem.*, 17, 394, 1969.
33. **Geuskens, M., Burglen, M. J., and Uriel, J.,** *In vitro* binding of ^{3}H-estradiol to eosinophil and neutrophil granulocytes in various tissues (normal and neoplastic) of newborn and adult rats, *Virchows Arch. B, Cell Pathol.*, 24, 67, 1977.
34. **Klebanoff, S. J.,** Estrogen binding by leukocytes during phagocytosis, *J. Exp. Med.*, 145, 983, 1977.
35. **Lee, S. H.,** Uterine epithelial and eosinophil estrogen receptors in rats during the estrous cycle, *Histochemistry*, 74, 443, 1982.
36. **Tchernitchin, A., Tchernitchin, X., Robel, P., and Baulieu, E. E.,** Liaison de l'oestradiol dans les leucocytes polinucléaires éosinophiles humains, *C. R. Acad. Sci. Paris*, 280 D, 1477, 1975.
37. **Tchernitchin, A. and Tchernitchin, X.,** Characterization of the estrogen receptors in the uterine and blood eosinophil leukocytes, *Experientia*, 32, 1240, 1976.
38. **Tchernitchin, A.,** Fine structure of rat uterine eosinophils and the possible role of eosinophils in the mechanism of estrogen action, *J. Steroid Biochem.*, 4, 277, 1973.
39. **Tchernitchin, A. N.,** Uterine eosinophils and their role in the mechanism of action of estrogens, in *Proc. 7th World Congr. Obstet. Gynecol., Moscow, Int. Congr. Ser. 279*, Excerpta Medica, Amsterdam, 1973, 12.
40. **Tchernitchin, A., Roorijck, J., Tchernitchin, X., Vandenhende, J., and Galand, P.,** Dramatic early increase in uterine eosinophils after oestrogen administration, *Nature (London)*, 248, 142, 1974.
41. **Tchernitchin, A., Tchernitchin, X., and Galand, P.,** New concepts on the action of oestrogens in the uterus and the role of the eosinophil receptor system, *Differentiation*, 5, 145, 1976.
42. **Tchernitchin, A.,** The role of eosinophil receptors in the non-genomic response to oestrogens in the uterus, *J. Steroid Biochem.*, 11, 417, 1979.
43. **Tchernitchin, A. N. and Galand, P.,** Dissociation of separate mechanisms of estrogen action by actinomycin D, *Experientia*, 38, 511, 1982.
44. **Tchernitchin, A. N.,** Eosinophil-mediated non-genomic parameters of estrogen stimulation: a separate group of responses mediated by an independent mechanism, *J. Steroid Biochem.*, 19, 95, 1983.
45. **Tchernitchin, X., Tchernitchin, A., and Galand, P.,** Dynamics of eosinophils in the uterus after oestrogen administration, *Differentiation*, 5, 151, 1976.
46. **Colburn, P. and Buonassisi, V.,** Estrogen binding sites in endothelial cell cultures, *Science*, 201, 817, 1978.
47. **Pietras, R. J. and Szego, C. M.,** Specific binding sites for oestrogen at the outer surface of isolated endometrial cells, *Nature (London)*, 265, 69, 1977.
48. **Müller, R. E., Johnston, T. C., and Wotiz, H. H.,** Binding of estradiol to purified uterine plasma membranes, *J. Biol. Chem.*, 254, 7895, 1979.
49. **Pietras, R. J. and Szego, C. M.,** Partial purification and characterization of oestrogen receptors in subfractions of hepatocyte plasma membranes, *Biochem. J.*, 191, 743, 1980.
50. **Nenci, I., Fabris, G., Marzola, A., and Marchetti, E.,** The plasma membrane as an additional level of steroid-cell interaction, *J. Steroid Biochem.*, 15, 231, 1981.
51. **Watson, G. H. and Muldoon, T. G.,** Microsomal estrogen receptors in rat uterus and anterior pituitary, *Fed. Proc. Fed. Am. Soc. Exp. Biol.*, 36, 912, 1977.
52. **Clark, J. H. and Peck, E. J., Jr., Eds.,** *Female Sex Steroids*, Springer-Verlag, Berlin, 1979.
53. **Sutherland, R. L., Murphy, L. C., Foo, M. S., Green, M. D., Whybourne, A. M., and Krozowski, Z. S.,** High-affinity anti-oestrogen binding site distinct from the oestrogen receptor, *Nature (London)*, 288, 273, 1980.
54. **Faye, J. C., Lasserre, B., and Bayard, F.,** Antiestrogen specific, high affinity saturable binding sites in rat uterine cytosol, *Biochem. Biophys. Res. Commun.*, 93, 1225, 1980.
55. **Murphy, L. C. and Sutherland, R. L.,** A high-affinity binding site for the antioestrogens, tamoxifen and CI 628, in the immature rat uterine cytosol which is distinct from the oestrogen receptor, *J. Endocrinol.*, 91, 155, 1981.

56. **Gulino, A. and Pasqualini, J. R.,** Heterogeneity of binding sites for tamoxifen and tamoxifen derivatives in estrogen target and nontarget fetal organs of guinea pig, *Cancer Res.*, 42, 1913, 1982.
57. **Flandroy, L. and Galand, P.,** Oestrogen-induced increase in uterine cGMP content: a true hormonal action?, *Mol. Cell. Endocrinol.*, 13, 281, 1979.
58. **Flandroy, L. and Galand, P.,** Changes in cGMP and cAMP content in the estrogen-stimulated rat uterus: temporal relationship with other parameters of hormonal stimulation, *J. Cyclic Nucleotide Res.*, 4, 145, 1978.
59. **Kuehl, F. A., Jr., Ham, E. A., Zanetti, M. E., Sanford, C. H., Nicol, S. E., and Goldberg, N. D.,** Estrogen-related increase in uterine guanosine 3′,5′-cyclic monophosphate level, *Proc. Natl. Acad. Sci. U.S.A.*, 71, 1866, 1974.
60. **Kang, Y. H., Sahai, A., Criss, W. E., and West, W. L.,** Ultracytochemical localization of estrogen-stimulated guanylate cyclase in rat uterus, *J. Histochem. Cytochem.*, 30, 331, 1982.
61. **Szego, C. M.,** Lysosomal function in nucleocytoplasmic communication, in *Lysosomes in Biology and Pathology*, Vol. 4, Dingle, J. T. and Dean, R. T., Eds., North-Holland, Amsterdam, 1975, 385.
62. **Hechter, O., Yoshinaga, K., Cohn, C., Dodd, P., and Halkerston, I. D. K.,** In vitro stimulatory effects of nucleotides and nucleosides on biosynthetic processes in castrated rat uterus, *Fed. Proc. Fed. Am. Soc. Exp. Biol.*, 24, 384, 1965.
63. **Kvinnsland, S.,** Estradiol-17β and cAMP: in vitro studies on the cervicovaginal epithelium of neonatal mice, *Cell Tissue Res.*, 175, 325, 1976.
64. **Resnik, R., Battaglia, F. C., Makowski, E. L., and Meschia, G.,** The effect of actinomycin D on estrogen-induced uterine blood flow, *Am. J. Obstet, Gynecol.*, 122, 273, 1975.
65. **Penney, L. L., Frederick, R. J., and Parker, G. W.,** 17β-Estradiol stimulation of uterine blood flow in oophorectomized rabbits with complete inhibition of uterine ribonucleic acid synthesis, *Endocrinology*, 109, 1672, 1981.
66. **Hechter, O. and Halkerston, I. D. K.,** On the action of mammalian hormones, in *The Hormones*, Vol. 5, Pincus, G., Thimann, K. V., and Astwood, E. A., Eds., Academic Press, New York, 1964, 697.
67. **Singhal, R. L. and Lafreniere, R.,** Induction of uterine phosphofruktokinase by cyclic 3′,5′-monophosphate, *Endocrinology*, 87, 1099, 1970.
68. **Singhal, R. L. and Lafreniere, R. T.,** Metabolic control mechanisms in mammalian systems. XV. Studies on the role of adenosine 3′,5′-monophosphate in estrogen action on the uterus, *J. Pharmacol. Exp. Ther.*, 180, 86, 1972.
69. **Singhal, R. L.,** Cyclic adenosine 3′,5′-monophosphate and estrogenic stimulation of uterine metabolism, *Adv. Pharmacol. Chemother.*, 11, 99, 1973.
70. **Kvinnsland, S.,** Effects of d-propranolol and estradiol on the cervicovaginal epithelium. Correlation with adenylate cyclase activity, *Cell Tissue Res.*, 173, 325, 1976.
71. **Tchernitchin, A., Tchernitchin, X., Rodríguez, A., Mena, M. A., Unda, C., Mairesse, N., and Galand, P.,** Effect of propranolol on various parameters of estrogen stimulation in the rat uterus, *Experientia*, 33, 1536, 1977.
72. **Resnik, R., Killam, A. P., Barton, M. D., Battaglia, F. C., Makowski, E. L., and Meschia, G.,** The effect of various vasoactive compounds upon the uterine vascular bed, *Am. J. Obstet. Gynecol.*, 125, 201, 1976.
73. **Phaily, S. and Senior, J.,** Modification of oestrogen-induced uterine hyperaemia by drugs in the ovariectomized rat, *J. Reprod. Fertil.*, 53, 91, 1978.
74. **Soto-Feine, N., Petersen, V., and Tchernitchin, A. N.,** Are prostaglandins involved in early estrogen action?, *Experientia*, 37, 1351, 1981.
75. **Katz, M. and Creasy, R. K.,** Uterine blood flow distribution after indomethacin infusion in the pregnant rabbit, *Am. J. Obstet. Gynecol.*, 140, 430, 1981.
76. **Tchernitchin, A. N. and Galand, P.,** Oestrogen levels in the blood, not in the uterus, determine uterine eosinophilia and oedema, *J. Endocrinol.*, 99, 123, 1983.
77. **Tchernitchin, A., Rooryck, J., Tchernitchin, X., Vandenhende, J., and Galand, P.,** Effects of cortisol on uterine eosinophilia and other oestrogenic responses, *Mol. Cell. Endocrinol.*, 2, 331, 1975.
78. **Tchernitchin, A. N., López-Solís, R. O., Cartes, R., Rodríguez, A., Mena, M. A., and Unda, C.,** Developmental changes of estrogenic responses in the rat uterus, *J. Steroid Biochem.*, 13, 1369, 1980.
79. **Steinsapir, J., Rojas, A. M., Alarcón, O., and Tchernitchin, A. N.,** Effect of insulin and epinephrine on some early oestrogenic responses in the rat uterus, *Acta Endocrinol.*, 99, 263, 1982.
80. **Steinsapir, J., Rojas, A. M., Mena, M., and Tchernitchin, A. N.,** Effects of thyroid hormone on some uterine responses to estrogen, *Endocrinology*, 110, 1773, 1982.
81. **Petersen, V., Soto-Feine, N., and Tchernitchin, A. N.,** Effect of indomethacin on estrogenic response, *IRCS Med. Sci.*, 8, 478, 1980.
82. **Ui, H. and Mueller, G.,** The role of RNA synthesis in early estrogen action, *Proc. Natl. Acad. Sci. U.S.A.*, 50, 256, 1963.

83. **Hamilton, T. H.,** Sequence of RNA and protein synthesis during early estrogen action, *Proc. Natl. Acad. Sci. U.S.A.*, 51, 83, 1964.
84. **Means, A. R. and Hamilton, T. H.,** Evidence for derepression of nuclear protein synthesis and concomitant stimulation of nuclear RNA synthesis during early estrogen action, *Proc. Natl. Acad. Sci. U.S.A.*, 56, 686, 1966.
85. **Singhal, R. L., Thomas, J. A., and Parukelar, M. R.,** Failure of β-adrenergic blocking agents to alter estrogenic induction of uterine enzymes, *Life Sci.*, 11(1), 255, 1972.
86. **Zor, U., Koch, Y., Lamprecht, S. A., Ausher, J., and Lindner, R. H.,** Mechanism of oestrogen action on the rat uterus: independence of cyclic AMP, Prostaglandin E_2 and β-adrenergic mediation, *J. Endocrinol.*, 58, 525, 1973.
87. **Katzenellenbogen, B. S. and Katzenellenbogen, J. A.,** Antiestrogens: studies using an in vitro estrogen-responsive uterine system, *Biochem. Biophys. Res. Commun.*, 50, 1152, 1973.
88. **Katzenellenbogen, B. S. and Ferguson, E. R.,** Antiestrogen action in the uterus: biological ineffectiveness of nuclear bound estradiol after antiestrogen, *Endocrinology*, 97, 1, 1975.
89. **Bichon, M. and Bayard, F.,** Dissociated effects of tamoxifen and oestradiol-17β on the uterus and liver functions, *J. Steroid Biochem.*, 10, 105, 1979.
90. **McCormack, S. A. and Glasser, S. R.,** Differential response of individual uterine cell types from immature rats treated with estradiol, *Endocrinology*, 106, 1634, 1980.
91. **Katzenellenbogen, B. S.,** Dynamics of steroid hormone receptor action, *Annu. Rev. Physiol.*, 42, 17, 1980.
92. **Katzenellenbogen, B. S., Bhakoo, H. S., Hayer, J. R., and Schmidt, W. N.,** Uterine estrogen-induced protein: an index of uterine sensitivity to hormones, in *Steroid Induced Uterine Proteins*, Beato, M., Ed., Elsevier/North-Holland, Amsterdam, 1980, 267.
93. **Eriksson, H. A., Hardin, J. W., Markaverich, B., Upchurch, S., and Clark, J. H.,** Estrogen binding in the rat uterus: heterogeneity of sites and relation to uterotrophic response, *J. Steroid Biochem.*, 12, 121, 1980.
94. **Korach, K. S. and Lamb, J. C., IV,** Estrogen action in the mouse uterus: differential nuclear localization of estradiol in uterine cell types, *Endocrinology*, 108, 1989, 1981.
95. **Black, L. J. and Goode, R. L.,** Evidence for biological action of the antiestrogen LY117018 and tamoxifen by different mechanism, *Endocrinology*, 109, 987, 1981.
96. **King, R. J. B., Townsend, P. T., Siddle, N., Whitehead, M. I., and Taylor, R. W.,** Regulation of estrogen and progesterone receptor levels in epithelium and stroma from pre- and postmenopausal endometria, *J. Steroid Biochem.*, 16, 21, 1982.
97. **Keefer, D. A.,** Dynamics of in situ estrogen uptake by nuclei of individual pituitary and uterine cell types, *Horm. Metab. Res.*, 14, 209, 1982.
98. **Martin, P. M. and Sheridan, P. J.,** Towards a new model for the mechanism of action of steroids, *J. Steroid Biochem.*, 16, 215, 1982.
99. **Funder, J. W. and Barlow, J. W.,** Heterogeneity of glucocorticoid receptors, *Cir. Res.*, 46(Suppl. I), I-83, 1980.
100. **Do, Y. S., Loose, D. S., and Feldman, D.,** Heterogeneity of glucocorticoid binders: a unique and a classical dexamethasone-binding site in bovine tissues, *Endocrinology*, 105, 1055, 1979.
101. **Melnykovych, G., Matthews, E., Gray, S., and Lopez, I.,** Inhibition of cholesterol biosynthesis in Hela cells by glucocorticoids, *Biochem. Biophys. Res. Commun.*, 71, 506, 1976.
102. **Tchernitchin, A. N., Tchernitchin, N., Anguita-Salas, J., Cañas-Kramarosky, M. M., Grunert, G., and Mena, M. A.,** Progesterone-dexamethasone antagonism in the rabbit eye, *IRCS Med. Sci.*, 10, 261, 1982.
103. **Epifanova, O. I.,** Mitotic cycles in estrogen-treated mice. A radioautographic study, *Exp. Cell Res.*, 42, 562, 1966.
104. **Martin, L. and Finn, C.,** Hormonal regulation of cell division in epithelial and connective tissues of the mouse uterus, *J. Endocrinol.*, 41, 363, 1968.
105. **Leroy, F., Galand, P., and Chretien, J.,** The mitogenic action of ovarian hormones on the uterine and the vaginal epithelium during the oestrous cycle of the rat. A radioautographic study, *J. Endocrinol.*, 45, 441, 1969.
106. **Smith, J. A., Martin, L., King, R. J. B., and Vertes, M.,** Effects of estradiol-17β and progesterone on total and nuclear-protein synthesis in epithelial and stromal tissues of the mouse uterus, and of progesterone on the ability of these tissues to bind estradiol-17β, *Biochem. J.*, 119, 773, 1970.
107. **Tachi, C., Tachi, S., and Lindner, H. R.,** Modification by progesterone of estradiol-induced cell proliferation, RNA synthesis and oestradiol distribution in the rat uterus, *J. Reprod. Fertil.*, 31, 59, 1972.
108. **Martel, D. and Psychoyos, A.,** Different responses of rat endometrial epithelium and stroma to induction of estradiol binding sites by progesterone, *J. Reprod. Fertil.*, 64, 387, 1982.

109. **Alberga, A. and Baulieu, E. E.,** Binding of estradiol in castrated rat endometrium in vivo and in vitro, *Mol. Pharmacol.*, 4, 311, 1968.
110. **Tseng, L. and Gurpide, E.,** Effect of estrone and progesterone on the nuclear uptake of estradiol by slices of human endometrium, *Endocrinology*, 93, 245, 1973.
111. **Pietras, R. J. and Szego, C. M.,** Steroid hormone-responsive, isolated endometrial cells, *Endocrinology*, 96, 946, 1975.
112. **Vladimirsky, F., Chen, L., Amsterdam, A., Zor, U., and Lindner, H. R.,** Differentiation of decidual cells in cultures of rat endometrium, *J. Reprod. Fertil.*, 49, 61, 1977.
113. **Williams, D. and Gorski, J.,** Preparation and characterization of free cell suspensions from the immature rat uterus, *Biochemistry*, 12, 297, 1973.
114. **Fleming, H., Namit, C., and Gurpide, E.,** Estrogen receptors in epithelial and stromal cells of human endometrium in culture, *J. Steroid Biochem.*, 12, 169, 1980.
115. **Fleming, H. and Gurpide, E.,** Rapid fluctuations in the levels of specific estradiol binding sites in endometrial cells in culture, *Endocrinology*, 108, 1744, 1981.
116. **Soto-Feine, N. and Tchernitchin, A. N.,** Glucocorticoid receptors in dental pulp: a preliminary report, *J. Endodontics*, 8, 136, 1982.
117. **Collao, C. and Tchernitchin, A.,** Effect of sulphydryl group blockage on estrogen binding by the receptors of the uterine eosinophils, *IRCS Med. Sci.*, 4, 87, 1976.
118. **Mena, M. A., Unda, C., Rodríguez, A., and Tchernitchin, A. N.,** Effect of urea on estrogen binding by rat uterine eosinophils in vitro, *IRCS Med. Sci.*, 9, 73, 1981.
119. **Mena, M. A., Unda, C., Rodríguez, A., and Tchernitchin, A. N.,** Impairment of estrogen binding by rat uterine eosinophils under the effect of tryptophan residues oxidation in vitro, *IRCS Med. Sci.*, 9, 74, 1981.
120. **Maurer, W. and Primbsch, E.,** Grösse der beta-Selbstabsorption bei der ^{3}H-Autoradiographie, *Exp. Cell Res.*, 33, 8, 1964.
121. **Kopriwa, B. M. and Leblond, C. P.,** Improvements in the coating technique of radioautography, *J. Histochem. Cytochem.*, 10, 269, 1962.
122. **Gersh, I.,** The Altmann technique for fixation by drying while freezing, *Anat. Rec.*, 53, 309, 1932.
123. **Tchernitchin, A., Wenk, E. J., Weinstein, B. I., Dunn, M. W., Gordon, G. G., and Southren, A. L.,** Radioautography of dexamethasone in rabbit eye tissues, *Metab. Pediatr. Ophthalmol.*, 3, 37, 1979.
124. **Tchernitchin, N., Anguita-Salas, J., Cañas-Kramarosky, M. M., Cartes, R., and Tchernitchin, A. N.,** Autoradiography of ^{3}H-dexamethasone following topical ophthalmic administration, *IRCS Med. Sci.*, 9, 887, 1981.
125. **Jansen, M. T.,** Testing the performance of a simple freeze-drying apparatus for histochemical purposes, *Exp. Cell Res.*, 7, 318, 1954.
126. **Baillie, A. H., Ferguson, M. M., and Hart, D. M., Eds.,** *Developments in Steroid Histochemistry*, Academic Press, London, 1966, chap. 1.
127. **Scatchard, G.,** The attractions of proteins for small molecules and ions, *Ann. N.Y. Acad. Sci.*, 51, 660, 1949.
128. **Garcia, M. and Rochefort, H.,** Evidence and characterization of the binding of two ^{3}H-labeled androgens to the estrogen receptor, *Endocrinology*, 104, 1797, 1979.
129. **Funder, J. W.,** Steroid antagonists: pharmacological effects and physiological roles?, *J. Steroid Biochem.*, 11, 87, 1979.
130. **Bullock, L. P., Bardin, C. W., and Ohno, S.,** The androgen insensitive mouse: absence of intranuclear androgen retention in the kidney, *Biochem. Biophys. Res. Commun.*, 44, 1537, 1971.
131. **Sibley, C. H. and Tomkins, G. M.,** Mechanisms of steroid resistence, *Cell*, 2, 221, 1974.
132. **De Paepe, J. C.,** Autoradiographic study of the distribution in mice of oestrogens labelled with carbon-14 and tritium, *Nature (London)*, 185, 264, 1960.
133. **Mobbs, B. G.,** Localization of tritiated oestradiol-17β within tissues by autoradiography, *J. Endocrinol.*, 27, 129, 1963.
134. **Ullberg, S. and Bengtsson, G.,** Autoradiographic distribution studies with natural oestrogens, *Acta Endocrinol.*, 43, 75, 1963.
135. **Inman, D. R., Banfield, R. E. J., and King, R. J. B.,** Autoradiographic localization of oestrogen in rat tissues, *J. Endocrinol.*, 32, 17, 1965.
136. **Rytomäa, T.,** Organ distribution and histochemical properties of eosinophil granulocytes in rat, *Acta Pathol. Microbiol. Scand.*, 50 (Suppl. 140), 1, 1960.
137. **Bjersing, L. and Borglin, N. E.,** Effect of hormones on incidence of uterine eosinophilia in rats, *Acta Pathol. Microbiol. Scand.*, 60, 27, 1964.
138. **Tchernitchin, A., Tchernitchin, X., and Galand, P.,** Correlation of estrogen-induced uterine eosinophilia with other parameters of estrogen stimulation, produced with estradiol-17β and estriol, *Experientia*, 31, 993, 1975.

139. **Anderson, W. A., DeSombre, E. R., and Kang, Y.-H.,** Estrogen-progesterone antagonism with respect to specific marker protein synthesis and growth by the uterine endometrium, *Biol. Reprod.*, 16, 409, 1977.
140. **Tchernitchin, A.,** Effect of estradiol-17β, estrone, estriol, equilin, progesterone, testosterone and corticosterone on the *in vitro* uptake of tritiated estradiol-17β, estrone and estriol by uterine eosinophils of the rat, *Proc. 5th Int. Congr. Pharmacol., San Francisco, Volunteer Abstr.*, 1972, 231.
141. **Tchernitchin, A.,** *In vitro* uptake and interaction of (^{125}I) thyroxine and estrogens by uterine eosinophils of the rat, *Biol. Reprod.*, 7, 124, 1972.
142. **Tchernitchin, A., Steinsapir, J., Unda, C., Mena, M. A., Rodríguez, A., Hernández, M. R., Soto, N., Tchernitchin, X., Bravo, L., and De la Lastra, M.,** Effect of aminophylline on estrogen binding by uterine eosinophils, *IRCS Med. Sci.*, 5, 456, 1977.
143. **Bergink, E. W.,** Oestriol receptor interactions: their biological importance and therapeutic implications, *Acta Endocrinol. Suppl.*, 233, 9, 1980.
143a. **Tchernitchin, A. N., Rodríguez, A., Mena, M. A., Maturana, M., Gutiérrez, C., Muñoz, P., Astorga, L. A., and Diharasarri, M. J.,** A new radiohistochemical procedure for the evaluation of the affinity of the uterine eosinophil estrogen receptors for estradiol, *IRCS Med. Sci.*, 12, 240, 1984.
144. **Jensen, E. V., Hurst, D. J., DeSombre, E. R., and Jungblut, P. W.,** Sulfhydryl groups and estrogen-receptor interaction, *Science*, 158, 385, 1967.
145. **Puca, G. A. and Bresciani, F.,** Binding activity of estrogen receptors destroyed by iodination, *Nature (London)*, 225, 1251, 1970.
146. **Mena, M. A., Diharasarri, M. J., and Tchernitchin, A. N.,** Effect of guanidinium chloride on estrogen binding by rat uterine eosinophils *in vitro*, *IRCS Med. Sci.*, 10, 521, 1982.
147. **Ruh, T. S., Katzenellenbogen, B. S., Katzenellenbogen, J. A., and Gorski, J.,** Estrone interaction with the rat uterus: in vitro response and nuclear uptake, *Endocrinology*, 92, 125, 1973.
148. **Arvidson, N. G. and Terenius, L.,** Oestrogens and anti-oestrogens show dissociation between early uterine vascular responses and uterotrophic effects in mice, *Acta Endocrinol.*, 100, 290, 1982.
149. **Hisaw, F. L., Jr.,** Comparative effectiveness of estrogens on fluid imbibition and growth of the rat's uterus, *Endocrinology*, 64, 276, 1959.
150. **Hechter, O. and Halkerston, I. D. K.,** Effects of steroid hormones on gene regulation and cell metabolism, *Ann. Rev. Physiol.*, 27, 133, 1965.
151. **Campbell, P. S., Newman, G. A., Loveless, G. C., Wilson, H. J., and Eley, M. H.,** Differential uterine responsivity to diethystilbestrol: apparent bases for contrasting estrogenic potency, *Biol. Reprod.*, 23, 78, 1980.
152. **Sullivan, D. A. and Wira, C. R.,** Hormonal regulation of immunoglobulins in the rat uterus: uterine response to a single estradiol treatment, *Endocrinology*, 112, 260, 1983.
153. **Shelesnyak, M. C.,** Histamine releasing activity of natural estrogens, *Proc. Soc. Exp. Biol. Med.*, 100, 739, 1959.
154. **Steinsapir, J., Rojas, A. M., and Tchernitchin, A. N.,** Theophylline-estrogen interaction in the rat uterus. Role of the ovary, *Am. J. Physiol.*, 242, E121, 1982.
155. **Nenci, I., Marchetti, E., Marzola, A., and Fabris, G.,** Affinity cytochemistry visualizes specific estrogen binding sites on the plasma membrane of breast cancer cells, *J. Steroid Biochem.*, 14, 1139, 1981.
156. **Clark, J. H., Hardin, J. W., Upchurch, S., and Eriksson, H.,** Heterogeneity of estrogen binding sites in the cytosol of the rat uterus, *J. Biol. Chem.*, 253, 7630, 1978.
157. **Markaverich, B. M., Upchurch, S., and Clark, J. H.,** Progesterone and dexamethasone antagonism of uterine growth: a role for a second nuclear binding site for estradiol in estrogen action, *J. Steroid Biochem.*, 14, 125, 1981.
158. **Clewell, W. H., Carson, B. A., and Meschia, G.,** Comparison of uterotrophic and vascular effects of estradiol-17β and estriol in the mature organism, *Am. J. Obstet. Gynecol.*, 129, 384, 1977.
159. **Flandroy, L. and Galand, P.,** In vitro stimulation by estrogen of guanosine 3′,5′-monophosphate accumulation in incubated rat uterus, *Endocrinology*, 106, 1187, 1980.
160. **Stone, G. M. and Bagget, B.,** The effect of some anti-metabolites on the uptake of tritriated estradiol by the vagina of the ovariectomized mouse, *Steroids*, 5, 495, 1965.
161. **Jensen, E. V.,** Mechanism of oestrogen action in relation to carcinogenesis, *Can. Cancer Conf.*, 6, 143, 1966.
162. **Jensen, E. V., Suzuki, T., Numata, M., Smith, S., and DeSombre, E. R.,** Estrogen-binding substances of target tissues, *Steroids*, 13, 417, 1969.
163. **Sarff, M. and Gorski, J.,** Control of estrogen-binding protein concentration under basal conditions and after estrogen administration, *Biochemistry*, 10, 2557, 1971.
164. **Mester, J. and Baulieu, E. E.,** Dynamics of oestrogen-receptor distribution between the cytosol and nuclear fractions of immature rat uterus after oestradiol administration, *Biochem. J.*, 146, 617, 1975.
165. **Horwitz, K. B. and McGuire, W. L.,** Actinomycin D prevents nuclear processing of estrogen receptor, *J. Biol. Chem.*, 253, 6319, 1978.

166. **Nicol, S. E. and Goldberg, N. D.,** Inhibition of estrogen-induced increase in uterine guanosine 3′,5′-cyclic monophosphate levels by inhibitor of protein and RNA synthesis, *Biochemistry*, 15, 5490, 1976.
167. **Flandroy, L. and Galand, P.,** Oestrogen-induced increase in uterine cGMP content in vitro: effects of inhibitors of protein and RNA synthesis, *Mol. Cell. Endocrinol.*, 25, 49, 1982.
168. **Rosenfeld, M. G. and O'Malley, B.,** Steroid hormones: effects on adenylcyclase activity and adenosine 3′5′-monophosphate in target tissues, *Science*, 168, 253, 1970.
169. **Szego, C. M. and Davis, J. S.,** Adenosine 3′5′-monophosphate in rat uterus: acute elevation by estrogen, *Proc. Natl. Acad. Sci. U.S.A.*, 58, 1711, 1967.
170. **Szego, C. M. and Davis, J. S.,** Inhibition of estrogen-induced elevation of cyclic 3′,5′-adenosine monophosphate in rat uterus. I. By β-adrenergic receptor blocking drugs, *Mol. Pharmacol.*, 5, 470, 1969.
171. **Dupont-Mairesse, N., Van Sande, J., Rooryck, J., Fastrez-Boute, N., and Galand, P.,** Mechanism of estrogen action — independence of several responses of the rat uterus from the early increase in adenosine 3′,5′-cyclic monophosphate, *J. Steroid. Biochem.*, 5, 173, 1974.
172. **Kano, T.,** Effects of estrogen and progesterone on adrenoceptors and cyclic nucleotides in rat uterus, *Jpn. J. Pharmacol.*, 32, 535, 1982.
173. **Brandon, J. M.,** Failure of indomethacin and D-2-bromolysergic acid diethylamide to modify the acute oedematous reaction to oestradiol in the uterus of the immature rat, *J. Endocrinol.*, 78, 291, 1978.
174. **Finlay, T. H., Katz, J., Kirsch, L., Levitz, M., Nathoo, S. A., and Seiler, S.,** Estrogen-stimulated uptake of plasminogen by the mouse uterus, *Endocrinology*, 112, 856, 1983.
175. **Ham, E. A., Cirillo, V. J., Zanetti, M. E., and Kuehl, F. A., Jr.,** Estrogen-directed synthesis of specific prostaglandin in uterus, *Proc. Natl. Acad. Sci. U.S.A.*, 72, 1420, 1975.
176. **Terenius, L.,** Uptake of radioactive oestradiol in some organs of immature mice. The capacity and structural specificity of uptake studied with several steroid hormones, *Acta Endocrinol.*, 50, 584, 1965.
177. **Grunert, G. and Tchernitchin, A. N.,** Effect of progesterone on the non-genomic response to oestrogen, *J. Endocrinol.*, 94, 307, 1982.
178. **Grunert, G. and Tchernitchin, A. N.,** Effect of progesterone on the fate of uterine eosinophils in estrogen pretreated rats, *IRCS Med. Sci.*, 9, 645, 1981.
179. **Castrillón, M. A., Grunert, G., López, M., and Tchernitchin, A. N.,** Effect of progesterone on spleen eosinophilia, *IRCS Med. Sci.*, 10, 955, 1982.
180. **Grunert, G., Neumann, G., Porcia, M., and Tchernitchin, A. N.,** Dose effect of progesterone on the degranulation of uterine eosinophils, *IRCS Med. Sci.*, 11, 193, 1983.
181. **Rinard, G. A. and Chew, C. S.,** Interacting effects of estrogen, progesterone and catecholamines on rat uterine cyclic AMP and glycogen phosphorilase, *Life Sci.*, 16, 1507, 1975.
182. **Paul, P. K. and Duttagupta, P. N.,** Inhibition of oestrogen-induced increase in hepatic and uterine glycogen by progesterone in the rat, *Acta Endocrinol.*, 72, 762, 1973.
183. **Greiss, F. C., Jr., and Anderson, S. G.,** Effect of ovarian hormones on the uterine vascular bed, *Am. J. Obstet. Gynecol.*, 107, 829, 1970.
184. **Garris, D. R.,** Temporal relationships between uterine weight, uterine blood flow, and serum progesterone and estradiol levels in the pseudopregnant rat, *Proc. Soc. Exp. Biol. Med.*, 169, 334, 1982.
185. **Szego, C. M.,** Role of histamine in mediation of hormone action, *Fed. Proc. Fed. Am. Soc. Exp. Biol.*, 24, 1343, 1965.
186. **Josefsson, B.,** Studies on eosinophil granulocytes. V. Evidence against the role of histamine as a mediator of eosinophilia in the uterus of the rat, *Acta Endocrinol.*, 58, 532, 1968.
187. **Clark, K. E., Farley, D. B., VanOrden, D. E., and Brody, M. J.,** Estrogen-induced uterine hyperemia and edema persist during histamine receptor blockade, *Proc. Soc. Exp. Biol. Med.*, 156, 411, 1977.
188. **Galand, P. and Dupont-Mairesse, N.,** Evaluation of the role of estrogen contamination on functional studies with uterine RNA from estradiol-treated rats, *Endocrinology*, 90, 936, 1972.
189. **Soto, N. and Tchernitchin, A.,** Colchicine-estrogen interactions, *Experientia*, 35, 558, 1979.
190. **Fujimoto, G. I. and Morrill, G. A.,** Effect of colchicine on estrogen action. I. Inhibition of 17β-estradiol-induced water and potassium uptake in the immature rat uterus, *Biochim. Biophys. Acta*, 538, 226, 1978.
191. **Kalimi, M. and Fujimoto, G. I.,** Effect of colchicine on estrogen action. II. Translocation of 17β-estradiol cytosol receptor complex into the nucleus, *Biochim. Biophys. Acta*, 538, 231, 1978.
192. **Clark, J. H. and Gorski, J.,** Ontogeny of the estrogen receptor during early uterine development, *Science*, 169, 76, 1070.
193. **Luck, I. N., Gschwendt, M., and Hamilton, T. H.,** Oestrogenic stimulation of macromolecular synthesis is correlated with nuclear hormone-receptor complex in neonatal rat uterus, *Nat. New Biol.*, 245, 24, 1973.
194. **Sabag, N., Castrillón, M. A., and Tchernitchin, A.,** Cortisol-induced migration of eosinophil leukocytes to lymphoid organs, *Experientia*, 34, 666, 1978.
195. **Szego, C. M. and Roberts, S.,** Steroid action and interaction in uterine metabolism, *Recent Prog. Horm. Res.*, 8, 419, 1953.
196. **Nicolette, J. A. and Gorski, J.,** Cortisol effects on the uterine response to estrogen, *Endocrinology*, 74, 955, 1964.

197. **Shelesnyak, M. C.,** Some experimental studies on the mechanism of ova-implantation in the rat, *Recent Prog. Horm. Res.*, 13, 269, 1959.
198. **Clark, J. H., Markaverich, B., Upchurch, S., Eriksson, H., Hardin, J. W., and Peck, E. J., Jr.,** Heterogeneity of estrogen binding sites: relationship to estrogen receptors and estrogen responses, *Recent Prog. Horm. Res.*, 36, 89, 1980.
199. **Szego, C. M. and Davis, J. S.,** Inhibition of estrogen-induced cyclic AMP elevation in the rat uterus. II. By glucocorticoids, *Life Sci.*, 8 (1), 1109, 1969.
200. **Lippe, B. M. and Szego, C. M.,** Participation of adrenocortical hyperactivity in the suppressive effect of systemic actinomycin D on uterine stimulation by oestrogen, *Nature (London)*, 207, 272, 1965.
201. **Keeping, H. S., Newcombe, A. M., and Jellinck, P. H.,** Modulation of estrogen-induced peroxidase activity in the rat uterus by thyroid hormones, *J. Steroid Biochem.*, 16, 45, 1982.
202. **Cidlowski, J. A. and Muldoon, T. G.,** Modulation by thyroid hormones of cytoplasmic estrogen receptor concentrations in reproductive tissues of the rat, *Endocrinology*, 97, 59, 1975.
203. **Gardner, R. M., Kirkland, J. L., Ireland, J. S., and Stancel, G. M.,** Regulation of uterine response to estrogen by thyroid hormone, *Endocrinology*, 103, 1164, 1978.
204. **Steinsapir, J., Rojas, A. M., Bruzzone, M. E., Alarcón, O., and Ampuero, R.,** About the mechanism by which insulin inhibits estrogen-induced uterine edema in the rat, *Biol. Reprod.*, 28(Suppl. 1), 42, 1983.
205. **Ruh, M. F., Ruh, T. S., and Klitgaard, H. M.,** Uptake and retention of estrogens by uteri from rats in various thyroid states, *Proc. Soc. Exp. Biol. Med.*, 134, 558, 1970.
206. **Miura, S. and Koide, S. S.,** Effect of insulin and growth hormone on rat uterine RNA synthesis, *Proc. Soc. Exp. Biol. Med.*, 133, 882, 1970.
207. **Rojas, A. M. and Steinsapir, J.,** Multiple mechanisms of regulation of estrogen action in the rat uterus: effects of insulin, *Endocrinology*, 112, 586, 1983.
208. **Harbon, S., Do Khac, L., and Vesin, M. F.,** Cyclic AMP binding to intracellular receptor proteins in rat myometrium. Effect of epinephrine and prostaglandin E_1, *Mol. Cell. Endocrinol.*, 6, 17, 1976.
209. **Lafreniere, R. T. and Singhal, R. L.,** Theophylline-induced potentiation of estrogen action, *Steroids*, 17, 323, 1971.
210. **Cho-Chung, Y. S.,** On the interaction of cyclic AMP-binding protein and estrogen receptor in growth control, *Life Sci.*, 24, 1231, 1979.
211. **Steinsapir, J., Rojas, A. M., Tchernitchin, A., and Pacheco, R.,** Inhibition by theophylline of the estrogen-induced uterine wet weight increase in the ovariectomized adult rat, *IRCS Med. Sci.*, 6, 512, 1978.
212. **Rogovine, V. V., Piruzian, L. A., and Muravjev, R. A.,** *Peroksidazosomy* (Russian), Nauka, Moscow, 1977.
213. **Archer, R. K., Ed.,** *The Eosinophil Leukocytes*, Blackwell Scientific, Oxford, 1963.
214. **Beeson, P. B. and Bass, D. A., Eds.,** *The Eosinophil*, W. B. Saunders, Philadelphia, 1977.
215. **Weller, P. F. and Goetzl, E. J.,** The regulatory and effector roles of eosinophils, *Adv. Immunol.*, 27, 339, 1979.
216. **Colley, D. G. and James, S. L.,** Participation of eosinophils in immunological systems, in *Cellular, Molecular and Clinical Aspects of Allergic Disorders*, Gupta, S. and Good, R. A., Eds., Plenum Press, New York, 1979, 55.
217. **Smith, J. A. and Goetzl, E. J.,** Cellular properties of eosinophils: regulatory, protective and potentially pathogenic roles in inflammatory states, in *The Cell Biology of Inflammation*, Weissmann, G., Ed., Elsevier/North-Holland, Amsterdam, 1980, 189.
218. **Tavassoli, M.,** Eosinophil, eosinophilia and eosinophilic disorders, *CRC Crit. Rev. Clin. Lab. Sci.*, 16, 35, 1981.
219. **Smith, J. A.,** Molecular and cellular properties of eosinophils, *Ric. Clin. Lab.*, 11, 181, 1981.
220. **Wurgaft, R. and Tchernitchin, A.,** Histochemical study of rat uterus under estrogen stimulation, *IRCS Med. Sci.*, 4, 85, 1976.
221. **Hibbs, M. S., Mainardi, C. L., and Kang, A. H.,** Type-specific collagen degradation by eosinophils, *Biochem. J.*, 207, 621, 1982.
222. **Mandell, M. S. and Sodek, J.,** Metabolism of collagen types I, III and V in the estradiol-stimulated uterus, *J. Biol. Chem.*, 257, 5268, 1982.
223. **Hernández, M. R., Wurgaft, R., and Tchernitchin, A.,** Effects of estrogen on uterine stroma, *IRCS Med. Sci.*, 5, 113, 1977.
224. **Hubscher, T.,** Role of eosinophil in the allergic reaction. II. Release of prostaglandins from human eosinophilic leukocytes, *J. Immunol.*, 114, 1389, 1975.
225. **Finlay, T. H., Katz, J., Rasums, A., Seiler, S., and Levitz, M.,** Estrogen-stimulated uptake of α1-protease inhibitor and other plasma proteins by the mouse uterus, *Endocrinology*, 108, 2129, 1981.
226. **Schumacher, G. F. B.,** Alpha-1-antitrypsin in genital secretions, *J. Reprod. Med.*, 5, 13, 1970.
227. **Weller, P. B. and Goetzl, E. J.,** The human eosinophil. Role in host defense and tissue injury, *Am. J. Pathol.*, 100, 793, 1980.

228. **Strickland, S., Reich, E., and Sherman, M. I.,** Plasminogen activator in early embryogenesis: enzyme production by trophoblast and parietal endoderm, *Cell*, 9, 231, 1976.
229. **Mullins, D. E., Brazer, F. W., and Roberts, R. M.,** Secretion of a progesterone-induced inhibitor of plasminogen activator in the porcine uterus, *Cell*, 20, 865, 1980.
230. **Padawer, J.,** The mast cell and immediate hypersensitivity, in *Modern Concepts and Developments in Immediate Hypersensitivity*, Vol. 7, Bach, M. K., Ed., Marcel Dekker, New York, 1978, 301.
231. **Henderson, W. R., Chi, E. Y., and Klebanoff, S. J.,** Eosinophil peroxidase-induced mast cell secretion, *J. Exp. Med.*, 152, 265, 1980.
232. **Pincus, S. H., DiNapoli, A. M., and Schooley, W. R.,** Superoxide production by eosinophils: activation by histamine, *J. Invest. Dermatol.*, 79, 53, 1982.
233. **Kater, L. A., Goetzl, E. J., and Austen, K. F.,** Isolation of human eosinophil phospholipase D, *J. Clin. Invest.*, 57, 1173, 1976.
234. **Orange, R. P., Murphy, R. C., and Austen, K. R.,** Inactivation of slow-reacting substance of anaphylaxis (SRS-A) by arylsulfatase, *J. Immunol.*, 113, 316, 1974.
235. **Wasserman, S. I., Goetzl, E. J., and Austen, K. F.,** Inactivation of slow reacting substance of anaphylaxis by human eosinophil arylsulfatase, *J. Immunol.*, 114, 645, 1975.
236. **Henderson, W. R., Jörg, A., and Klebanoff, S. J.,** Eosinophil peroxidase-mediated inactivation of leukotrienes B_4, C_4, and D_4, *J. Immunol.*, 128, 2609, 1982.
237. **Gleich, G. J.,** The eosinophil. New aspects of structure and function, *J. Allergy Clin. Immunol.*, 60, 73, 1977.
238. **Zipper, J., Ferrando, G., Sáez, G., and Tchernitchin, A.,** Intrauterine grafting in uterus of autologous and homologous adult rat skin, *Am. J. Obstet. Gynecol.*, 94, 1056, 1966.
239. **Tchernitchin, A., Tchernitchin, X., Bongiovanni, A. M., and Chandross, R.,** Effect of hydrogen peroxide on estrogen binding by uterine eosinophils, *in vitro*, *J. Steroid Biochem.*, 5, 693, 1974.
240. **Castrillón, M. A., Sabag, N., and Tchernitchin, A. N.,** Effect of glucocorticoids in lymphoid tissue. An electron microscopical study, *IRCS Med. Sci.*, 9, 552, 1981.
241. **Peterson, A. P., Altman, L. C., Hill, J. S., Gosney, K., and Kadin, M. E.,** Glucocorticoid receptors in normal human eosinophils: comparison with neutrophils, *J. Allergy Clin. Immunol.*, 68, 212, 1981.
242. **Archer, R. K.,** Eosinophil leukocyte-attracting effect of histamine in skin, *Nature (London)*, 187, 155, 1960.
243. **Colley, D. G.,** Eosinophils and immune mechanisms. I. Eosinophil stimulation promoter (ESP): a lymphokine induced by specific antigens or phytohemagglutinin, *J. Immunol.*, 110, 1419, 1973.
244. **Parish, W. E.,** Substances that attract eosinophils *in vitro* and *in vivo*, and that elicit blood eosinophilia, *Antibiot. Chemother.*, 19, 233, 1974.
245. **Roux, M. E., McWilliams, M., Phillips-Quagliata, J. M., Weiz-Carrington, P., and Lamm, M. E.,** Origin of IgA-secreting plasma cells in the mammary gland, *J. Exp. Med.*, 146, 1311, 1977.
246. **McDermott, M. R. and Bienenstock, J.,** Evidence for a common mucosa immunologic system. I. Migration of B immunoblasts into intestinal, respiratory and genital tissues, *J. Immunol.*, 122, 1892, 1979.
247. **Seelig, L. L., Jr. and Beer, A. E.,** Transepithelial migration of leukocytes in the mammary gland of lactating rats, *Biol. Reprod.*, 18, 736, 1978.
248. **Seelig, L. L., Jr.,** Dynamics of leukocytes in rat mammary epithelium during pregnancy and lactation, *Biol. Reprod.*, 22, 1211, 1980.

Chapter 3

FLUORESCEINATED ESTRONE BINDING BY NORMAL AND NEOPLASTIC CELLS OF HUMAN AND ANIMAL ORIGIN

Nurten Gunduz, Elizabeth Saffer, and Bernard Fisher

TABLE OF CONTENTS

I. Introduction 40

II. Methodology 40
- A. Ligand Investigated 40
- B. Cells Examined 40
- C. Preparation of Cells for 17-FE-ER-Determination 41
- D. Characterization of Cells Examined 41

III. Specificity of 17-FE Binding 42
- A. Effect of Ligand Concentration 42
- B. Variation of Incubation Time and Temperature 42
- C. In Vivo and In Vitro Competitive Binding 44
- D. Correlation of Radioactive Ligand Binding with 17-FE Binding 45

IV. The Application of 17-FE for Biological Investigations 47
- A. Interrelation between Tumor Cell Proliferation and 17-FE Binding 48
- B. Relation of Age to ERI and LI in Intact Livers 50

V. Findings from Human Mammary Tumors 51
- A. Variation of ERI 51
- B. Correlation of ERI with Cytosol ER (fmol) 52
- C. 17-FE Binding by Cells Obtained by Needle Aspiration 54

VI. Comment 57

References 58

I. INTRODUCTION

Currently accepted methods of estrogen receptor determination measure the binding of [^{3}H] estradiol or a close [^{3}H] analog with receptor proteins in the cytosol of tissue homogenates.[1] Despite widespread use, there are disadvantages associated with biochemical methods for ER determination. The cytosol method requires a relatively large amount of tumor, includes cytosol from nontumor cells, is time consuming, and, most importantly, it fails to provide information regarding the proportion of cells containing the receptor. These limitations have prompted development of other methods for determining the tumor ER status. Recent reports describe immunochemical[2-4] and histochemical[5-7] procedures which have been used in an effort to overcome these disadvantages. For the past 4 years we have been employing 1-(*N*)-fluoresceinyl-estrone-thiosemicarbazone which consists of a fluorescein moiety coupled to the 17 β-position of estrone (17-FE),[8,9] to determine ER in various tissues.[10-12] This ligand can pass through the cell membrane, bind to receptors, and can translocate into the nucleus of the intact cells. The method provides a unique opportunity to follow the action of hormone receptor complex (HRC) visually under the microscope, and also enables determination of the proportion of cells with ER. That value can be expressed as the estrogen receptor index (ERI).

This report presents an overview of our experience with the use of a methodology which permits the detection of ER in intact viable single cells in less than 2 hr. It also presents some of the biological information which we have obtained indicating the credibility of and justification for the use of the method.

II. METHODOLOGY

A. Ligand Investigated

The ligand used is 1-(*N*)-fluoresceinyl-estrone-thiosemicarbazone which consists of a fluorescein moiety coupled to the 17 β-position of estrone (17-FE). All ligand was prepared and supplied by W. B. Dandliker, Ph.D., University Research Foundation, San Diego, Calif. and C. Y. Meyers, Department of Chemistry, S. Illinois University, Carbondale, Ill.

Initial experiments regarding the specificity of binding of 17-FE by ER and the lack of nonspecific binding at the concentrations needed to bind ER have been reported by Dandliker and associates.[8] To compare [^{3}H] labeled E_2 to 17-FE for determination of ER, a fluorometric dextran-coated charcoal (DCC) assay with 17-FE showed 210 fmol of specific binding sites per milligram protein as compared to 170 fmol of specific sites per milligram protein obtained with the use of [^{3}H]E_2 in rabbit uterus. The 17-FE ligand is approximately 100-fold smaller in size than various comparable fluoresceinated estrogens prepared in other laboratories; it can pass through the cell membrane, bind to receptors, and can translocate into the nucleus of the intact cells.

B. Cells Examined

Our investigation have used cells from mammary tumors and normal tissues. A transplanted mouse tumor arising in a C3H/HeJ female mouse and maintained in C3HeB/FeJ mice has been extensively studied. The growth characteristics and receptor content of these transplanted tumors have been determined repeatedly in our laboratory.[13-15] Several other mammary tumors, derived from female mice and rats, were also evaluated relative to their 17-FE-ER: (1) spontaneous tumors in the C3H/HeJ hosts, (2) spontaneous $CD8F_1$ tumors in the $CD8F_1$ hosts, (3) transplanted urethan-induced MXT tumor in C57BL × DBA/2f F_1 mice, and (4) DMBA-induced 13762NF tumor in Fischer 344 Mai rats. Human breast cancers were obtained at the time of excision for both DCC assay of ER and 17-FE binding. The normal tissues which provided cells for the determination of ERI included mouse uterus and

Table 1
PROPORTION OF CELLS FROM ANIMAL MAMMARY TUMOR DISPLAYING 17-FE BINDING (ERI)

Tumor	Host	No. tumors	Mean ± SE	Range
C3H[a]	Mouse	14	29.7 ± 0.15	26.8—32.0
C3H/HeJ[b]	Mouse	11	11.4 ± 3.13	2.2—36.0
$CD8F_1$[b]	Mouse	11	5.2 ± 1.24	1.1—16.4
MXT[a]	Mouse	16	27.7 ± 2.10	13.6—39.5
13762NF[a]				
14 days	Rat	10	8.5 ± 2.30	1.8—22.8
28 days	Rat	25	14.4 ± 1.20	3.6—22.4

[a] Transplanted.
[b] Spontaneous.

liver, as well as rat vaginal epithelium and liver. Cells which served as negative controls were derived from muscle, spleen, kidney, lymphocytes, and erythrocytes.

C. Preparation of Cells for 17-FE-ER Determination

For optimal 17-FE-ER determination slight modifications in the preparation of single-cell suspensions are required depending upon the tissue being examined. Suspensions of mouse tumor cells were prepared in McCoy's medium by mincing with scissors. Human breast tumor cells, after removing attached fat, were similarly prepared, but in RPMI 1640 medium. Both media were supplemented with 20% calf serum. The brei obtained was filtered through a stainless steel sieve. Liver samples were rinsed in cold PBS to remove blood and cut into small pieces. The cells were gently teased from the cut surfaces by triturating with cold PBS. Vaginal epithelial cells were obtained from rat vagina by irrigating with 0.5 mℓ of PBS. No additional preparation was needed.

All single-cell suspensions were washed with PBS and incubated with 17-FE. Following incubation, the cells were washed twice (5 min each) with PBS at 300 g centrifugation. The ligand concentrations employed ranged from 10^{-5} to 10^{-10} M and the time of incubation of cells ranged from 15 to 180 min. The standard incubation is 1 hr at 37°C with 2×10^{-7} M 17-FE ligand. The suspensions were washed with cold PBS to remove excess ligand. Tubes containing washed cells were maintained in a light-free ice bath. A drop of the suspension to be studied was placed on a slide and a cover slip applied. The cells were examined with a 40× phase contrast objective on a Zeiss® epi-fluorescence microscope having a halogen light source with a BG-23-exciter filter and a K520 barrier filter.

D. Characterization of Cells Examined

All suspensions derived from mouse, rat, and human mammary cancers, rat liver, and vaginal epithelial cells contained two populations of cells: those binding 17-FE, as evidenced by their fluorescence, and others which failed to display such binding. Five mouse and rat tumor types examined for 17-FE binding demonstrated variation in the proportion of ER-positive cells (Table 1). While this variation was prominent among spontaneous tumors, the transplanted tumors showed lesser variation among tumors of the same type. This variation indicates that the 17-FE was binding specifically to ER since if it were nonspecific it would be expected that all of the tumors would demonstrate the same proportion of 17-FE-positive cells despite their heterogeneous origin. The proportion of ER-positive cells was higher in the transplanted mouse tumors than in the spontaneous ones and the range of values was narrower. The rat tumor differed from the mouse tumors in that the ER content increased with duration of tumor growth.

The distribution of fluorescence within the cell varied in both mouse mammary tumors (Table 2). The greatest proportion of fluorescent tumor cells displayed fluorescence of both cytoplasm and nucleus. An occasional ER-negative tumor cell showed a very faint fluorescence of nucleolus. In our studies, tumor cells which displayed brilliant nuclear and/or cytoplasmic fluorescence are considered to be 17-FE-ER-positive and all other types are ER negative.

Between 50 and 60% of hepatocytes from normal intact female rat livers demonstrated fluorescence which was, in all but an occasional cell, confined to the cytoplasm. Parenchymal cells from male rat livers contain a lower proportion of cytoplasmic fluorescent cells (~40%). Mouse hepatocytes contain 40 to 50% ligand binding cells.

The proportion of rat vaginal epithelial cells displaying 17-FE binding varied according to the estrus cycle (Table 3). In general, para-basal cells showed fluorescence in both cytoplasm *and* nucleus. As the cycle progressed more cells displayed fluorescence of the cytoplasm or of the nucleus only.

III. SPECIFICITY OF 17-FE BINDING

A. Effect of Ligand Concentration

Varying ligand concentrations over a wide range failed to increase the proportion of fluorescent cells in populations obtained from mouse or human mammary tumors. However, as the concentration of ligand was increased, the brightness of the fluorescence within the cells also increased despite no change in the proportion demonstrating fluorescence. When the concentration ranged between 3×10^{-10} to 1×10^{-5} M, the percentage of mouse tumor cells which displayed fluorescence remained constant (Table 4). When the concentration was less than 3×10^{-10} M, the fluorescence was too evanescent either to permit determination of the proportion of cells displaying fluorescence or the location of the fluorescence within cells. The use of ligand in increasing concentrations from 10^{-10} to 10^{-5} M failed to increase the proportion of fluorescent cells in populations obtained from human tumors demonstrating relatively few cells with fluorescence (Table 4).

These findings indicate the specificity of the binding. They show that type II binding does not occur in a cell independently of type I binding in that cell.[16] Since our investigations utilized a concentration of $\leqslant 10^{-7}$ M, type III binding would not have been expected to occur. If type III binding did take place when a ligand concentration of 10^{-5} M was used, it presumably occurred only in cells demonstrating types I and II binding, since no increase in 17-FE-ER-positive cells resulted. Thus, it would seem that nonspecific binding is limited, should it occur, only to cells which contain type I ER.

B. Variation of Incubation Time and Temperature

When mouse tumor cells were incubated with 17-FE for 15 to 60 min at 23°C or between 30 and 120 min at 37°C, the percentage of cells displaying fluorescence was essentially the same providing that the time employed was equal to or greater than 30 min (Table 5). More prolonged incubation failed to increase the population of cells displaying fluorescence. Incubation of human breast cancer cells for 2 hr at 37°C failed to increase the proportion of fluorescent cells beyond that observed following 1 hr of incubation. With an incubation time of 60 min, variation of temperature (4 or 37°C) failed to alter the proportion of ligand-binding cells. These findings also indicate that a cell population contains a certain proportion of differentiated cells which synthesize ER, and prolonged incubation with 17-FE does not increase the 17-FE-ER-positive cells. Moreover, it would have been anticipated that the opportunity for nonspecific binding to occur would have increased with prolonged incubation

Table 2
DISTRIBUTION OF TUMOR CELL TYPES RELATIVE TO THEIR FLUORESCENCE

		Fluorescent-positive cells			Fluorescent-negative cells	
		Total cell	Nucleus	Cytoplasm	Nucleolus	None
Mouse mammary tumors (15)[a]	a[b]	28.1 ± 0.67[c]	2.2 ± 0.48	1.0 ± 0.51	14.1 ± 4.20	54.2 ± 4.53
	b	88.6 ± 2.28	6.8 ± 1.44	2.9 ± 1.28	20.9 ± 6.31	79.0 ± 6.31

[a] Number in parentheses, number of tumors.
[b] (a)Percentage of all cells; (b) percentage of fluorescent-positive or -negative cells.
[c] Mean ± SE.

From Fisher, B., Gunduz, N., Zheng, S., and Saffer, E. A., *Cancer Res.*, 42, 540, 1982. With permission.

Table 3
DISTRIBUTION OF ESTROGEN RECEPTORS IN VAGINAL CELLS OF THE RAT DURING THE ESTRUS CYCLE

Cycle	17-FE-ER % positive
Diestrus 1	24.3 ± 2.8[a] (7)[b]
Diestrus 2	30.1 ± 2.9 (11)
Proestrus	17.5 ± 2.2 (9)
Estrus	4.2 ± 0.8 (23)

[a] Mean ± SE.
[b] Number of specimens.

Table 4
EFFECT OF LIGAND CONCENTRATION ON FLUORESCENCE WITH 1 HR INCUBATION AT 37°C

	17-FE (mol ligand concentration)	Fluorescent-positive cells (%)		No. of tumors
		Av	Range	
Mouse	3×10^{-10}	35.8	33.6—38.0	2
	2×10^{-9}	33.9	33.4—34.3	2
	2×10^{-8}	31.8	30.6—33.5	4
	1×10^{-7}	31.4	27.5—36.6	4
	2×10^{-7}	33.5	28.2—42.9	12
	1×10^{-6}	26.7	25.0—30.0	4
	1×10^{-5}	34.0	32.9—35.1	2
Human				
191[a]	2×10^{-9}	7.3		
	2×10^{-7}	8.6		
	2×10^{-5}	8.5		
195[a]	2×10^{-10}	5.2		
	2×10^{-7}	3.8		
	2×10^{-5}	5.2		
194[a]	2×10^{-10}	8.9		
	2×10^{-7}	7.7		
	2×10^{-5}	7.0		

[a] Patient number.

From Fisher, B., Gunduz, N., Zheng, S., and Saffer, E. A., *Cancer Res.*, 42, 540, 1982. With permission.

times and that the proportion of cells displaying fluorescence would have become greater. That such did not occur argues for specificity of the binding.

C. In Vivo and In Vitro Competitive Binding

To further determine the specificity of 17-FE binding, three agents were used for evaluating competitive binding: (1) estradiol [(estratrien-3, 17β-diol)-1,3,5(10)] (Steraloids, Inc.), (2) tamoxifen [ICI46474(1-*p*-β-dimethylaminoethoxyphenyl-*trans*-1,2-diphenylbut-1-ene)],

Table 5
EFFECT OF INCUBATION TIME AND TEMPERATURE ON FLUORESCENCE

	Time of incubation (min)	ER-positive cells (%)		No. of tumors
		Mean	Range	
Mouse tumor				
At 23°C	15	21.8	21.6—21.9	2
	30	32.2	30.2—34.2	2
	45	30.4	30.2—30.5	2
	60	32.5	31.3—35.6	2
At 37°C	30	30.2	27.4—33.0	2
	60	31.8	30.8—32.7	3
	120	30.1	24.9—35.1	2
Human tumor				
At 37°C	60	23.5	4.0, 15.4, 51.1	3
	120	20.0	3.7, 15.0, 41.3	3
For 60 min	4°	17.2	3.6—32.1	5
	37°	18.1	4.0—35.3	5

From Fisher, B., Gunduz, N., Zheng, S., and Saffer, E. A., *Cancer Res.*, 42, 540, 1982. With permission.

Nolvadex® (Stuart Pharmaceuticals Division of ICI Americas, Inc.), and (3) estrone [Δ 1,3,4,(10-estratrien-3-ol-17-one)] (Sigma). If 17-FE were binding to ER the presence of a competitor for the same binding site should decrease the proportion of fluorescent-positive cells and/or the intensity of the fluorescence.

When normal mouse uterine cells were incubated with tamoxifen for 60 min at 37°C prior to 17-FE incubation, there was a complete abrogation of fluorescence. Without tamoxifen between 25 and 41% of cells were 17-FE-ER positive (Table 6). Tamoxifen also inhibited 17-FE binding of mouse tumor cells. The use of increased concentrations of that agent decreased the percentage of such cells demonstrating fluorescence. The decrease was similar whether the time of incubation of the cells with tamoxifen was 30, 45, or 60 min. Exposure of mouse tumor cells to increasing amounts of estradiol also resulted in an increasing reduction in the number of fluorescent cells. Following in vitro incubation or in vivo treatment of liver cells with various competitors such as estrone, estradiol, and tamoxifen prior to 17-FE incubation, a decrease in 17-FE binding was observed (Table 7). The overall intensity was much lower in all cells displaying fluorescence when they had been incubated with those competitors. The decrease in the proportion of 17-FE-ER-positive cells was remarkably similar in cell populations treated with competitors in vivo or in vitro.

Failure to observe complete blockage of 17-FE binding in all cells by competitors may be considered to indicate that nonspecific binding took place. It is important to reiterate that in those remaining cells demonstrating fluorescence there was a decrease in intensity, indicating that blockage of ER sites had occurred. Since the unit of measurement in this method differs from that when ER is determined in cytosol preparations, it is not possible to indicate whether competitive binding is greater or less in one or the other method of analysis.

D. Correlation of Radioactive Ligand Binding with 17-FE Binding

Since 17-FE binding is a measure of the proportion of ER-positive cells and cannot quantitate the amount of receptor per cell, whereas the cytosol method quantitates the amount of receptor within a tissue sample but not the proportion of cells with ER, there can be no direct quantitative comparison between the two methods. Only if all marker-positive cells,

Table 6
IN VITRO INHIBITION OF 17-FE BINDING

Species	Tissue	Temp (°C)	Time (min)	Competitor (mol)	17-FE (mol)	Fluorescent cells (%) Av	Range	No. of tumors
Mouse	Uterus	37	60	None	10^{-7}	31.1	25.0—40.9	4
				Tamoxifen (10^{-5})	10^{-7}	0	0—0	
Mouse	Tumor (C3H)	23	30	None	10^{-6}	32.2	30.2—34.2	2
				Tamoxifen (10^{-6})	10^{-6}	21.1	20.9—21.3	2
				Tamoxifen (10^{-5})	10^{-6}	10.2	9.3—11.1	2
		23	45	None	10^{-6}	30.4	30.2—30.5	2
				Tamoxifen (10^{-6})	10^{-6}	19.0	18.2—19.8	2
				Tamoxifen (10^{-5})	10^{-6}	9.9	8.8—11.4	2
		23	60	None	10^{-6}	32.5	31.3—35.6	2
				Tamoxifen (10^{-6})	10^{-6}	14.1	12.9—15.2	2
				Tamoxifen (10^{-5})	10^{-6}	8.5	72—9.8	2
		37	60	None	10^{-7}	28.5	27.0—30.0	2
				Estradiol (10^{-7})	10^{-7}	16.1	14.8—17.3	2
				Estradiol (10^{-6})	10^{-7}	13.8	13.2—14.2	2
				Estradiol (10^{-5})	10^{-7}	7.0	6.1—7.9	2
Human	Tumor							
	219[a]	37	60	None	10^{-7}	31.3		
				Tamoxifen (10^{-5})	10^{-7}	8.9		
	226[a]	37	60	None	10^{-7}	12.7		
				Tamoxifen (10^{-5})	10^{-7}	5.3		
	231[a]	37	60	None	10^{-7}	45.6		
				Tamoxifen (10^{-5})	10^{-7}	13.8		

[a] Patient number.

From Fisher, B., Gunduz, N., Zheng, S., and Saffer, E. A., *Cancer Res.*, 42, 540, 1982. With permission.

Table 7
IN VITRO AND IN VIVO INHIBITION OF 17-FE OF RAT LIVERS

Treatment	Temp (°C)	Time	Competitor (mol)	17-FE (mol)	Fluorescent cells (%) Mean ± SEM	No. of rats
In vitro						
	37	60 min	None	10^{-7}	59.9 ± 2.5	6
	37	60 min	Estrone (10^{-5})	10^{-7}	36.4 ± 4.5	6
	37	60 min	Estradiol (10^{-5})	10^{-7}	37.5 ± 4.1	6
	37	60 min	Tamoxifen (10^{-5})	10^{-7}	37.6 ± 3.8	6
	37	180 min	None	10^{-7}	60.6 ± 1.6	6
	37	180 min	Estrone (10^{-5})	10^{-7}	32.5 ± 4.9	6
	37	180 min	Estradiol (10^{-5})	10^{-7}	37.4 ± 2.8	6
	37	180 min	Tamoxifen (10^{-5})	10^{-7}	42.4 ± 3.7	6
In vivo	—	14 days[a]	None	10^{-7}	50.0 ± 1.7	12
	—	14 days	Estrone	10^{-7}	37.2 ± 3.2	3
	—	14 days	Estradiol	10^{-7}	37.6 ± 2.6	22
	—	14 days	Tamoxifen	10^{-7}	26.3 ± 1.0	18
	—	21 days	Estradiol	10^{-7}	51.5 ± 1.2	3

[a] Each competitor is administered at 0, 1, and 7 days and animals were sacrificed at 14 and 21 days.

Table 8
CORRELATION OF RADIOACTIVE LIGAND BINDING WITH 17-FE BINDING (MOUSE TUMORS)

	Cytosol protein (fmol/mg)	Fluorescent cells (% positive)
Untreated	8	28.0
	9	20.1
	13	28.7
	10	26.9
	10	27.5
	12	27.3
	6	26.5
	10.0 ± 0.7[a]	26.1 ± 3.1
4 Days following cyclophosphamide (240 mg/kg × 1)	19	57.3
	15	45.0
	17	47.4
	17.0 ± 1.2	49.7 ± 3.7
7 Days following tamoxifen (2.5 mg × 1)	5	9.8
	7	15.0
	0	12.6
	4.0 ± 2.1	12.5 ± 1.5

[a] Mean ± SE.

From Fisher, B., Gunduz, N., Zheng, S., and Saffer, E. A., *Cancer Res.*, 42, 540, 1982. With permission.

not only within a tumor but in all tumors, possessed the same number of ER binding sites would a correlation between the two methods be possible. It is unfortunate that critics of the 17-FE method consider that lack of such numerical correlation prevents acceptance of the use of 17-FE binding as a measure of ER. Investigations conducted by us indicate that while such a numerical relationship is often lacking, there is a remarkable correlation between the two in identifying the direction of change which occurs when receptor content is altered. The following experiments exemplify the concordance.

When the ER content of tumors from untreated mice was determined by the dextran-coated charcoal assay[17] and by 17-FE binding, findings were similar from tumor to tumor (Table 8). Following a single dose of cyclophosphamide, the proportion of fluorescent cells was found to be twice that of cells from tumors in untreated hosts. Similarly, the ER determined by radioactive ligand was doubled. When a single dose of tamoxifen was administered to animals, both the ER determined by dextran-coated charcoal and the percentage of fluorescent cells decreased proportionally.

In another experiment (Table 9) the proportion of cells binding 17-FE decreased in livers following partial hepatectomy (PH). Livers were examined for ER content at the time of and at intervals following PH by means of the incorporation of $[^3H]E_2$ using the DCC method. Findings were concordant with those employing 17-FE.

The findings noted in the above experiments would have been extremely unlikely if the binding of the 17-FE ligand was random and nonspecific, while that of the $[^3H]E_2$ ligand was specific.

IV. THE APPLICATION OF 17-FE FOR BIOLOGICAL INVESTIGATIONS

Studies have been and are currently being carried out by us which provide additional

Table 9
EFFECT OF PARTIAL HEPATECTOMY ON THE 17-FE-ER AND THE LEVEL OF CYTOSOL ER IN THE LIVER REMNANT

Time after PH	17-FE-ER		Cytosol ER	
(hr)	Livers (#)	%	Livers (#)	fmol
0	122	60.6 ± 0.95[a]	25	20.6 ± 1.64
1	4	52.8 ± 1.40	—	—
3	3	41.9 ± 2.35	—	—
6	4	45.8 ± 7.49	4	26.8 ± 0.32
18	10	50.6 ± 2.99	8	3.5 ± 0.57
24	8	41.3 ± 1.65	6	15.5 ± 2.02
48	16	40.7 ± 3.62	6	14.4 ± 1.35
72	36	32.2 ± 1.40	—	—
120	4	32.3 ± 1.90	—	—
168	21	55.3 ± 2.76	6	19.2 ± 2.70

[a] Mean ± SE.

support for the validity of the methodology. They also demonstrate its usefulness for obtaining biological information regarding ER which conventional biochemical analyses will not provide. The following are a few examples.

A. Interrelation Between Tumor Cell Proliferation and 17-FE Binding

Investigations by us[14] have demonstrated that within 24 hr following removal of a primary C3H mammary tumor, changes occurred in the kinetics of cells in a distant tumor focus. There was an increase in the LI which persisted for between 7 and 10 days. A measurable increase in tumor size became apparent about a week following tumor removal. Recently, we have assessed the proportion of single intact viable mouse mammary tumor cells containing ER as determined by 17-FE and the proportion labeled with [^{3}H]TdR (LI) in the same cell population after either primary tumor removal, radiation, cyclophosphamide (CY), or tamoxifen (T) administration.[12]

We observed that following tumor removal the proportion of cells demonstrating 17-FE binding decreased as the LI increased, but returned to the level present prior to tumor removal at the time that the proportion of labeled cells receded to its initial level (Figure 1). Conversely, when a decrease in tumor LI was observed following radiation (Figure 2) or CY administration (Figure 3), there was a concomitant elevation in the proportion of fluorescent cells. The prolonged depression in LI following radiation was associated with a sustained increase in the proportion of ER-containing cells. Findings following CY administration were dose related. The greater the dose of drug the greater the decrease in LI, and the greater the concomitant increase in ER-containing cells. Following tamoxifen administration a decrease occurred in the proportion of fluorescing cells due to competitive binding. There was no alteration in LI (Figure 4).

These findings, revealing that there is an inverse relationship between the proportion of DNA synthesizing cells and the proportion of cells demonstrating fluorescence, indicate that in the mouse mammary tumor the proliferating population of cells contains fewer (or no) ER-containing cells than does the nonproliferating population. Since cells comprising the growth fraction of tumors are more sensitive to radiation and to alkylating agents (CY), it is likely that in these studies there is a selective destruction of proliferating cells following the use of those therapeutic modalities. Only if the cells destroyed do not express ER to the same extent as do cells in the G_0 and G_1 phase would their deletion result in an increase in

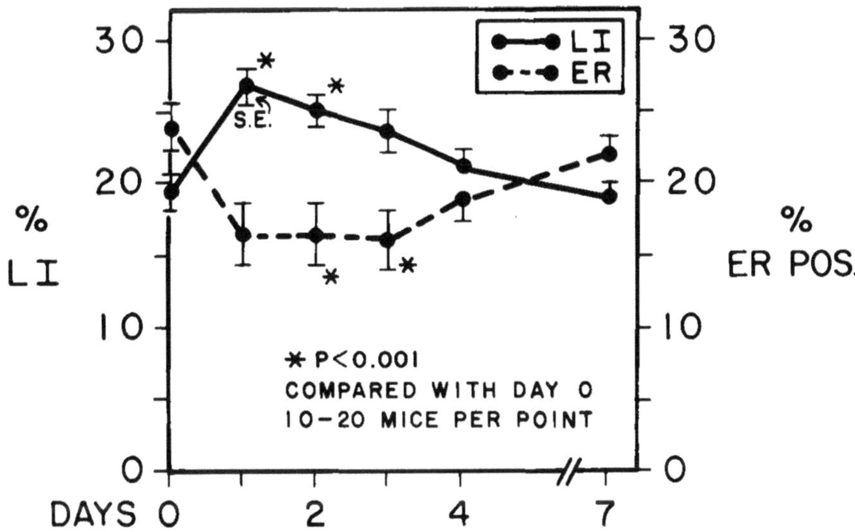

FIGURE 1. Relation between ER and LI metastases following primary tumor removal. (From Fisher, B., Gunduz, N., and Saffer, E. A., *Cancer Res.*, 43, 5244, 1983. With permission.)

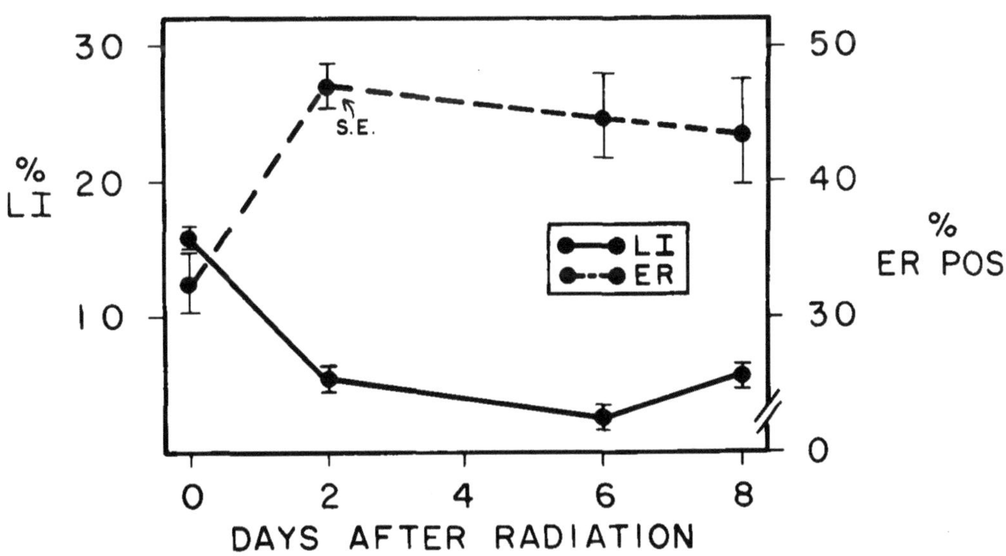

FIGURE 2. Relation of labeling index (LI) to estrogen receptor (ER) following tumor radiation. (From Fisher, B., Gunduz, N., and Saffer, E. A., *Cancer Res.*, 43, 5244, 1983. With permission.)

the proportion of ER-positive cells. Similarly, a decrease in ER-positive cells would be likely to occur only if the proportion of proliferating cells which increased following primary tumor removal contains fewer cells with ER than does the nonproliferating cell pool. Thus, it would seem that the change in the proportion of ER-containing cells is the result of an alteration of the proliferating cell population by the therapies employed.

The findings following the administration of tamoxifen indicating no change in LI but a decrease in the proportion ER-containing cells are not to be considered in conflict with the thesis formulated from the other observations that there is a correlation betwen ER and LI values. The tamoxifen, when administered in vivo or when incubated with tumor cells in

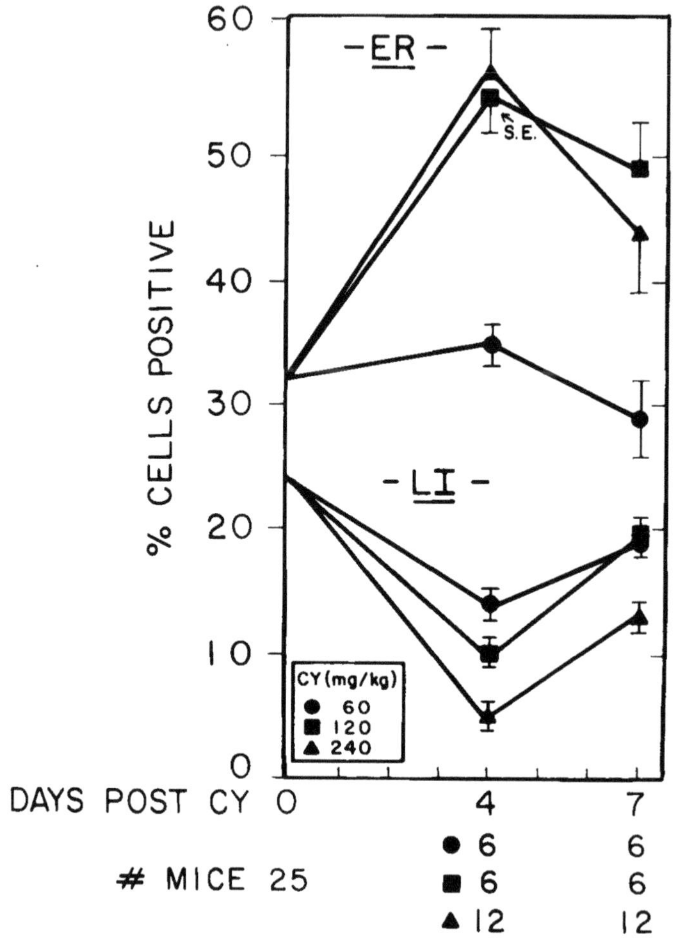

FIGURE 3. Effect of cyclophosphamide (CY) on estrogen receptor (ER) and labeling index (LI). (From Fisher, B., Gunduz, N., and Saffer, E. A., *Cancer Res.*, 43, 5244, 1983. With permission.)

vitro prior to 17-FE incubation, competitively binds with ER resulting in a decrease in the percentage of cells displaying fluorescence. The mechanism of reduction in detected ER is, thus, entirely different from that occurring following primary tumor removal. It has no relationship to changes in LI. These studies provide additional support to the credibility of the use of 17-FE for the determination of ER in individual tumor cells. It is extremely unlikely that the specific relationship between the proportion of LI and fluorescing cells could have been so clearly defined unless the ligand was specifically binding to ER. Since the relationships described are in keeping with current concepts regarding ER and tumor growth, it is entirely reasonable to conclude that the method is, indeed, depicting ER in individual tumor cells.

B. Relation of Age to ERI and LI in Intact Livers

The proportion of ER-positive cells in livers increased rapidly during the growth of animals (Table 10). At 5 days of life ~10% of cells demonstrated 17-FE binding. By 35 days, the ERI was equivalent to that demonstrated by mature (4-month-old) adult animals (~60%). The proportion of DNA synthesizing cells (LI) was inversely related to the age of animals and the percent of parenchymal cells with ER, decreasing from 30% at 5 days after birth to 2% at maturity. In virtually all cells, the 17-FE-ER binding was confined to the cytoplasm.

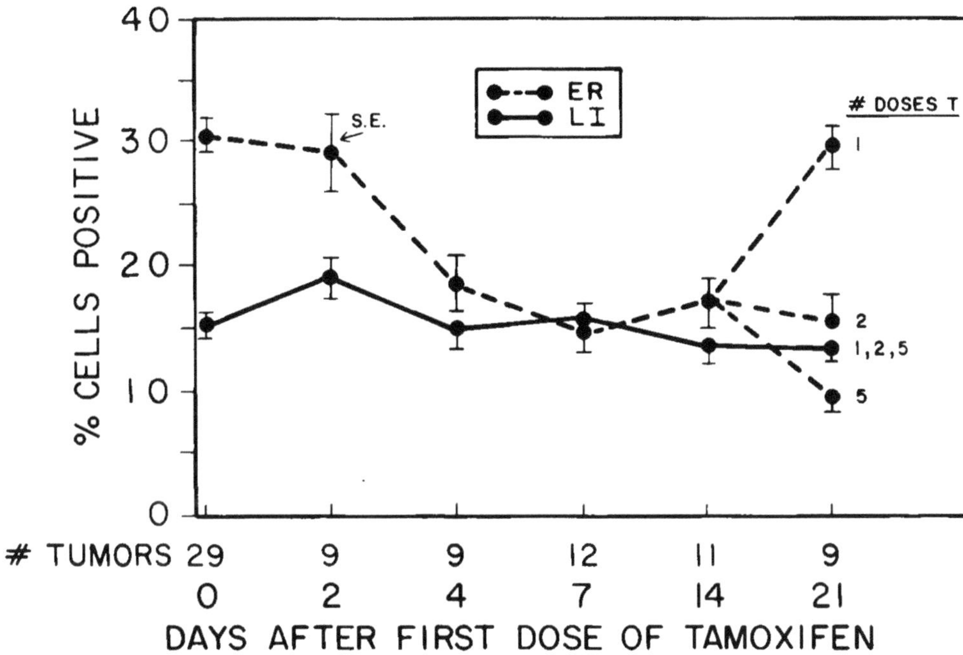

FIGURE 4. Effect of tamoxifen (T) on estrogen receptor (ER) and labeling index (LI). (From Fisher, B., Gunduz, N., and Saffer, E. A., *Cancer Res.*, 43, 5244, 1983. With permission.)

Table 10
RELATION OF AGE OF RATS TO ESTROGEN RECEPTOR INDEX (ERI) AND LABELING INDEX (LI) IN PARENCHYMAL CELLS OF INTACT LIVERS

	Age (days)				
	5	**14**	**28**	**35**	**112**
ERI	9.4 ± 0.7[a]	34.6 ± 0.9	42.9 ± 1.2	64.8 ± 0.9	60.6 ± 1.0
	(8)	(5)	(11)	(3)	(122)
LI	30.4 ± 1.1	3.7 ± 0.5	3.9 ± 0.8	4.6 ± 0.6	2.4 ± 0.1
	(8)	(5)	(11)	(3)	(46)

[a] % — Mean ± SE; in parentheses, number of livers.

These experiments provide additional support to those obtained in tumor, which indicate that in rapidly proliferating tissue an elevated LI is associated with a decreased ER. Our observation that the ERI of 5-day-old rats was only 10% and that it increased with maturity, reaching adult levels at 1 month is completely consistent with those who noted low concentrations of ER in supernates of liver from immature rats.[12,19,31] Once again, it is unlikely that findings such as these could have occurred by chance with 17-FE binding to cells nonspecifically.

V. FINDINGS FROM HUMAN MAMMARY TUMORS

A. Variation of ERI

While human tumors have a greater proportion of ER-positive cells than do spontaneous

Table 11
DISTRIBUTION OF TUMOR CELL TYPES RELATIVE TO THEIR FLUORESCENCE

		Total cell (type A)	Nucleus (type B)	Cytoplasm (type C)	Nucleolus (type D)	None (type E)
Human breast	a[a]	12.1 ± 1.70[b]	1.4 ± 0.34	4.2 ± 1.15	18.4 ± 1.68	63.8 ± 2.55
Cancer (41)[c]	b	68.4 ± 5.10	7.8 ± 4.89	23.7 ± 4.89	22.4 ± 3.14	77.6 ± 3.17

[a] (a) Percentage of all cells; (b) percentage of fluorescent-positive or -negative cells.
[b] Mean ± SE.
[c] Numbers in parentheses, number of tumors.

Table 12
COMPARISON OF ER WITH MARKER-POSITIVE CELLS IN HUMAN BREAST CANCER

Patient no.	Age	ER (fmol)	ERI	Agreement
154	65	0	35.4	−
143	47	0	42.4	−
153	57	0	20.3	−
156	55	0	43.0	−
159	51	0	51.1	−
177	54	0	9.1	+
184	36	0	8.5	+
195	60	0	3.3	+
228	33	0	25.3	−
221	67	1	6.5	+
207	73	2	57.5	−
168	51	3	13.9	−
185	63	6	4.3	+
152	47	6	26.8	−
200	64	7	17.8	−
213	56	8	8.5	+
162	52	8	16.8	−
194	60	9	8.6	+

Note: ER poor, <10 fmol.

From Fisher, B., Gunduz, N., Zheng, S., and Saffer, E. A., *Cancer Res.*, 42, 540, 1982. With permission.

mouse tumors, they show the same wide range of values. The ERI of 58 tumors was 20.0 ± 1.77 SE with a range of 2.8 to 57.5%. As in the animals, the greatest proportion of fluorescent cells displayed both cytoplasmic and nuclear fluorescence (Table 11). The wide range in ERI was similar to the variation observed from tumor to tumor when ER was determined by the DCC assay (Tables 12, 13, and 14). It is of interest that no human tumor in this series, or in a greatly expanded one, has ever demonstrated a complete absence of fluorescent cells.

B. Correlation of ERI with Cytosol ER (fmol)

A number of investigators employing a variety of histochemical methods have attempted to correlate their findings with those obtained by the biochemical method.[5-7,18] Any attempt to conclude from the reports of their findings that there is or is not correlation between the

Table 13
COMPARISON OF ER WITH 17-FE-POSITIVE CELLS IN HUMAN BREAST CANCER

Patient no.	Age	ER (fmol)	ERI	Agreement
171	61	10	3.8	−
175	45	10	28.6	+
179	50	11	30.3	+
176	54	13	10.7	+
198	71	14	11.9	+
146	45	17	22.1	+
150	75	18	28.4	+
157	70	18	14.7	+
138	64	20	34.8	+
212	54	22	36.7	+
155	50	23	15.8	+
148	66	25	34.0	+
174	61	25	4.0	−
180	48	31	9.5	−
151	78	33	12.6	+
190	56	33	18.0	+
137	44	34	35.0	+
139	71	36	34.6	+
172	70	36	14.5	+
189	45	39	4.5	−
149	54	48	31.8	+
226	43	49	12.7	+
163	67	52	7.3	−
222	63	61	28.6	+
186	48	61	14.0	+
167	58	64	13.0	+
188	76	73	11.3	+
141	72	77	20.6	+
173	63	81	15.3	+
204	72	97	3.7	−
225	56	95	18.6	+

Note: ER positive, 10 to 99 fmol.

From Fisher, B., Gunduz, N., Zheng, S., and Saffer, E. A., *Cancer Res.*, 42, 540, 1982. With permission.

histo- and biochemical methods seems perilous. The variations in technique used (temperature, incubation times, differences in ligand, and the methods of scoring) in these studies preclude combining findings and making a summary conclusion. To reach a conclusion from any one of the studies which would be definitive in resolving the issue as to whether a correlation between the two methods exists would be equally erroneous.

Our own results,[10] overall, fail to indicate either a qualitative or quantitative correlation between the histo- and biochemical methods. A closer evaluation indicates that, although there was no direct quantitative relationship, 81% of tumors having 10 to 99 fmol (Table 13) were associated with >10% 17-FE-positive cells. Conversely, 81% of tumors with 10 to 39% 17-FE-positive cells were found to have >10 fmol ER. The poorest association involved tumors with negative or borderline ER determined from cytosol (Table 12). A substantial number of those tumors displayed a significant proportion of 17-FE-positive cells. In six of nine tumors with 0 fmol, the 17-FE-positive cells ranged between 20 and 51%.

Table 14
COMPARISON OF ER WITH MARKER-POSITIVE CELLS IN HUMAN BREAST CANCER

Patient no.	Age	ER(fmol)	Cells with marker (%)	Agreement
191		108	8.6	−
219	56	113	36.0	+
140	71	146	30.1	+
142	55	158	43.1	+
147	73	369	13.2	+
158	53	412	7.2	−
161	68	467	2.8	−
183	84	>700	3.9	−
145	84	717	34.3	+

Note: ER positive $\geq$100 fmol.

From Fisher, B., Gunduz, N., Zheng, S., and Saffer, E. A., *Cancer Res.*, 42, 540, 1982. With permission.

When discrepancies of this magnitude occur, the correctness of the findings from either of the two methods may be suspect. The precision of the biochemical method is subject to so many variables that its use as the standard against which all other methods are to be judged is of concern.[19] The possibility cannot be ignored that those in ER-negative tumors represent false-negative biochemical results rather than false-positive cytochemical findings. The poor correlation at the other end of the spectrum of ER values is equally disconcerting (Table 14). The finding that tumors with ER reported as 412, 467, and >700 fmol, for example, had ERI poor tumors is difficult to explain. It is not implied that the false-negative or false-positive determinations are the results of erroneous analyses. The cytochemical and biochemical findings were derived from different portions of the same tumor, and heterogeneity may have been influential. Only if the two analyses could be carried out on the same cells could there be proof that the two methods do or do not agree.

Failure to detect a direct correlation in quantitative measurements obtained by the two methods is not necessarily an indication that one or the other is inappropriate. Since the biochemical method measures the ER content in the cytosol of all cells in the sample without indicating the proportion of the cells in which the sites reside, the cytochemical evaluation determines the proportion of cells with ligand-binding sites but fails to provide precise information regarding the number of sites per cell, and there is little reason to anticipate that a correlation would be demonstrated. Only if all 17-FE cells, not only within a tumor but in all tumors, possessed the same number of ER binding sites would a direct correlation between the methods result.

Our finding of a lack of quantitative correlation would seem to indicate that, indeed, all cells do not possess the same degree of ER. There is the possibility that, when the ER of a tumor as determined biochemically is relatively high and the proportion of 17-FE-positive cells is low, the number of receptor sites in the few positive cells may be great. On the other hand, when there is a high proportion of 17-FE-positive cells but a low ER, the number of sites per cell may be low. Whether the number of cells displaying ER is more significant in predicting response to treatment than is the amount of receptor, or vice versa, is unknown.

C. 17-FE Binding by Cells Obtained by Needle Aspiration

Since there is increasing emphasis on earlier detection of primary breast cancers, there is

Table 15
COMPARISON OF CYTOSOL ER WITH 17-FE-POSITIVE CELLS: ER POSITIVE (≥10 fmol)

			$[^3H]E_2$ (fmol)	17-FE Positive (%)		Agreement		
			SS[a]	NA[b]	SS[a]			
	Pt.#	Age	(A)	(B)	(C)	A vs. B	A vs. C	B vs. C
1	210	30	12	18.7	46.7	+	+	+
2	223	52	56	10.3	9.0	+	−	−
3	226	34	49	18.2	11.8	+	+	+
4	249	47	42	27.4	8.6	+	−	−
5	146	45	17	27.5	22.1	+	+	+
6	265	58	88	40.2	28.5	+	+	+
7	272	42	28	18.8	10.5	+	+	+
8	277	44	13	14.3	12.8	+	+	+
9	186	48	61	16.7	14.0	+	+	+
10	219	56	113	15.8	10.4	+	+	+
11	116	68	467	21.2	2.8	+	−	−
12	155	50	23	44.0	15.8	+	+	+
13	280	44	13	31.0	29.5	+	+	+
14	280	44	13	20.4	21.7	+	+	+
15	302	49	139	45.5	24.4	+	+	+
Agreement (%)						100	80	80

[a] SS: surgical specimen.
[b] NA: needle aspirate.

From Gunduz, N., Zheng, S., and Fisher, B., *Cancer*, 52, 1251, 1983. With permission.

clearly a need for a method which requires only small numbers of tumor cells. A variety of other clinical situations exist in which the availability of a technique that could supply information regarding tumor ER from a small cell population would be helpful. We have reported[11] our experience with 17-FE binding for detecting the proportion of estrone binding cells in samples obtained by fine-needle aspiration of tumors prior to their removal. Just as in our previous investigation, while there was no correlation between the quantitative results of the cyto- and biochemical methods, there was a strong relationship between the two methods. All (100%) tumors having ER levels ≥10 fmol displayed fluorescence in ≥10% of cells obtained by needle aspiration (Table 15). There was, however, a poorer association between the two methods when the ER level determined from tumor cytosol was negative or borderline (<10 fmol) (Table 16). Concordance with the cytochemical method using cells obtained by needle aspiration was then only 54%. In the 5 of 11 with disagreement, there was good agreement in all between the results obtained when cells for 17-FE binding came from the needle aspirate or from the excised tumor leading to question the findings from the biochemical method.

When similar comparisons were made between results obtained by the biochemical method and that obtained by 17-FE binding using cells obtained from pieces of removed tumor, the results were almost identical to those from our previous study.[10] The concordance in the prior study when tumors had ER levels between 10 and 99 fmol was 81% (25 of 31), and in the present one 83% (10 of 12). When tumor ER values were <10 fmol the concordance (<10% of cells fluorescent) in the two studies was 39 and 36%, respectively. That the discrepant findings relative to ER-negative tumors in both studies may represent false-negative biochemical results rather than false-positive cytochemical findings must be considered.

Table 16
COMPARISON OF CYTOSOL ER WITH 17-FE-POSITIVE CELLS: ER POOR (<10 fmol)

	Pt.#	Age	$[^3H]E_2$ (fmol) SS[a] (A)	17-FE Positive % NA[b] (B)	17-FE Positive % SS[a] (C)	Agreement A vs. B	Agreement A vs. C	Agreement B vs. C
1	185	63	6	9.9	4.3	+	+	+
2	194	60	0	3.2	8.6	+	+	+
3	195	60	0	3.5	3.8	+	+	+
4	200	64	7	8.7	17.8	+	−	−
5	207	73	2	20.2	57.5	−	−	+
6	262	84	4	3.2	4.1	+	+	+
7	266	57	5	8.2	17.9	+	−	−
8	279	51	0	26.4	16.8	−	−	+
9	279	65	9	26.1	29.7	−	−	+
10	296	45	9	15.2	12.3	−	−	+
11	299	63	0	28.0	14.2	−	−	+
12	214	63	ND[c]	4.3	2.6	ND[c]	ND[c]	+
Agreement (%)						54	36	83

[a] SS: surgical specimen.
[b] NA: needle aspirate.
[c] Not determined.

From Gunduz, N., Zheng, S., and Fisher, B., *Cancer*, 52, 1251, 1983. With permission.

The lack of concordance in 17-FE binding by the two cell suspensions, i.e., those from needle aspiration and those from the removed tumor, may be related to the method of sampling. The heterogeneity of the cellular composition of a tumor as regards ER content is well appreciated. When aspiration was performed the needle was manipulated so that cells were obtained from different planes in the tumor. Sampling from the resected tumor was restricted to cells from the small fragment of tissue available and was, therefore, apt to be less representative. The better agreement between the results of the biochemical method and ligand binding by cells from the aspirate than by the cells from the resected specimen may also be related to sampling. The larger amount of tumor used for the biochemical analysis is likely to be more representative of the tumor than is the small fragment available for preparation of a cell suspension for 17-FE binding. Only if all analyses could be carried out on the same cells would there be absolute justification for conclusions regarding the agreement or lack of agreement of the various methods.

There are a variety of clinical situations in which the availability of a technique that can supply information regarding tumor ER from a small cell population would be helpful. The primary breast cancer detected only by mammography, which is too small for ER assay by the standard biochemical method, early metastases in the skin of the operative site or recurrences in sites which are not accessible for removal such as in the liver or lung, are all particularly suitable for the use of the cytochemical technique using small numbers of cells. The detection of ER in cells aspirated from pleural effusions or ascitic fluid obtained by aspiration and from lesions where the diagnosis is questionable, such as the solitary lung nodule in patients having a previously removed primary breast cancer, may aid in differentiating between a primary lung cancer and a recurrence of the breast cancer. Similarly, the finding of ER in cells aspirated from a clinically positive axillary node in the absence of evidence of a primary cancer may indicate that tumor in the node is derived from an occult breast cancer. In patients with advanced disease, ER determination at intervals in cells obtained by needle aspiration may provide information regarding the effect of therapy. Aliquots of cells obtained by needle aspiration for ER measurement may be used for determination of labeling index, permitting correlation of the two parameters. As a result of our experience to date, with the use of 17-FE for the determination of ER-positive cells, we believe that the method is worthy of greater consideration for clinical use.

Increasing evidence indicates the value of *both* estrogen and progesterone (PR) levels of breast cancer. Consequently, there is urgent need for the development of a ligand suitable for determining PR in a fashion similar to that presently described for recognition of cells containing ER.

VI. COMMENT

A direct approach for validating the cytochemical technique in ER determination is to relate the finding of marker-positive cells with patient response to therapy as has been done in establishing the credibility of the biochemical method. Such an evaluation at this time may be difficult to pursue since the application of hormone therapy is directed by the biochemical assay. In order to obtain appropriate information, all patients, either ER positive or negative or marker positive or negative, must be administered hormonal therapy. Information might, thus, be obtained which would provide insight as to why only ≅60% of ER-positive patients respond to hormonal manipulation. Perhaps the proportion of marker-positive cells is the determinant; i.e., those with ER-positive but marker-poor tumors fail to respond.

The use of multimodality therapy for treatment is so pervasive that to ascertain whether a particular response is the result of hormonal perturbation alone or to the other agents used becomes virtually impossible. Only the use of a standardized hormonal therapy alone in a series of comparable patients is likely to provide convincing information regarding the comparability of the biochemical and cytochemical methods if patient response is to be used to indicate the efficacy of the cytochemical method.

As a result of our experience with the use of 17-FE for the determination of marker-positive cells, we conclude that the method is worthy of greater consideration than it has been given. The analyses, as carried out in these studies, can be performed rapidly on very few cells with good reproducibility and correlation between observers once they have become familiar with it. The method eliminates one of the major shortcomings of the cytosol technique in that it is possible to recognize the cells being evaluated as the test is being performed. Since the cytochemical technique is measuring the proportion of cells with ER, and since the biochemical method is measuring the total concentration of ER in the sample, it is inappropriate, particularly in light of the present results, to consider findings with the former as meaningless because of lack of concordance with the latter. The information provided by the cytochemical method may, indeed, add to the accuracy of the biochemical results for the prediction of responses to hormonal therapy or may even prove to provide more accuracy in its own right. Further evaluation is required.

REFERENCES

1. **Wittliff, J. L., Hilf, R., Brooks, W. F., Jr., Savlov, E. D., Hall, T. C., and Orlando, R. A.,** Specific estrogen-binding capacity of the cytoplasmic receptor in normal and neoplastic breast tissues of humans, *Cancer Res.*, 32, 1992, 1972.
2. **Ghosh, L., Ghosh, B. C., and Das Gupta, T. K.,** Immunocytological localization of estrogen in human mammary carcinoma cells by horseradish-antihorseradish peroxidase complex, *J. Surg. Oncol.*, 10, 221, 1978.
3. **Pertschuk, L. P., Tobin, E. H., Brigati, D. J., Kim, D. S., Bloom, N. D., Gaetjens, E., Berman, P. J., Carter, A. C., and Degenschein, G. A.,** Immunofluorescent detection of estrogen receptors in breast cancer: comparison with dextran-coated charcoal and sucrose gradient assays, *Cancer*, 41, 907, 1978.
4. **Mercer, W. D., Lippman, M. E., Wahl, T. M., Carlson, C. A., Wahl, D. A., Lezotte, D., and Teague, P. O.,** The use of immunocytochemical techniques for the detection of steroid hormones in breast cancer cells, *Cancer*, 46, 2859, 1980.
5. **Barrows, G. H., Stroupe, S. B., and Riehm, J. D.,** Nuclear uptake of a 17β-estradiol-fluorescein derivative as a marker of estrogen dependence, *Am. J. Clin. Pathol.*, 73, 330, 1980.
6. **Pertschuk, L. P., Tobin, E. H., Tanapat, P., Gaetjens, E., Carter, A. C., Bloom, N. D., Macchia, R. H., and Eisenberg, K. B.,** Histochemical analyses of steroid hormone receptors in breast and prostatic carcinoma, *J. Histochem. Cytochem.*, 28, 799, 1980.
7. **Lee, S. H.,** Cancer cell estrogen receptor of human mammary carcinoma, *Cancer*, 44, 1, 1979.
8. **Dandliker, W. B., Brawn, R. J., Hsu, M.-L., Brawn, P. N., Levin, J., Meyers, C. Y., and Kolb, V. M.,** Investigation of hormone-receptor interactions by means of fluorescence labeling, *Cancer Res.*, 38, 4212, 1978.
9. **Nenci, I., Beccati, M. D., Piffanelli, A., and Lanza, G.,** Detection and dynamic localization of estradiol-receptor complexes in intact target cells by immunofluorescence technique, *J. Steroid Biochem.*, 7, 505, 1976.
10. **Fisher, B., Gunduz, N., Zheng, S., and Saffer, E. A.,** Fluoresceinated estrone binding by human and mouse breast cancer cells, *Cancer Res.*, 42, 540, 1982.
11. **Gunduz, N., Zheng, S., and Fisher, B.,** Fluoresceinated estrone binding by cells from human breast cancer obtained by needle aspiration, *Cancer*, 52, 1251, 1983.
12. **Fisher, B., Gunduz, N., and Saffer, E. A.,** The interrelation between tumor cell proliferation and 17-fluoresceinated estrone binding following primary tumor removal, radiation, cyclophosphamide or tamoxifen, *Cancer Res.*, 43, 5244, 1983.
13. **Fisher, B. and Gunduz, N.,** Further observations on the inhibition of tumor growth by corynebacterium parvum with cyclophosphamide. X. Effect of treatment on tumor cell kinetics in mice, *J. Natl. Cancer Inst.*, 62, 1545, 1979.
14. **Gunduz, N., Fisher, B., and Saffer, E. A.,** Effect of surgical removal on the growth and kinetics of residual tumor, *Cancer Res.*, 39, 3861, 1979.
15. **Fisher, B., Gunduz, N., and Saffer, E. A.,** Influence of the interval between primary tumor removal and chemotherapy on kinetics and growth of metastases, *Cancer Res.*, 43, 1488, 1983.
16. **Hilf, R.,** personal communication.
17. **Eagon, P. K., Fisher, S. E., Imhoff, A. F., Porter, L. E., Stewart, R. R., Van Thiel, D. H., and Lester, R.,** Estrogen-binding proteins of male rat liver: influences of hormonal change, *Arch. Biochem. Biophys.*, 201, 486, 1980.
18. **McCarty, K. D., Jr., Woodard, B. H., Nichols, D. E., Wilkinson, W., and McCarty, K. S., Sr.,** Comparison of biochemical and histochemical techniques for estrogen receptor analyses in mammary carcinoma, *Cancer*, 46, 2842, 1980.
19. **Poulsen, H. S.,** Oestrogen receptor assay — limitation of the method, *Eur. J. Cancer*, 17, 495, 1981.

Chapter 4

HISTOCHEMICAL STUDY OF ESTROGEN RECEPTORS IN THE RAT UTERUS WITH A HYDROPHILIC FLUORESCENT ESTRADIOL CONJUGATE

Sin Hang Lee

TABLE OF CONTENTS

I. Introduction 60

II. Cellular Estrogen Receptors 60
A. Estrogen Receptors in Cryostat Frozen Sections 61
B. In Vitro Conditions in Histochemical Assay 61

III. Fluorescent Steroid Histochemical Technique 62
A. Technical Procedure 62
B. Interpretation and Limitations 63
C. In Vitro Criteria of Estrogen Receptors 63

IV. Estrogen Receptors in the Rat Uterus 65

V. Effects of Estrogens and Antiestrogens 74

VI. Concluding Remarks 79

Acknowledgments 81

References 81

I. INTRODUCTION

Biological response of the female reproductive system, especially the uterus and the vagina of the rodents, has been used as a yardstick to measure estrogenic and antiestrogenic activity for many decades.[1-4] Since the discovery that [^{3}H]-estradiol injected into rats was preferentially retained by certain target tissues,[5] much attention has been focused on the study of the estrogen binding macromolecules or estrogen receptor (ER) proteins of the rat uterus. Numerous attempts have been made to purify and characterize these binding proteins, and to define their exact intracellular distribution in the uterus under different physiological and experimental conditions.[6-11] Based on quantitation of the specific estrogen binding activity in the cytosolic and the nuclear fraction of tissue homogenates, the now widely accepted general scheme of steroid-receptor interaction was developed.

While the biochemical research carried out with tissue homogenates has made a major contribution to the study of steroid hormone action, its methodology has not taken into account the fact that the uterus is composed of many cell types. Each of these cell types may respond to hormone stimulation with different sensitivity, and may even contain a different amount of cytoplasmic receptors which may not be all soluble for their recovery in the cytosol. In addition, with the harsh mechanical treatment that is often required to pulverize the frozen uterine tissue for extraction, cytosolic contamination by soluble components derived from other intracellular organelles becomes inevitable. Besides, there is practically no published evidence to confirm the purity of the nuclear fractions employed in the biochemical studies. Therefore, the only reliable information provided by the protein assays is the number of estrogen binding sites in the supernatant of the tissue homogenate of the whole uterus. Whether all macromolecules that are found to bind estradiol in the cytosols with a high affinity and hormonal specificity would be capable of being translocated into the nucleus of the intact cells upon in vivo hormonal stimulation is open to question.

Radioautography provides an alternative to localize estrogen binding in different cell types and in different compartments of a target cell. After injection of radiolabeled estradiol into animals, the nuclei of various cell types of the uterus seem to have the highest concentration of radioactivity,[12,13] indicating that the nucleus of the target cell is probably the final destination of much of the injected estradiol or its metabolites. This finding is consistent with the hypothesis of the two-step cytosol-nuclear interaction. However, the radioautographic technique is less effective in demonstrating the primary binding sites in the cell, for example, those in the cytoplasm. According to the biochemical data, most of the estrogen binding proteins of the cells should be located in the cytoplasm. Another drawback for radioautography is that it is difficult to use this technique to study a large number of animals under different experimental conditions because its extreme sensitivity and the variable time intervals required for an optimum development of a radioautogram make a comparative semiquantitative study almost impossible.

This chapter introduces an in vitro histochemical technique for localizing the intracellular estrogen binding sites in cryostat frozen sections of the rat uterus, using a macromolecular, hydrophilic, fluorescent estradiol conjugate as a histochemical reagent. The technique was originally developed for observation of the human breast cancer cells rich in cytoplasmic estrogen receptors,[14,15] and the results have been found reproducible in many independent laboratories.[16-24] It has the advantage of providing a means to survey estrogen binding activity of numerous potential target cells in a large number of tissue sections with little effort, and thus facilitates comparative study of this functional change in a specific cell type under different physiological and experimental conditions.

II. CELLULAR ESTROGEN RECEPTORS

A major difference between the histochemical and the biochemical approach to studying estrogen receptors is that the former is concerned with receptors in relatively intact cellular or tissue fragments, and the latter with free receptor protein molecules. Since most of our information on steroid-receptor interaction has been derived from research work on soluble receptor proteins, the use of tissue-bound receptors as a model for in vitro study is sometimes regarded unorthodox. Therefore, two general remarks pertinent to the demonstration of receptor sites in histochemical tissue preparations must be made before any technical procedure can be considered. These remarks are directed toward answering two important questions. First, are there any receptors left in the cells whose cytoplasmic membranes have been disrupted? Second, in what way do the receptors bound to the cells differ from the receptor proteins in cytosols in terms of ligand binding characteristics?

A. Estrogen Receptors in Cryostat Frozen Sections

Some biochemists who are primarily accustomed to assay ER proteins in tissue cytosols have questioned the feasibility of using cryostat sections as suitable materials for study.[25-27] According to the traditional biochemists' concept, a cell is merely a bag of fluid with a nucleus in it. Once the bag was ruptured, its contents in the fluid would have leaked out freely; there would be nothing left in the cytoplasm of the cells in a cryostat frozen section for estrogen binding study. However, more recent findings in the field of cell biology research have produced convincing evidence that this is not the case. To the contrary, the cytoplasm of a cell is a highly organized compartment. Many physiologically active molecules, possibly including enzymes and specific receptors, may be bound to the cytoplasmic cytoskeleton which is, in turn, formed by a complex network of three major filament systems, the microtubules, microfilaments, and intermediate filaments plus an interconnecting filigree of thin, filamentous bridges.[28-30] Therefore, it is quite probable that the methods employed in the biochemical assay can only detect a small fraction of the cytoplasmic estrogen binding substance that is easily extracted into the supernatant of the tissue homogenate. Cryostat frozen sections of the rat uterus incubated in solutions containing [^{3}H]-estradiol have been shown to bind the ligand with a high affinity and hormonal specificity, indicating that there is a considerable amount of nondiffusible estrogen receptors retained in the section.[31,32] It has been reported that the microsomal fraction of the homogenate of target tissues, indeed, contains a significant amount of specific estrogen receptors.[33,34]

B. In Vitro Conditions in Histochemical Assay

Unlike the biochemical procedures which are designed to measure the binding activity of soluble ER proteins in dilute solutions, a histochemical method is concerned with detecting the binding activity of estrogen receptors bound to relatively intact microsegments of the cells. It is natural to expect that the number of estrogen binding sites per volume in the cytoplasm of a target cell is much higher than that found in an environment customarily employed for competitive binding protein analysis. For example, a 40-fold dilution of tissue homogenates based on the wet weight of rat uteri has been used for cytosol assays.[35] It has been well known that there is a highly uneven distribution of ER proteins among different types of target tissue, [36] presumably, the higher the intracellular receptor content there is, the more sensitive are the cells to hormonal manipulation. Therefore, the concentration of binding sites in the cytoplasm of a receptor-rich target cell may be many times higher than that of a receptor-poor target cell in the same rat uterus, and may be hundreds of times that in the protein-ligand incubation mixture used for the sucrose density gradient or the dextran-coated charcoal cytosol assays.

In a cytosol assay system, the concentration of the ligands bound to receptor proteins can never exceed the concentration of the free ligand initially added to the incubation mixture. However, in a histochemical or cytochemical assay setting, the concentration of ligand in a target cell may well exceed that in solution many times. This can be confirmed by suspending frozen and thawed MCF-7 tumor cells in a low concentration of histochemical reagent with constant gentle rotation. At the end of incubation when equilibrium is reached, most of the added free fluorescent steroid molecules are found to be bound by the tumor cells, so that the concentration of fluorescent estradiol conjugate in the tumor cell pellet can be more than tenfold higher than the original in solution.

There is a difference not only in concentration, but in physical characteristics between the receptors in the cell and the receptors in cytosol. Unlike those soluble in supernatant of a tissue homogenate, the receptor sites demonstrated histochemically are bound to cellular structures. The situation is reminiscent of an immobilized enzyme system. In comparative studies, it has been shown that because immobilized enzymes are less accessible to substrate and are subject to partitioning effects leading to a modified microenvironment in the domain of the immobilized enzyme particle, great differences are expected to exist between these fixed enzymes and the soluble enzymes, as reflected in the values of their kinetic parameters.[37] Therefore, the kinetic parameters used to characterize the binding behavior of ER sites in tissue sections may well differ from those based on soluble receptor proteins.

In addition to the potential difference in physical properties between the soluble receptors and the tissue-bound receptors, there may be chemical differences between the two as well. While it has been widely reported that ER proteins are very labile in solution, the intracellular receptors appear to be relatively stable in cryostat frozen sections that have been kept for several hours at 4°C. This finding indicates that there is at least a different rate of chemical denaturation process between these two forms of receptors. Similar differences have been previously demonstrated between soluble and cell-bound enzymes. For example, soluble or purified aspartate aminotransferase is almost instantaneously irreversibly inactivated by a 1% glutaraldehyde solution, whereas the activity of the same enzyme when still bound to the cells of the rat liver tissue even in cryostat sections is resistant to the same treatment.[38,39]

Depending on experimental conditions, a cryostat frozen section may measure somewhere between 5 and 14 μm in thickness. At the very surface of the tissue section in direct contact with the ligand solution, there is probably a free exchange between the bound and unbound ligand, and the concentration of unbound ligand should approximately equal that in the solution. However, in the deeper layers of the section toward the glass slide, the concentration of free ligand decreases progressively. This can be confirmed by incubating two superimposed cryostat frozen sections put on one microscopic glass slide. The target cells in the section sandwiched between the top section and the glass slide often fail to show any positive ligand binding activity at all, indicating that the diffusion of a histochemical reagent is quite limited, especially when mechanical mixing or stirring is avoided, like in almost all histochemical staining procedures.

III. FLUORESCENT STEROID HISTOCHEMICAL TECHNIQUE

A. Technical Procedure

The principle of the fluorescent steroid histochemical techniques is based on the assumption that steroid hormone molecules are capable of binding to their receptors, soluble or insoluble, even when the hormone is covalently coupled to a dye or a carrier as long as its physiologically active determinants or radicals are exposed. Such a possibility was initially demonstrated with fluoresceinated 6-carboxymethyl oxime estradiol-17β.[40] However, because of its low solubility and hydrophobic nature that often cause poor differential staining in tissue sections, this compound has not been widely tested.

In order to bring the steroid hormones into aqueous solution, a hydrophilic molecule, such as bovine serum albumin (BSA), may be used as a carrier. 17β-Estradiol, acting as a hapten coupled to BSA via a chemical handle at its 6 position, has been shown to be capable of eliciting production of specific antibodies against the free steroid hormone. For the histochemical study, fluoresceinated BSA is used for the coupling so that each mole of BSA carries about 4 mol of fluorescein isothiocyanate and 26 mol of 17β-estradiol-6-carboxymethyl oxime radicals. The technical aspect concerning the synthesis of this histochemical estrogen receptor marker has been dealt with previously in detail.[14] Most conveniently, this compound is diluted to a concentration containing about 2×10^{-4} *M* of bound steroid for histochemical staining of frozen sections, although much lower concentrations have been used successfully.

For the staining procedure, a cryostat frozen section of fresh tissue specimen of about 10-μm thick on a clean microscopic glass slide is air-dried at 4°C for about 1 hr. The section is then rehydrated with phosphate-buffered saline (PBS) which is drained and wiped off a few seconds later. The rehydrated tissue section is covered with a drop of fluorescent steroid conjugate, and incubated in a humid chamber for 2 hr at room temperature. At the end of the incubation the conjugate is drained off, and the section is first rinsed very gently with PBS and then immersed in PBS again for 1 to 24 hr with at least one change of solution. Finally, the section is air-dried at room temperature and mounted in buffered glycerin.

A similar approach has been used to detect progesterone receptors using 11α-hydroxyprogesterone hemisuccinate-BSA-tetramethylrhodamine isothiocyanate as a progesterone receptor tracer.[15]

B. Interpretation and Limitations

Histochemical techniques are generally designed to localize a cell type or an intracellular organelle which is rich in a specific function or rich in a product of such a function against a background composed of other cells or cell products. If every cell and every component part of each cell of a tissue exhibits the same degree of a certain functional activity, the latter cannot be convincingly demonstrated histochemically because there is no built-in background which is needed for interpretation of results in a histochemical preparation just as a blank is needed in a biochemical assay. In a frozen section which has been stained for ER binding activity, the background varies according to the organs being studied. For example, the intensity of fluorescence in the background of a section of the spleen is much lower than that of a uterus, presumably, because the cells in the uterus, in general, do bind more fluorescent steroid conjugate than those in the spleen. But within the uterus, again, only those cells containing exceptionally high numbers of binding sites can be recognized by this histochemical technique.

Therefore, the observer should be aware of the background of the kind of tissue being studied. In view of the well-known physiological effects of estrogens with influence on the activity of so many organs and tissues, estrogen target cells must be widely distributed with various degrees of responsiveness to hormonal stimulation, not only between different cell types, but also between cells of the same histological type in different locations or different organs. Overextrapolation of established criteria in an uncharted system is dangerous and must be avoided.

C. In Vitro Criteria of Estrogen Receptors

Because of the physicochemical characteristics of tissue receptors, special in vitro experimental conditions were used to show that the binding between cryostat sections and the estrogen ligands, indeed, involves specific receptors. Several 10-μm-thick cryostat cross sections of 16 to 20 segments of proestrous rat uteri were cut and air-dried at 4°C, incubated in a solution of [^{3}H]-estradiol with or without a competitor, rinsed or washed for different periods of time, air-dried, and extracted with ethanol. The ethanolic extract was assayed for radioactivity which is considered to represent the amount of bound ligand in the system at equilibrium.

High-affinity binding — Using a low concentration of [^{3}H]-estradiol in solution, it is difficult to saturate all of the high-affinity binding sites in cryostat frozen sections within a few hours. One must increase the concentration of the radiolabeled ligand to 10^{-7} *M* or higher to give a maximum binding of these sites, then equilibrate the incubated sections in a ligand-free buffer for 16 hr. Suppose that the concentration of free ligand in the system was about 10^{-10} *M* at equilibrium; then, theoretically, no more than 10% of the ligand bound to the sections would be due to low-affinity binding sites with a K_d for estradiol on the order of 10^{-9} *M* or above. Therefore, even when a concentration of hydrophilic fluorescent estradiol conjugate considered unusually high by biochemical standards is used in the staining solution for histochemical binding study, the amount of residual fluorescent ligand eventually remaining bound to the cells after an extended period of washing in ligand-free solutions can be regarded representing the result of a high-affinity binding. A similar principle has been used for the sucrose density gradient analysis of ER proteins and to differentiate the type I and type II binding sites in cytosols.[41,42]

Limited binding capacity — The second generally accepted in vitro criterion for steroid receptor is the requirement to show that the number of binding sites in the tissue is limited. In order to do so, the frozen sections of the rat uterus were first incubated in a solution containing 10^{-6} or 10^{-5} *M* of [^{3}H]-estradiol, then were allowed to equilibrate in solutions with decreasing concentrations of free [^{3}H]-estradiol. When equilibrium is reached in the system, the amount of [^{3}H]-estradiol that is firmly bound to the sections were found to remain relatively constant over a wide range of low concentrations of free ligand, for example, from lower than 10^{-12} to 10^{-9} *M*, indicating that the number of the high-affinity binding sites in the sections is limited.[43] In a similar manner, increasing the concentration of the fluorescent estradiol conjugate in the histochemical staining solution does not change the pattern of the ligand distribution in tissue sections at all, if adequate post-incubation washings are provided to remove the loosely bound ligand from the low-affinity binding sites.

Steroid hormone specificity — Although the binding may not be absolutely specific, generally one type of steroid receptor can only bind its own corresponding hormone. Once the binding sites of the receptor are occupied by the hormone, no more ligand can be taken up by the receptor. To prove that a histochemical reagent is, in fact, bound by the binding sites of a receptor, a competitive blocking experiment with the natural hormone or a known hormone competitor is necessary. Initially, great difficulties were encountered in an attempt to block the histochemical staining of the target cells by estradiol or DES when a coincubation system was employed.[14,40] Subsequently, it was found that it is necessary to preincubate the sections in a high concentration of competitor prior to the addition of the histochemical reagent in order to achieve a complete blocking of staining.[31] A partial blockage is difficult to recognize because a steroid or its competitor at high intracellular concentrations may cause an increase rather than a decrease in the intensity of fluorescence due to a peculiar phenomenon which may involve light scattering effects or energy transfer between closely placed molecules[44] (see next chapter). In addition, the incorporation of 10% glycerin in the incubation mixture has increased the solubility of the competitors. With these subsequent modifications in procedure, staining of the estrogen receptor sites in target cells by the histochemical reagent has been shown to be successfully blocked by a 100-fold excess of DES. In a coincubation system, the estradiol derivative covalently coupled to fluoresceinated BSA is found to be at least as effective as DES in preventing the binding of [^{3}H]-estradiol to the frozen sections of the rat uterus.[31]

Tissue specificity — A rigid categorization of all organs into target tissues and nontarget tissues of estrogen is probably impossible in view of the wide spectrum of physiological effects that may be induced by this class of hormones. However, according to biochemical studies, the uterus seems to contain the highest amount of ER proteins, and the lowest amount is found in the spleen.[36] Therefore, it appears safe to consider the uterus as a standard target organ and the spleen an example of a nontarget organ. Since no significant binding of the fluorescent steroid reagent by the cells in the frozen sections of the spleen is detected, and since the fluorescent staining is consistently observed in the epithelial cells of the target organs, such as the uterus and the mammary glands, it may be concluded that the staining by this histochemical reagent is at least specific for certain epithelial cells in the target organs of estrogen.[43]

IV. ESTROGEN RECEPTORS IN THE RAT UTERUS

Initially, it was hoped that the uterus of the adult female rat can be used as a positive tissue control for histochemical ER assay of human mammary cancers since the epithelial cells of human mammary glands and of human endometrium received in a clinical pathology laboratory do not always have a high ER level. However, it soon became clear that even the epithelial cells in the adult rat uterus varied tremendously in their cytoplasmic estrogen binding activity, ostensibly related to estrous cycles. The luminal epithelial cells of the uterus usually show the highest degree of binding at the late stage of proestrus when the uterus is markedly edematous and distended with fluid. But even in the proestrous stage, there is only a background fluorescence in the glandular epithelial cells situated deeply in the endometrial stroma, indicating a functional difference in the epithelial cells in two locations of the same organ. It is also of great interest to note that practically all of the cells with estrogen receptor activity were found to be positive for progesterone binding as well. This consistent coexistence of two receptors in the ER-rich epithelial cells has been repeatedly observed throughout all experiments described, subsequently, in this chapter.

In addition to the epithelial cells, eosinophilic leukocytes which vary greatly in number in the rat uterus during the estrous cycle were also stained selectively by the fluorescent estradiol conjugate.[43] However, the staining characteristics of eosinophils differ from those of the luminal epithelial cells in several aspects. First of all, the eosinophils are stained by dilute solutions of the fluorescein dye and fluoresceinated BSA without the estradiol moiety, while the epithelial cells are not selectively stained by these dyes. Secondly, not every batch of fluorescent estradiol conjugate can stain the eosinophilic leukocytes. Only when there are fluorescent compounds with a molecular size smaller than that of the intact BSA molecules in the conjugate (often as a by-product of the synthetic process), and only then can the eosinophils be stained. The highly purified conjugates composed of only the full-sized fluoresceinated BSA molecules carrying at least 24 radicals of estradiol derivative per mole of BSA do not stain the eosinophils. Certainly, the intracellular binding activity of the eosinophils does not fluctuate during the estrous cycle as in the epithelial cells. Finally, the eosinophils apparently have no progesterone receptors since they do not bind the hydrophilic fluorescent progesterone histochemical reagent. At one point it was thought that the staining of eosinophils was entirely due to their eosinophilic characteristics because of the molecular similarity between fluorescein and eosin. However, their staining by the fluorescent estradiol conjugate was found to be successfully blocked by a 100-fold of DES. These data collectively appear to support the claim that there is an ER system in the eosinophilic leukocytes.[45] Perhaps, the full-sized BSA steroid molecules failed to bind to the receptor sites firmly in these leukocytes because of steric hindrance. Whether this really is the case must await additional investigation.

In order to study the cyclic changes of the cytoplasmic ER in the luminal epithelial cells of the uterus, young adult female rats exhibiting a regular 4-day estrous cycle were selected. All animals were housed in an air-conditioned room, lighted from 5 a.m. to 7 p.m. The stages of the cycle were followed by daily vaginal smears for at least three cycles. Sometimes, two or three smears per day had to be made to identify the rats with a uterus whose luminal epithelial ER was at a peak level. This usually coincided with a time when all leukocytes in the vaginal smears had disappeared and about a half of the exfoliated epithelial cells still retained their nuclei. As soon as the cells became cornified and anuclear, the level of the luminal epithelial cytoplasmic ER dropped precipitously.

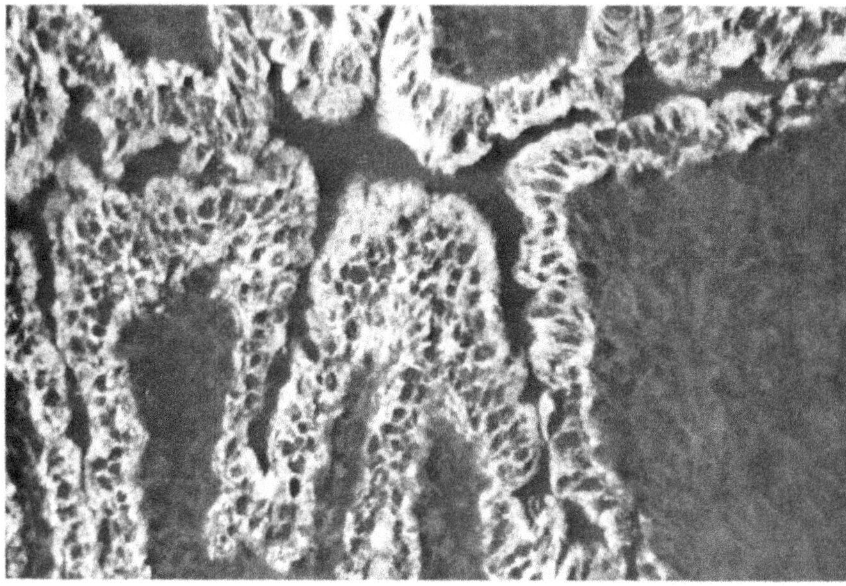

FIGURE 1. Uterus of a young adult female rat in late proestrus, showing strong estrogen binding activity covering the entire extra nuclear region in the luminal epithelial cells. (Magnification × 300.) (Reprinted with permission of Springer-Verlag, Heidelberg, from The Histochemistry of Estrogen Receptors, Lee, S. H., *Histochemistry*, 74, 443—452, 1982.)

The regular wax-and-wane cyclic changes of the intracellular ER activity of the luminal cells can be summarized as follows.

During the proestrous stage, the population of estrogen binding sites increases rapidly in density in the cytoplasm. At the end of this stage, the entire extranuclear region shows a coarsely granular bright fluorescence after staining (Figure 1*). This observation is best made on the proximal third of the uterine horn. The luminal cells in the distal uterine segments near the oviduct, though following the same pattern of cylic changes in ER contents, tend to show a less uniform and a less intense degree of estrogen binding. It is of interest to note that the distribution of ER-rich epithelial cells only extends to the end of the uterine horn, and vanishes abruptly at the junction where it meets the oviduct whose luminal epithelial cells appear to contain very low levels of estrogen receptors, if any (Figures 2 and 3). However, the lining cells of the fimbriated end of the oviduct (Figure 4) and the hypertrophic germinal epithelial cells covering the graafian follicles of the ovary (Figure 5) appear to contain high levels of cytoplasmic estrogen receptors in proestrus.

As soon as the cycle enters the estrous stage, whch is characterized by an abundance of cornified epithelial cells in the vaginal smear, uterine edema begins to subside and the estrogen binding activity of the luminal epithelial cells also drops. Toward the late stage of estrus, although the exfoliative cells in the vaginal smear at this time are still primarily composed of cornified epithelial cells, estrogen binding activity seems only confined to the base of the luminal epithelial cells and along the cell borders (Figure 6).

The lowest number of intracellular binding sites is observed in metestrus. At this stage, there is only occasional faint cytoplasmic fluorescence near the free cytoplasmic border of the luminal cells (Figure 7). The estrogen binding activity begins to increase in the diestrous stage when repopulation of the cytoplasm with binding sites is first evident at the base and

* All figures are photomicrographs of 10-μm-thick cryostat frozen sections which were stained with 17β-estradiol-6-carboxymethyl oxime-BSA-FITC solution and exhaustively washed in ligand-free buffered saline. Deposits of the fluorescent conjugate indicate the sites of estrogen binding activity in high density.

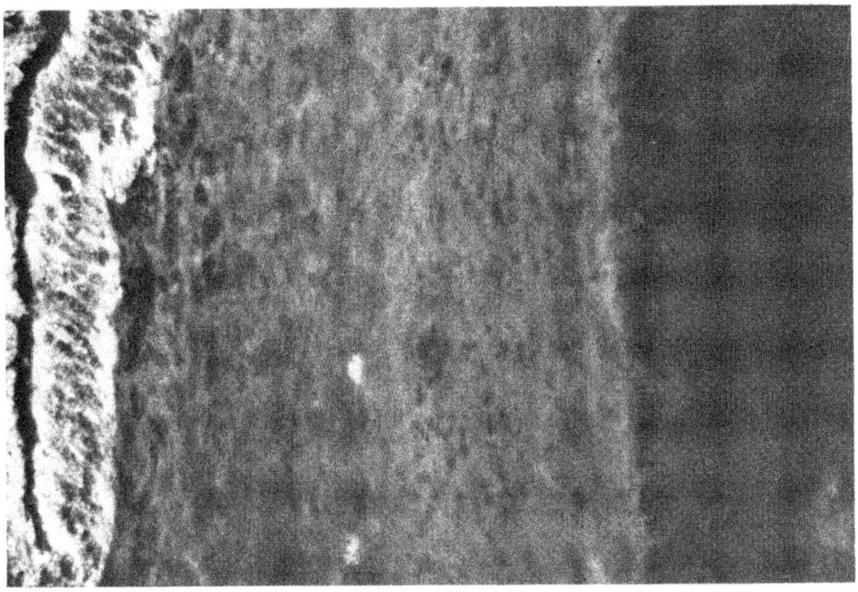

FIGURE 2. A section of a proestrous rat uterus showing luminal epithelial cells in the distal segment of a uterine horn (left) and the cells lining the lumen of the oviduct (right). Note the absence of estrogen binding activity in the epithelial cells of the oviduct in contrast to the strong cytoplasmic fluorescence in the luminal cells of the uterine horn. (Magnification × 300.)

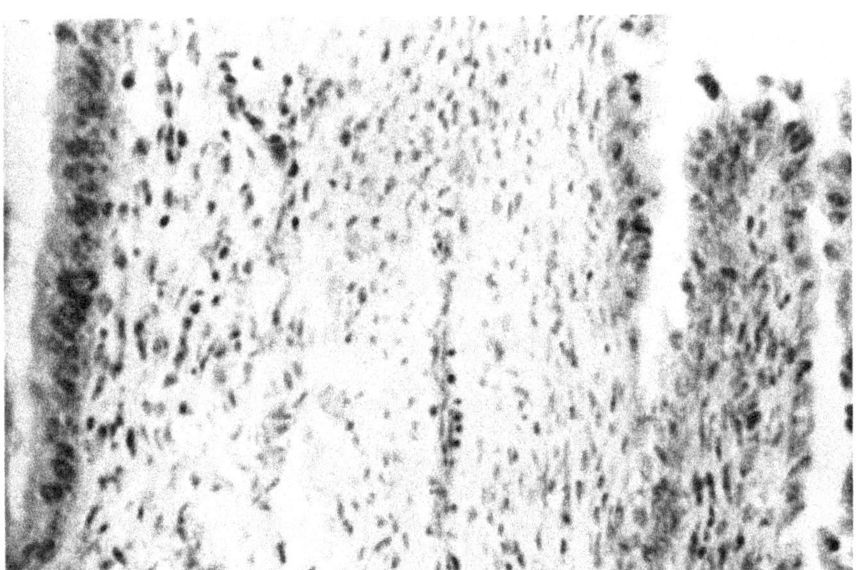

FIGURE 3. A serial section adjacent to that shown in Figure 2. Hematoxylin and eosin stain to show the morphological similarities between the luminal cells in the uterine horn (left) and in the oviduct (right). (Magnification × 300.)

near the free cytoplasmic borders of the cells in some segments of the luminal epithelium (Figure 8). This process continues into the proestrous stage and reaches its peak in late proestrus.

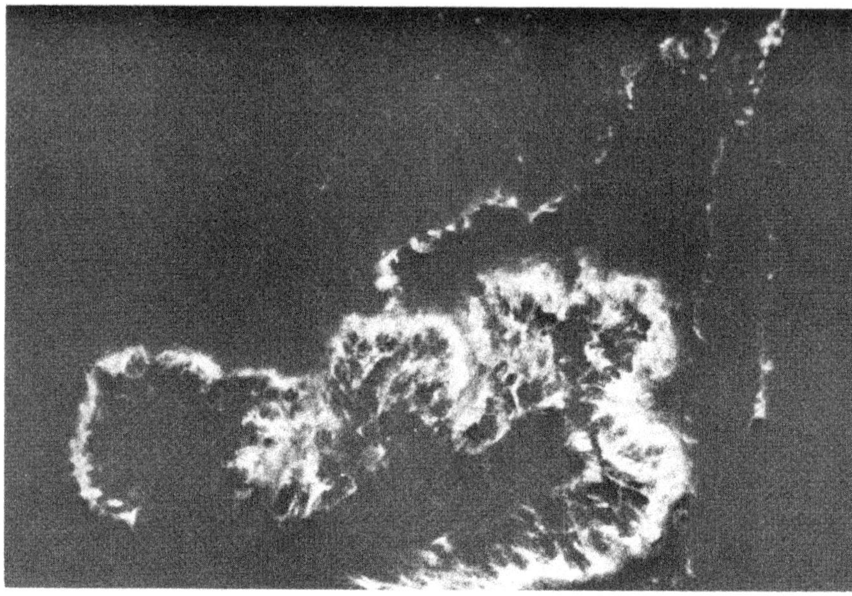

FIGURE 4.

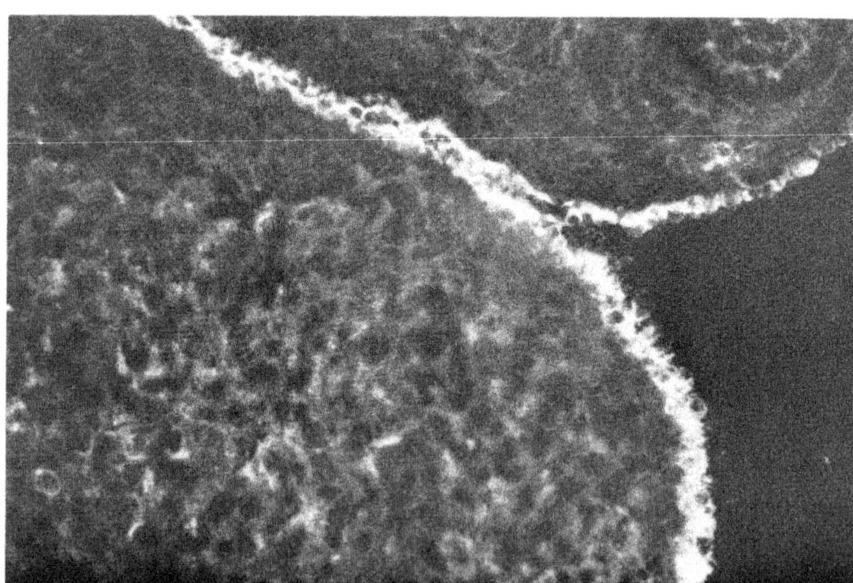

FIGURE 5.

FIGURES 4 and 5. Epithelial cells of the fimbriated end of the oviduct (Figure 4) and the germinal epithelial cells of the ovary (Figure 5) at the end of proestrous stage. Note strong cytoplasmic estrogen binding activity in both.(Magnification × 300.)

The regular cyclic fluctuation of ER contents in the cytoplasm of the luminal epithelial cells during the estrous cycle is easily observed. Similar changes may occur in other cell types of the uterus, but cannot be detected by this histochemical technique; for example, in order to demonstrate an increase in ER in the smooth muscle cells of the myometrium, an

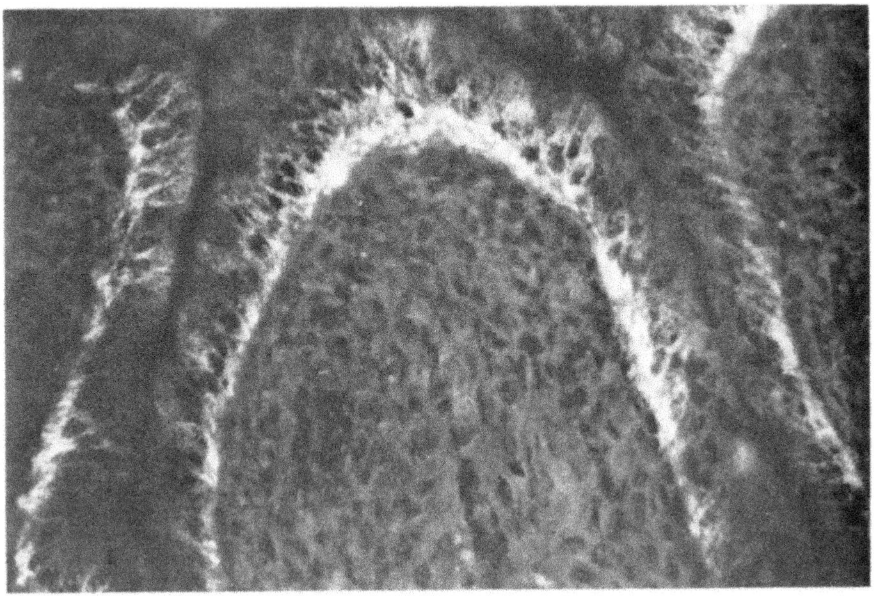

FIGURE 6. Adult rat uterus in late estrus. Estrogen binding activity is confined to the base and along the borders of the luminal epithelial cells. (Magnification × 300.) (Reprinted with permission of Springer-Verlag, Heidelberg, from The Histochemistry of Estrogen Receptors, Lee, S. H., *Histochemistry*, 74, 443—452, 1982.)

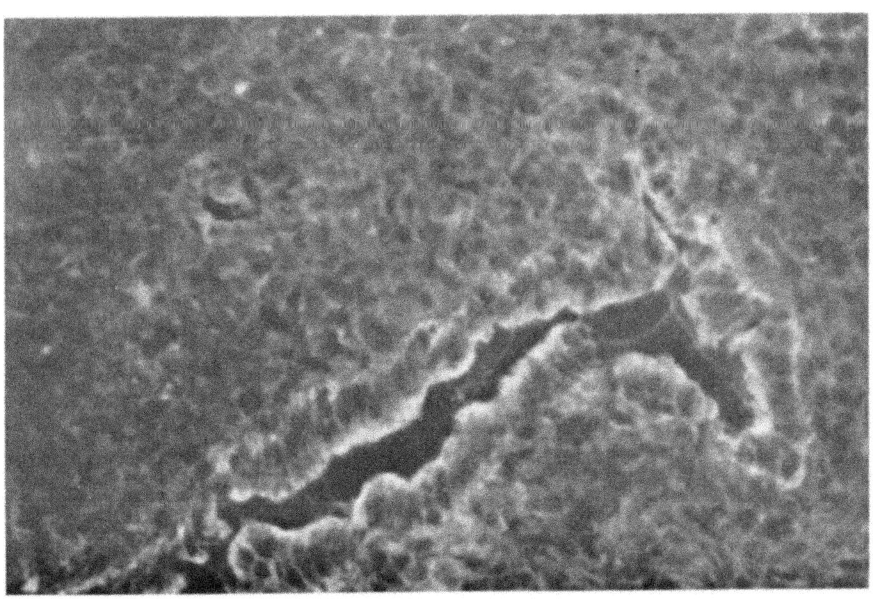

FIGURE 7. Rat uterus in metestrus. Little estrogen binding activity is noted except occasional weak fluorescence near the free border of the luminal cells. (Magnification × 300.) (Reprinted with permission of Springer-Verlag, Heidelberg, from The Histochemistry of Estrogen Receptors, Lee, S. H., *Histochemistry*, 74, 443—452, 1982.)

ovariectomized rat must be first prepared with multiple injections of large doses of estradiol (see below).

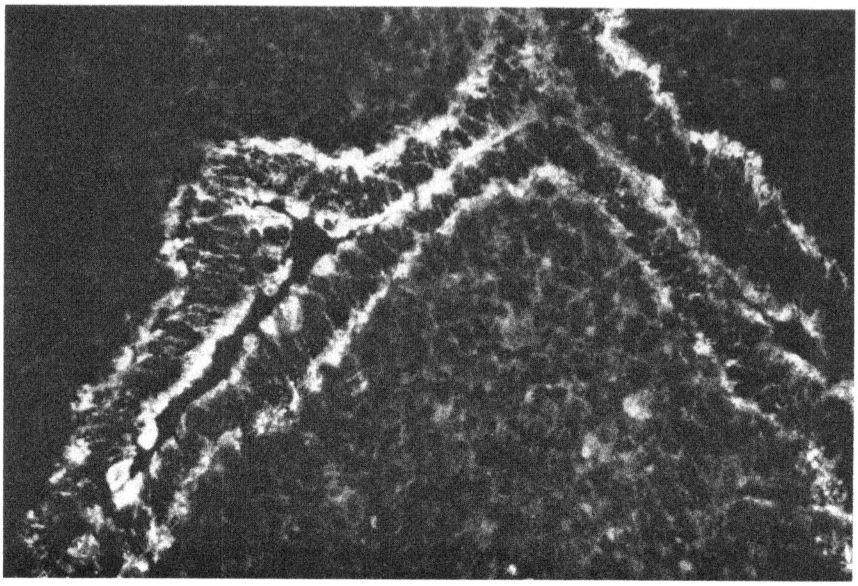

FIGURE 8. Rat uterus in diestrus. Focal reappearance of estrogen binding activity is noted near the free border and at the base of the luminal cells. (Magnification × 300.) (Reprinted with permission of Springer-Verlag, Heidelberg, from The Histochemistry of Estrogen Receptors, Lee, S. H., *Histochemistry*, 74, 443—452, 1982.)

The fluctuating pattern of ER proteins in the uterus of young adult female rats at different stages of the estrous cycle has been widely studied. However, the maximum concentration of ER proteins has been reportedly observed by various investigators at proestrus,[46] estrus,[47] and late diestrus.[48] Since ER protein assays measure only the extractable binding sites in the whole uterus, it is difficult to compare these data with the histochemical findings described in this chapter. The observation that the intracellular ER level peaks in proestrus when the unopposed effect of endogenous estradiol is expected at its highest point[49] is most consistent with the hypothesis of replenishment regulation by progesterone.[50] According to the latter hypothesis, the synthesis of uterine ER proteins is augmented by estrogens and suppressed by progesterone through its interference with the replenishment mechanism. In order to test whether this mechanism can be observed at the cellular level, immature female rats of 22 days old or ovariectomized adult female rats 3 weeks post-castration were injected subcutaneously standard doses of 17β-estradiol E_2 commonly used for priming.[50,51] The animals were killed randomly from 8 hr to 5 days after the initial steroid injection, and the uteri examined for intracellular estrogen binding activity. The results showed that three daily injections of 2.5 μg of E_2 (or as little as 0.08 μg) in the immature rats induced synthesis of ER in a quantity adequate to cover the entire cytoplasm of the luminal epithelial cells (Figure 9 and 10). For the ovariectomized rats, three daily injections of 10 μg of E_2 were used with similiar effects (Figure 11 and 12). Estrogen binding activity usually began to appear in these cells 24 hr after the initial injection, reaching its peak level in 48 to 72 hr and maintaining this high level for the remaining period of the experiment as long as the rats received daily injections of E_2. A single large dose of 2.5 mg of progesterone injected subcutaneously in immature rats (Figure 13) or 6.5 mg of progesterone in the ovariectomized adult rats 24 hr before sacrifice caused a massive loss of the estrogen binding sites in these cells. However, replacing progesterone with hydrocortisone in the same experiment produced no suppressive effects on the estrogen-induced augmentation of ER (Figure 14).

In order to test whether large doses of estrogen can block the intracellular estrogen binding

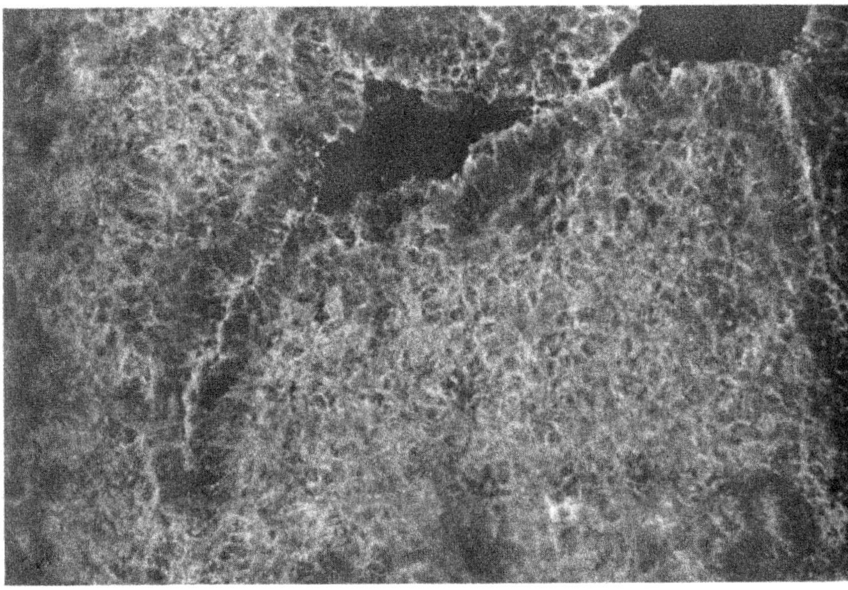

FIGURE 9. Uterus of a 25-day-old control immature rat which had received three daily injections of 0.5 mℓ of ethanol-saline. There is only a background fluorescence in the epithelial cells. (Magnification × 300.) (Reprinted with permission of Elsevier Science Publishing Company, Inc., New York, from Validity of a Histochemical Estrogen Receptor Assay, Lee, S. H., *The Journal of Histochemistry and Cytochemistry*, 32, 305—310. Copyright 1984 by The Histochemical Society, Inc.)

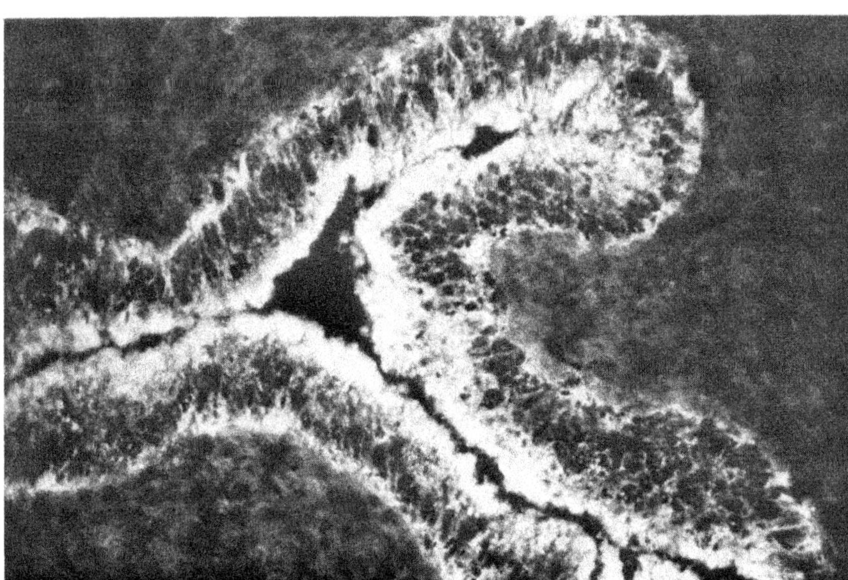

FIGURE 10. Uterus of a 25-day-old immature rat which had been given three daily subcutaneous injections of 2.5 μg of estradiol on day 22, 23, and 24. Note the strong estrogen binding activity in the cytoplasm of the luminal epithelial cells. Similar effects were induced with three daily injections of 0.08 μg of estradiol. (Magnification × 300.) (Reprinted with permission of Elsevier Science Publishing Company, Inc., New York, from Validity of a Histochemical Estrogen Receptor Assay, Lee, S. H., *The Journal of Histochemistry and Cytochemistry*, 32, 305—310. Copyright 1984 by The Histochemical Society, Inc.)

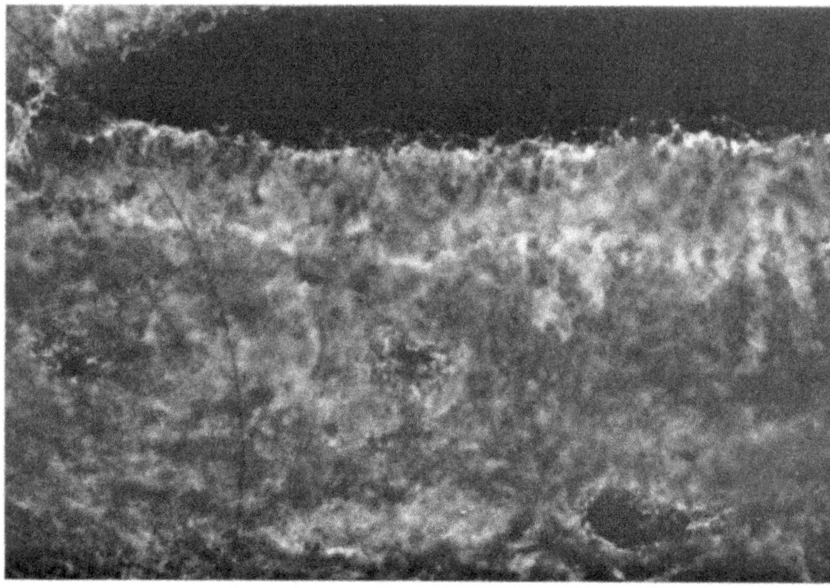

FIGURE 11. Atrophic uterus of an ovariectomized adult female rat 3 weeks after castration, showing a flat luminal epithelium with only a background fluorescence. (Magnification × 300.) (Reprinted with permission of Elsevier Science Publishing Compny, Inc., New York, from Validity of a Histochemical Estrogen Receptor Assay, Lee, S. H., *The Journal of Histochemistry and Cytochemistry*, 32, 305—310. Copyright 1984 by The Histochemical Society, Inc.)

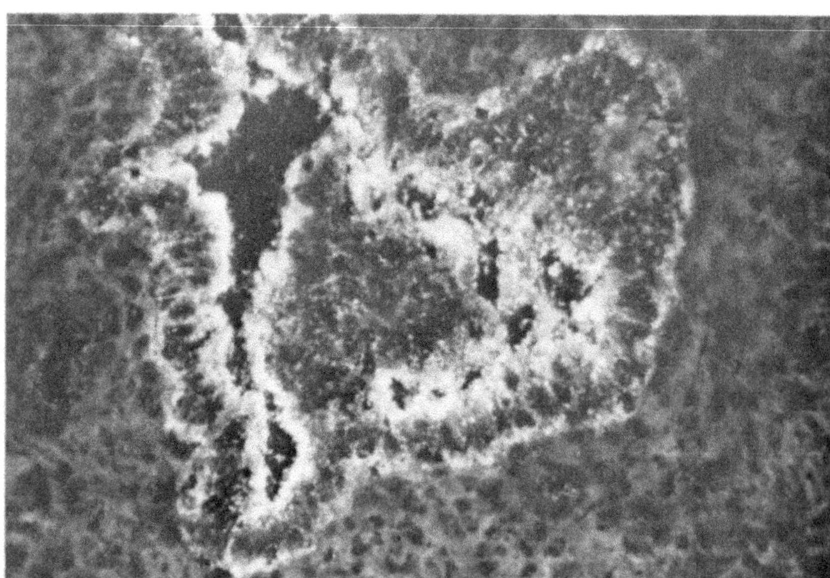

FIGURE 12. Uterus of an ovariectomized adult female rat which had received three daily injections of 10 μg of estradiol and was sacrificed 24 hr after the last injection. Note the strong estrogen binding activity in the cytoplasm of the hypertrophic and hyperplastic luminal epithelial cells. (Magnification × 300.)

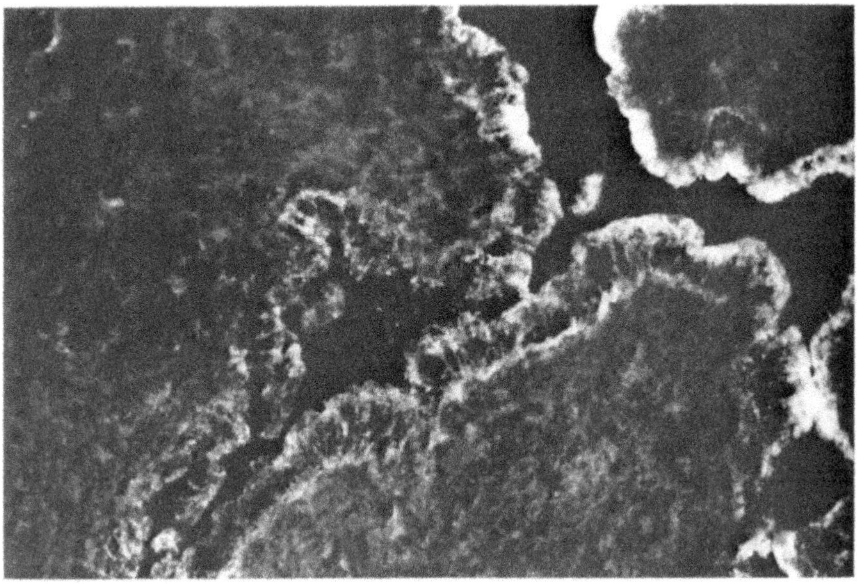

FIGURE 13. Uterus of a 25-day-old immature rat primed with three injections of estradiol (2.5 µg per day) was given a concomitant dose of 2.5 mg of progesterone with the last injection 24 hr before sacrifice. Note the massive losses of cytoplasmic estrogen binding activity in the epithelial cells compared to those treated with estradiol only (Figure 10). (Magnification × 300.) (Reprinted with permission of Elsevier Science Publishing Company, Inc., New York, from Validity of a Histochemical Estrogen Receptor Assay, Lee, S. H., *The Journal of Histochemistry and Cytochemistry*, 32, 305—310. Copyright 1984 by The Histochemical Society, Inc.)

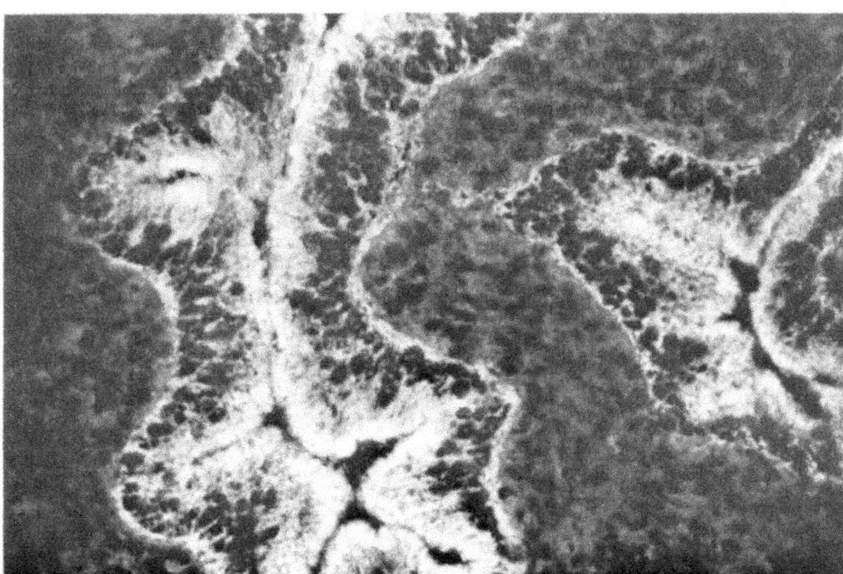

FIGURE 14. Uterus of an immature rat which had been primed with steroids in the same manner as described for the animal shown in Figure 13 except that 2.5 mg of hydrocortisone was used instead of progesterone. The hyperplastic and hypertrophic epithelial cells retain their strong estrogen binding activity, as those shown in Figure 10. (Magnification × 300.) (Reprinted with permission of Elsevier Science Publishing Company, Inc., New York, from Validity of a Histochemical Estrogen Receptor Assay, Lee, S. H., *The Journal of Histochemistry and Cytochemistry*, 32, 305—310. Copyright 1984 by The Histochemical Society, Inc.)

activity in vivo, five daily subcutaneous injections of 50 μg of E_2 in 1 mℓ of sesame oil were given to each ovariectomized adult female rat. The rats were sacrificed 24 hr after the last injection. All uteri were found to be markedly distended by fluid accumulated in the lumen. Interestingly enough, the hypertrophic luminal epithelial cells still showed a strong estrogen binding activity, although many of them flattened in shape, perhaps due to a rapid increase in intraluminal pressure (Figure 15). It also became clear that only under the condition of hyperestrogenic stimulation the estrogen binding activity in the smooth muscle cells of the myometrium could reach a level high enough for histochemical visualization (Figure 16).

During the estrous cycle, the intracellular epithelial estrogen binding activity appears to increase with a simultaneous synthesis of progesterone receptors as the result of an unopposed estrogenic stimulation. After ovulation, however, as the circulating progesterone reaches the target cells, its inhibitory action on the replenishment of estrogen receptor takes effect, causing massive losses of intracellular estrogen binding activity. This mechanism of control can be demonstrated at least in the luminal cells.[43,51] It is not known whether it also exists in other cell types of the uterus. Hydrocortisone as a nonsex steroid behaves like a neutral agent; it neither stimulates nor suppresses the synthesis of intracellular estrogen receptors.

V. EFFECTS OF ESTROGENS AND ANTIESTROGENS

The estrogenic and antiestrogenic effects of nonsteroidal antiestrogens have been extensively studied.[3,52,53] They all seem to manifest an initial estrogenic activity after administration in animals, and then render certain target tissues refractory to further estrogenic stimulation.[54,55] The estrogenic effect of these compounds is thought to be initiated by their binding to cytoplasmic estrogen receptors. The complex is then translocated to the nucleus where stimulation of the synthesis of various macromolecular substances is triggered, including the production of more cytoplasmic ER proteins. Thus, in their agonistic actions, the antiestrogens seem to follow the same sequence of the initial steps as that generally accepted for estrogens.[56-59] Their antagonistic effects on the estrogen action have been attributed to a failure to stimulate the replenishment mechanism which appears to be necessary for the maintenance of a high level of cytoplasmic estrogen receptors.[60-63] Since the uterine epithelial cells are known to be the target cells for both estrogens and antiestrogen,[64,65] they can be used as a potential in vivo cellular model for the study of estrogenic and antiestrogenic actions.

Immature 22-day-old female Sprague-Dawley rats were used. All steroidal hormones and nonsteroidal chemicals were dissolved in ethanol and diluted with normal saline immediately before use so that the final diluent contained 2% ethanol. For the convenience of studying the comparative effects of different chemical agents, the doses given to the rats were calculated on the basis of molarity.

To test a potential estrogenic activity, the rats were each given a daily subcutaneous injection of 44 nmol of E_2, DES, hydrocortisone, tamoxifen, nafoxidine, or nitromifene in 0.5 mℓ of diluent for five consecutive days. The animals were selected randomly and killed at 8 hr after the initial injection and then daily thereafter from day 23 to day 27 after birth, at least four animals in each group. Cryostat sections of the uterine horns were cut at 10 μm thick and processed for histochemical staining.

To test a potential antiestrogenic activity, the rats were injected with tamoxifen, nafoxidine, nitromifene, or hydrocortisone for five consecutive days as described above. From day 3 to day 5, an additional daily subcutaneous injection of 10 nmol of E_2 was also given at a different site. The control rats received three daily injections of E_2 and five daily injections of ethanol-saline solution. All animals were killed 24 hr after the last injection and the cryostat sections of the uterine horns stained for estrogen binding activity.

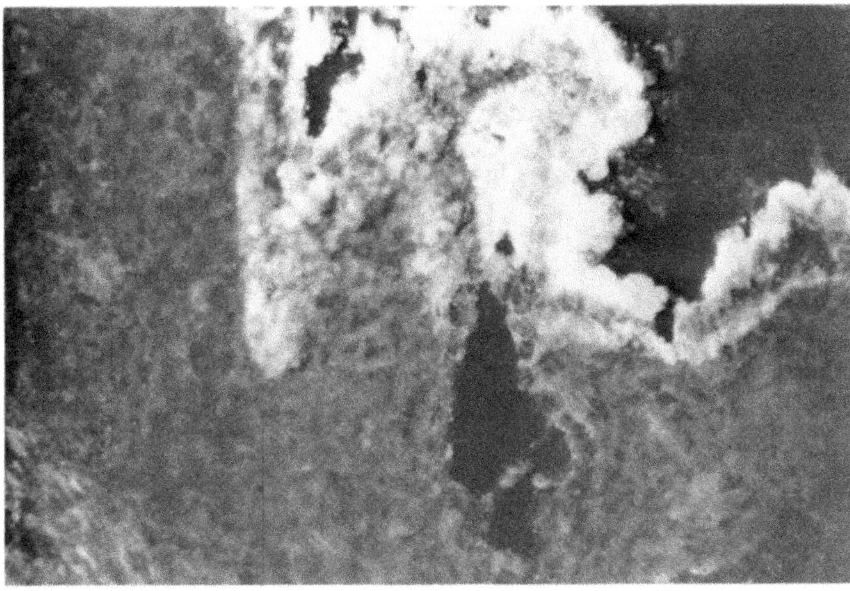

FIGURE 15.

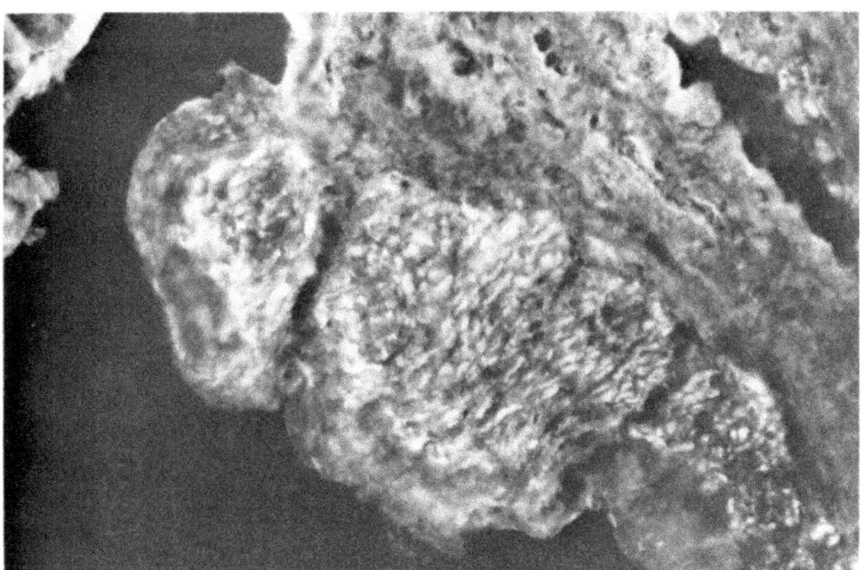

FIGURE 16.

FIGURES 15 and 16. Uterus of an ovariectomized adult female rat which had received five consecutive daily subcutaneous injections of 50 μg of estradiol and was sacrificed 24 hr after the last injection. The luminal epithelial cells (Figure 15) and the smooth muscle cells in the myometrium (Figure 16) show a strong cytoplasmic estrogen binding activity. (Magnification × 300.)

At 24 hr after injection of E_2, DES, tamoxifen, nafoxidine, or nitromifene, all rats began to show focal cytoplasmic fluorescence in some of the moderately hypertrophic uterine luminal epithelial cells, indicating an increase in estrogen binding activity. This binding

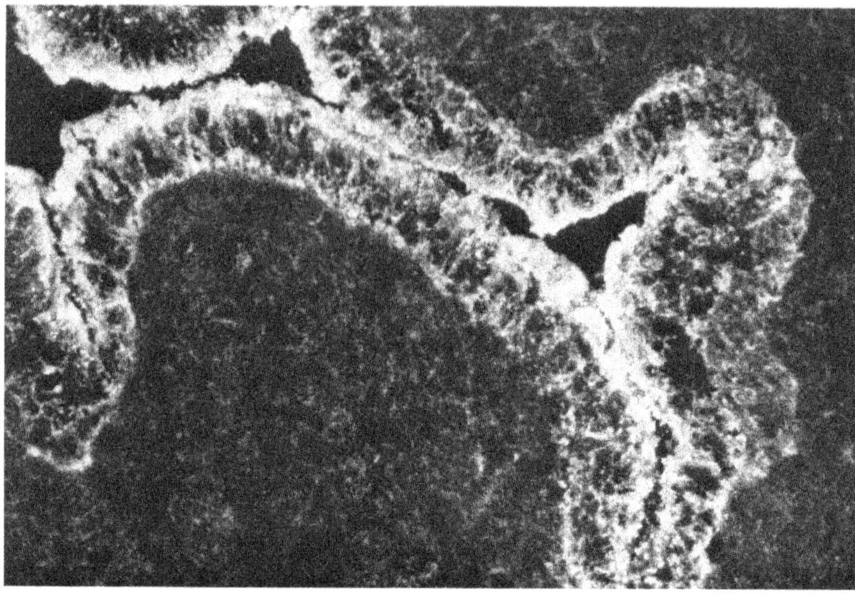

FIGURE 17. Uterus of an immature rat which was given consecutive daily subcutaneous injections of 44 nmol of tamoxifen. The rat was sacrificed 24 hr after the second injection. Note the strong estrogen binding activity in the cytoplasm of the luminal epithelial cells. Similar results were obtained with injections of equimolar doses of estradiol, DES, nitromifene, and nafoxidine, although in the nafoxidine treated rats, this high level of binding activity appeared a day or two later. (Magnification × 300.)

activity appeared to reach its peak level in another 24 hr in the animals treated wth E_2, DES, tamoxifen, or nitromifene, when practically the entire luminal epithelium showed a strong cytoplasmic fluorescence covering the extranuclear area of every cell (Figure 17), but not in the nafoxidine group in which the peak level of binding activity was not observed until 3 to 4 days after the initial injection. This high level of binding activity persisted throughout the entire experimental period in the animals treated with E_2 or DES thereafter as a plateau from day 24 to day 27, but dropped precipitously to a nondetectable level in the rats receiving injections of nonsteroidal antiestrogens in 24 hr after the peak level was reached (Figure 18). The endometrial epithelial cells retained their hypertrophic appearance even at the end of the experiment in all five groups treated with either estrogens or antiestrogens. No increased estrogen binding activity was observed in the uteri of the rats injected with hydrocortisone.

In the control rats receiving three daily injections of 10 nmol of E_2 from day 24 to day 26 after birth, there was a strong estrogen binding activity in the cytoplasm of the uterine epithelial cells (Figure 19). The E_2-induced augmentation of estrogen binding activity was not affected by the injections of hydrocortisone, but was completely neutralized by any of the three antiestrogens tested (Figure 20). This phenomenon was best observed after five daily injections of antiestrogens when the estrogen binding activity of the uterine luminal cells was so depressed that the intensity of cytoplasmic fluorescence exhibited by these cells, after the sections were stained in the histochemical reagent, was often even weaker than that observed with uteri of the untreated control animals.

The factual observations presented above can be recapitulated as follows.

Consecutive daily injections of estrogens or antiestrogens in the immature rats cause the cytoplasmic estrogen binding activity in the uterine luminal epithelial cells to increase greatly above the background level. This high level of binding activity is most conspicuously

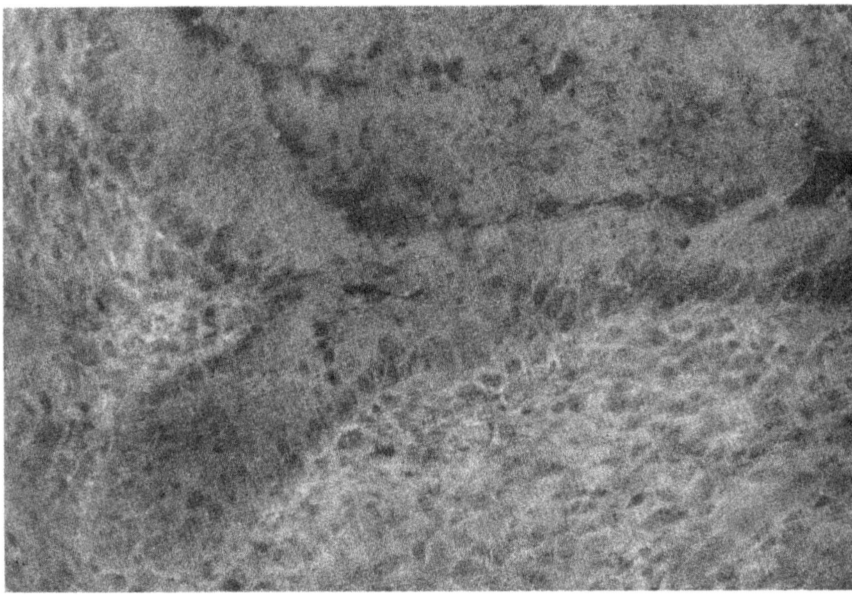

FIGURE 18. Uterus of an immature rat receiving five daily injections of 44 nmol of tamoxifen. The rat was sacrificed 24 hr after the last injection. The luminal epithelial cells still retain hypertrophic appearance, but have lost practically all of their cytoplasmic estrogen binding activity. In the animals which had been given all five daily injections of tamoxifen, nitromifene, or nafoxidene, the estrogen binding activity in these cells dropped to an undetectable level, usually in 24 hr after the peak level was reached. (Magnification × 300.)

demonstrated about 48 hr after the initial injection of E_2, DES, tamoxifen, and nitromifene, but may take 96 hr to develop under treatment of nafoxidine. While in the E_2 or DES groups the high level of cytoplasmic binding activity is maintained through the remainder of the observation period, the activity drops precipitously to a background or below-background level about 24 hr after reaching its peak in the tamoxifen, nitromifene, and nafoxidine groups, signaling the beginning of a refractory period during which time even additional injections of E_2 no longer increase the estrogen binding activity in these cells.

The most logical explanation for these phenomena is that the estrogen binding substance in the cytoplasm of the luminal epithelial cells as visualized by this histochemical technique represents part of the estrogen receptor system, since it behaves just like the soluble uterine ER proteins when the immature rats are treated with estrogens and antiestrogens.

One may argue that the repeated injections of antiestrogens may have simply saturated the ER sites already present in the epithelial cells; and the cytoplasmic ER saturated by the competitors could no longer be demonstrated histochemically. This argument will not hold because all estrogens and antiestrogens tested have caused an initial increase in the cytoplasmic estrogen binding activity, a factual observation not consistent with a simple action of physical blocking. Furthermore, continuous daily administration of equimolar doses of E_2 or DES did not lead to depletion of cytoplasmic estrogen binding activity.

The intracellular binding sites of the ER-rich target cells, especially those firmly bound to the cytoskeleton, probably can never be saturated in vivo by estrogens or antiestrogens injected subcutaneously in a location so remote from the uterus, because the concentration of these chemicals in the vicinity of the target cells will be determined by the carriers in the interstitial fluid and in the plasma, not by the absolute amount injected. In cytosol assays, as tissues are homogenized, estrogen previously in interstitial space or in plasma and unavailable for interaction with the cytoplasmic receptors is free to occupy the previously

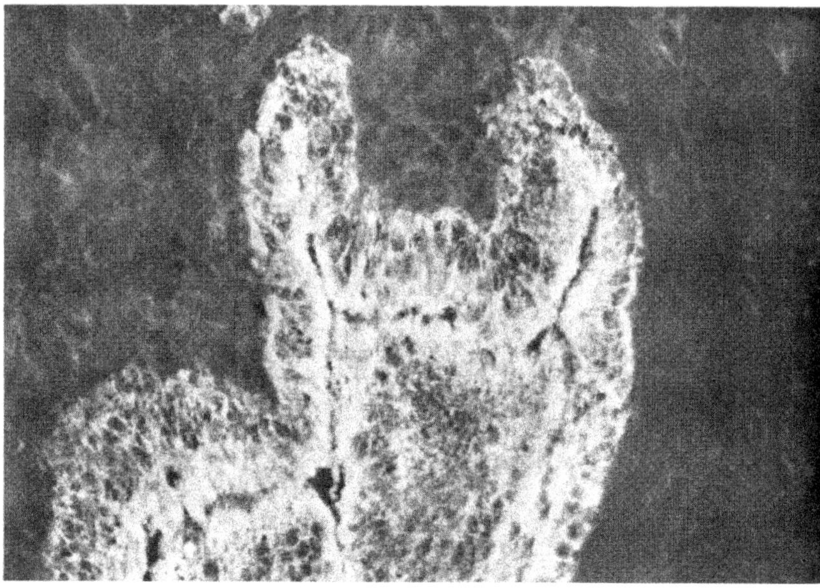

FIGURE 19.

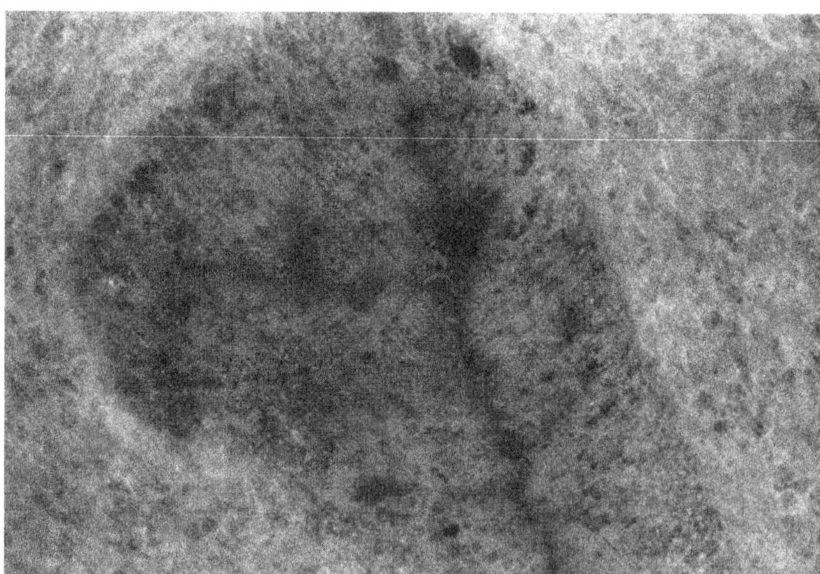

FIGURE 20.

FIGURES 19 and 20. Uterus of a rat given five daily injections of ethanol-saline and a concomitant injection of 10 nmol of estradiol during the last 3 days of injection (Figure 19), and a rat given five daily injections of 44 nmol of tamoxifen and a concomitant dose of 10 nmol of estradiol during the last 3 days of injection (Figure 20). The rats were sacrificed 24 hr after the last injection. Note the high estrogen binding activity in the cytoplasm of the uterine luminal cells in the control animal treated with estradiol, but not in that treated with estradiol plus tamoxifen. Similar inhibitory effects on the E_2-induced augmentation of cytoplasmic estrogen binding activity were observed with nitromifene and nafoxidine, but not with hydrocortisone. (Magnification × 300.)

unoccupied sites which are not in solution.[36] The final results may give an artificially high estimate of occupied cytoplasmic sites, even to the degree of seemingly saturating the total soluble binding sites.

The second question which may be raised is whether the estrogen binding sites induced or suppressed by these hormones or antihormones could be nonspecific binding proteins. To answer this question, in vitro competitive blocking experiments were carried out with a 100-fold molar excess of DES as the competitor, using experimental conditions referred to above in this chapter. The results have shown that the binding of the fluorescent estradiol ligand by these epithelial cells can be totally competitively blocked and resist prolonged washings in ligand-free solutions. Therefore, it is concluded that these are in fact, cytoplasmic high-affinity binding sites with high specificity for estrogen. However, whether all these binding sites demonstrated could be classified as the so-called type I receptors[35] and could be translocated to the nucleus in vivo is extremely difficult to prove or disapprove at this time.

VI. CONCLUDING REMARKS

The aim of this chapter was to present available in vitro and in vivo data, based on experiments using rat uterus as the target tissue, to show that the binding between the hydrophilic macromolecular fluorescent estradiol conjugate, which has been widely used as a histochemical reagent, and the estrogen receptors of the cells does have a high affinity, a limited binding capacity, hormonal specificity, and tissue specificity. The intracellular estrogen binding sites visualized in the luminal epithelial cells of the rat uterus respond to the physiological or nonphysiological estrogens and antiestrogens as expected of the soluble ER protein of the rat uterus. The evidence supports the claim that this histochemical technique is capable of identifying the cells rich in cytoplasmic estrogen receptors.

Nuclear estrogen receptors have not been convincingly demonstrated histochemically. It is possible that all estrogen binding sites in the nuclei are already occupied by intrinsic estrogens; or that the number of binding sites in the nucleus is too few for visualization.

In the same organ, different cell types may respond to estrogenic stimulation with different degrees of sensitivity. In the human uterus, for example, the cyclic changes in the histology of the endometrium are well known during the menstrual cycle, as the result of an interaction between the two female sex steroid hormones; but the morphological changes in the myometrium are hardly discernible. Experimental data from other investigators have also suggested that the luminal epithelial cells of the rat uterus may be more sensitive to estrogenic stimulation than the stroma and the myometrium.[64,65] A histochemical technique would be particularly useful if the function of a highly steroid-sensitive cell type is the subject of study.

The classic test for estrogenic activity is to observe the degrees of vaginal cornification in ovariectomized adult female rats and to measure the uterine weight increase in immature female rats after the injection of the test substance. For antiestrogenic activity, the ability of the test substance to inhibit the estrogen-induced vaginal cornification and to inhibit the estrogen-induced increase in uterine weight is measured. Using these tests, the antiestrogenic effects of many nonsteroidal antiestrogens can be demonstrated.[3,4] In the experiments presented in this chapter, the cytoplasmic estrogen receptors of the uterine luminal cells of the immature rats are chosen as an indicator. With this approach, even as high as 10 nmol (2.7 μg) of estradiol have been shown to be suppressed by 44 nmol of antiestrogens (e.g., 25 μg of tamoxifen citrate). If one used the uterine weight increase or the morphological changes of the epithelial cells in the reproductive tract as the criteria for interpretation, a different conclusion might have been reached. Both estradiol and antiestrogens induce cellular hypertrophy and hyperplasia in epithelium.[65,66] The histochemical technique may provide a more sensitive or an alternative approach to evaluate estrogenic or antiestrogenic activity, because the intracellular estrogen receptors of the luminal epithelial cells appear to be more sensitive to the action of sex steroid hormones than the morphological changes of various cell types of the rat uterus.

A repeated objection to using fluorescent histochemical methods for localizing estrogen receptors is the high concentration of ligands commonly used in the fluorescent histochemical staining solutions.[25-27] However, those who raise this objection often choose to ignore an important step shared by all histochemical techniques — that after having been stained in a ligand solution, the sections being studied are always exhaustively washed in saline, which instantaneously reduces the concentration of unbound ligands to a negligible level. Thus, at the end of the washing step, when the ligand bound to the tissue section is allowed to equilibrate with an essentially ligand-free saline solution that should contain only traces of estradiol derivative, conceivably at a concentration far below 10^{-9} *M* (well within the range of dissociation constants of the classical ER proteins), the low- and high-affinity binding can be readily distinguished. In fact, a similar principle has been used to separate type I and type II binding proteins in sucrose density gradients; estradiol dissociates itself from the type II sites rapidly when the free ligand in solution is reduced to a negligible concentration.[35,41,42]

Using fluorescent estradiol conjugates prepared in their own laboratories and comparing them with free estradiol, a group of authors reported that they have found a low relative binding affinity of the macromolecular estradiol compounds for ER proteins, and stated that these reagents are, therefore, not suitable for the histochemical localization of estrogen receptors.[67] However, others have shown the relative binding affinity of these reagents to be within an acceptable range under similar experimental conditions.[68] When cryostat frozen sections instead of cytosol have been used, the relative binding affinity of the macromolecular reagent with an average steroid-to-protein molecular ratio of greater than 20:1 might have been in the same range of that of DES.[31] The mechanism by which the macromolecular estradiol conjugates compete more effectively with [^{3}H]-estradiol binding to the receptor sites in tissue sections than with binding to the free receptor protein is unclear. Because of their special steric, hydrophilic, and electrostatic characteristics, the individual estradiol radicals covalently coupled to a protein molecule may not be able to bind as tightly to a single receptor site in solution as might free estradiol, but might fit readily between groups of binding sites when the latter were in close aggregates in cytoplasm, sterically hindering further diffusion of free radiolabeled estradiol into the section.

When the apparent conflicting points in kinetic parameters between the histochemical and the biochemical assay system are analyzed, much of them may be attributed to the difference in receptor conformation and concentration in the two different types of tissue preparations. It is possible that the cytosol assays are only capable of measuring the soluble ER proteins which can be, in turn, categorized into type I, type II, type III, etc. according to their in vitro ligand binding behaviors.[25] The histochemical assays, on the other hand, may localize the cellular estrogen receptors in a relatively insoluble form and in high density. A target cell may have to contain a large number of high-affinity binding sites, such as those in the luminal epithelial cells of the rat uterus and the epithelial cells of the human mammary ducts, before it can be visualized histochemically and recognized against a background of other concomitant cells. At least one type of high-affinity specific estrogen binding sites has been found to be associated with the membranes of the MCF-7 tumor cells.[69] Such binding sites may remain undetected by the customary cytosol assay. In view of the recently recognized heterogeneous nature of the ER proteins[35,42,70,71] which may be composed of several subclasses different from one another not only in function, but also in molecular conformation, the discrepancies between histochemical and biochemical findings related to such receptors may eventually complement each other in the further understanding of the effects of estrogen at the cellular level.

ACKNOWLEDGMENTS

The author wishes to express his appreciation to Ms. Doris Barclay for skilled photographic assistance, and to Miss Marilyn Weed for secretarial assistance.

REFERENCES

1. **Long, J. A. and Evans, H. M.,** The oestrous cycle in the rat and its associated phenomena, *Mem. Univ. Calif.*, 6, 1, 1922.
2. **Astwood, E. B.,** Changes in the weight and water content of the uterus of the normal adult rat, *Am. J. Physiol.*, 126, 162, 1939.
3. **Harper, M. J. K. and Walpole, A. L.,** A new derivative of triphenylethylene: effect on implantation and mode of action in rats, *J. Reprod. Fertil.*, 13, 101, 1967.
4. **Emmens, C. W.,** Early work on antioestrogens, in *Non-Steroidal Antioestrogens*, Sutherland, R. L. and Jordan, V. C., Eds., Academic Press, New York, 1981, 17.
5. **Jensen, E. V. and Jacobson, H. I.,** Basic guides to the mechanism of estrogen action, *Recent Prog. Horm. Res.*, 18, 387, 1962.
6. **Gorski, J., Toft, D., Shyamala, G., Smith, D., and Notides, A.,** Hormone receptors: studies on the interaction of estrogen with the uterus, *Recent Prog. Horm. Res.*, 24, 45, 1968.
7. **Jensen, E. V. and DeSombre, E. R.,** Mechanism of action of the female sex hormones, *Annu. Rev. Biochem.*, 41, 203, 1972.
8. **Jensen, E. V., Mohla, S., Gorell, T. A., and DeSombre, E. R.,** The role of estrophilin in estrogen action, *Vitam. Horm.*, 32, 89, 1974.
9. **Gorski, J. and Gannon, F.,** Current models of steroid hormone action: a critique, *Annu. Rev. Physiol.*, 38, 425, 1976.
10. **Greene, G. L., Gloss, L. E., Fleming, H., and DeSombre, E. R.,** Antibodies to estrogen receptor: immunochemical similarity of estrophilin from various mammalian species, *Proc. Natl. Acad. Sci. U.S.A.*, 74, 3681, 1977.
11. **Hubert, P., Mester, J., Dellacherie, E., Neel, J., and Baulieu, E. -E.,** Soluble biospecific macromolecule for purification of estrogen receptor, *Proc. Natl. Acad. Sci. U.S.A.*, 75, 3143, 1978.
12. **Stumpf, W. E. and Roth, L. J.,** High resolution autoradiography with dry mounted freeze-dried frozen sections. Comparative study of six methods using two diffusible compounds [^{3}H]-estradiol and [^{3}H]-mesobilirubinogen, *J. Histochem. Cytochem.*, 14, 274, 1966.
13. **Tchernitchin, A.,** Autoradiographic study of (6, 7-^{3}H) oestradiol — 17β incorporation into rat uterus, *Steroids*, 10, 661, 1967.
14. **Lee, S. H.,** Cytochemical study of estrogen receptor in human mammary cancer, *Am. J. Clin. Pathol.*, 70, 197, 1978.
15. **Lee, S. H.,** Cellular estrogen and progesterone receptors in mammary carcinoma, *Am. J. Clin. Pathol.*, 73, 323, 1980.
16. **Fetissof, F., Lansac, J., and Arbeilli-Brassart, B.,** Mise en évidence des récepteurs de l'oestradiol et de la progestérone sur préparations histologigues, *Ann. Anat. Pathol. (Paris)*, 25, 201, 1980.
17. **Paulsen, S. M., Johansen, P., Rasmussen, K. S., and Mygind, H.,** Histochemical demonstration of oestrogen receptors in cancer of the breast, *Ugeskr. Laeg.*, 143, 3119, 1981.
18. **Tominaga, T., Kitamura, M., Saito, T., Itoh, I., and Takikawa, H.,** Comparative histochemical and biochemical assays of estrogen receptors in breast cancer patients, *Gann*, 72, 60, 1981.
19. **Van Marle, J., Lindeman, J., Ariëns, A. Th., Labruyere, W., and Van Weeren-Kramer, J.,** Estrogen receptors in human breast cancer. I. Specificity of the histochemical localization of estrogen receptors using an estrogen-albumin FITC complex, *Virchows Arch. (Cell Pathol.)*, 40, 17, 1982.
20. **Eusebi, V., Cerasoli, P. T., Guidelli-Guidi, S., Grilli, S., Bussolati, G., and Azzopardi, J. G.,** A two-stage immunocytochemical method for oestrogen receptor analysis: correlation with morphological parameters of breast carcinomas, *Tumori*, 67, 315, 1981.
21. **Hanna, W., Ryder, D. E., and Mobbs, B. G.,** Cellular localization of estrogen binding sites in human breast cancer, *Am. J. Clin. Pathol.*, 77, 391, 1982.
22. **Jacobs, S. R., Wolfson, W. L., Cheng, L., and Lewin, K. J.,** Cytochemical and competitive protein binding assays for estrogen receptor in breast disease, *Cancer*, 51, 1621, 1983.

23. **O'Connell, M. D. and Said, J. W.,** Estrogen receptors in carcinoma of the breast. A comparison of the dextran-coated charcoal, immunofluorescent, and immunoperoxidase technique, *Am. J. Clin. Pathol.*, 80, 1, 1983.
24. **Yao, X., Meng, X., Chen, P., and Mei, Z.,** Histochemical estrogen receptor assay for selecting stage III breast cancers for combined chemohormonal therapy, *Lab. Invest.*, 48, 95, 1983.
25. **Chamness, G. C., Mercer, W. D., and McGuire, W. L.,** Are histochemical methods for estrogen receptor valid?, *J. Histochem. Cytochem.*, 28, 792, 1980.
26. **McCarty, K. S., Jr., Woodard, B. H., Nichols, D. E., Wilkinson, W., and McCarty, K. S., Sr.,** Comparison of biochemical and histochemical techniques for estrogen receptor analyses in mammary carcinoma, *Cancer*, 46, 2842, 1980.
27. **Chamness, G. C. and McGuire, W. L.,** Questions about histochemical methods for steroid receptors, *Arch. Pathol. Lab. Med.*, 106, 53, 1982.
28. **Weber, K. and Osborn, M.,** Cytoskeleton: definition, structure and gene regulation, *Pathol. Res. Pract.*, 175, 128, 1982.
29. **Walsh, T. P., Winzor, D. J., Clarke, F. M., Masters, C. J., and Morton, D. J.,** Binding of aldolase to actin-containing filaments, evidence of interaction with the regulatory proteins of skeletal muscle, *Biochem. J.*, 186, 89, 1980.
30. **Kammer, G. M., Smith, J. A., and Mitchell, R.,** Capping of human T cell specific determinants: kinetics of capping and receptor re-expression and regulation by the cytoskeleton, *J. Immunol.*, 130, 38, 1983.
31. **Lee, S. H.,** The histochemistry of estrogen receptors, *Histochemistry*, 71, 491, 1981.
32. **DeGoeij, A. F. P. M., Volleberg, M. P. W., Hondius, G. E., Frederik, P. M., and Bosman, F. T.,** Quantitative estrogen receptor assay on frozen sections using radiolabeled estradiol, *J. Steroid Biochem.*, 19, 100S, 1983.
33. **Jungblut, P. W., Gaues, J., Hughes, A., Kallweit, E., Sierralta, W., Szendro, P., and Wagner, R. R.,** Activation of transcription-regulating proteins by steroids, *J. Steroid Biochem.*, 7, 1109, 1976.
34. **Skinner, L. G., Barnes, D. M., and Ribeiro, G. G.,** The clinical value of multiple steroid receptor assays in breast cancer management, *Cancer*, 46, 2939, 1980.
35. **Clark, J. H., Markaverich, B., Upchurch, S., Eriksson, H., Hardin, J. W., and Peck, E. J., Jr.,** Heterogeneity of estrogen binding sites: relationship to estrogen receptors and estrogen responses, *Recent Prog. Horm. Res.*, 36, 89, 1980.
36. **Clark, J. H. and Peck, E. J., Jr.,** Steroid hormone receptors: basic principles and measurement, in *Receptors and Hormone Action*, Vol. 1, O'Malley, B. W. and Birnbaumer, L., Eds., Academic Press, New York, 1977, 383.
37. **Goldstein, L.,** Kinetic behavior of immobilized enzyme systems, *Methods Enzymol.*, 44, 397, 1976.
38. **Lee, S. H. and Torrack, R. M.,** Effects of lead and fixatives on activity of glutamic oxalacetic transaminase, *J. Histochem. Cytochem.*, 16, 181, 1968.
39. **Lee, S. H. and Torack, R. M.,** A biochemical and histochemical study of glutamic oxalacetic transaminase activity of rat hepatic mitochondria fixed *in situ* and *in vitro*, *J. Cell Biol.*, 39, 725, 1968.
40. **Dandliker, W. B., Levison, S. A., and Brawn, R. J.,** Hormone binding by cells and cell fragments as visualized by fluorescence microscopy, *Res. Commun. Chem. Pathol. Pharmacol.*, 14, 103, 1976.
41. **Clark, J. H., Hardin, J. W., Upchurch, S., and Eriksson, H.,** Heterogeneity of estrogen binding sites in the cytosol of the rat uterus, *J. Biol. Chem.*, 253, 7630, 1978.
42. **Eriksson, H. A., Hardin, J. W., Markaverich, B., Upchurch, S., and Clark, J. H.,** Estrogen binding in the rat uterus: heterogeneity of sites and relation to uterotrophic response, *J. Steroid Biochem.*, 12, 121, 1980.
43. **Lee, S. H.,** Uterine epithelial and eosinophil estrogen receptors in rats during the estrous cycle, *Histochemistry*, 74, 443, 1982.
44. **Undenfriend, S., Ed.,** *Fluorescence Assay in Biology and Medicine*, Academic Press, New York, 1962, 192.
45. **Tchernitchin, A.,** The role of eosinophil receptors in the nongenomic response to oestrogens in the uterus, *J. Steroid Biochem.*, 11, 417, 1979.
46. **Feherty, P., Robertson, D. M., Waynforth, H. B., and Kellie, A. E.,** Changes in the concentration of high-affinity oestradiol receptors in rat uterine supernatant preparations during the oestrous cycle, pseudopregnancy, pregnancy, maturation, and after ovariectomy, *Biochem. J.*, 120, 837, 1970.
47. **Lee, C. and Jacobson, H. I.,** Uterine estrogen receptor in rats during pubescence and the estrous cycle, *Endocrinology*, 88, 596, 1971.
48. **Kielhorn, J. and Hughes, A.,** Variations in uterine cytosolic oestrogen and progesterone receptor levels during the rat estrous cycle, *Acta Endocrinol.*, 86, 842, 1977.
49. **Yoshinaga, K., Hawkins, R. A., and Stocker, J. F.,** Estrogen secretion by the rat ovary *in vivo* during the estrous cycle and pregnancy, *Endocrinology*, 85, 103, 1969.

50. **Clark, J. H., Hsueh, A. J. W., and Peck, E. G., Jr.,** Regulation of estrogen receptor replenishment by progesterone, *Ann. N.Y. Acad. Sci.*, 286, 161, 1977.
51. **Lee, S. H.,** Validity of a histochemical estrogen receptor assay, supported by the observation of a cellular response to steroid manipulation, *J. Histochem. Cytochem.*, 32, 305, 1984.
52. **Dix, C. J. and Jordan, V. C.,** Modulation of rat uterine steroid hormone receptors by estrogen and antiestrogens, *Endocrinology*, 107, 2011, 1980.
53. **Sutherland, R. L. and Jordan, V. C., Eds.,** *Non-Steroidal Antioestrogens*, Academic Press, New York, 1981.
54. **Emmens, C. W.,** Compounds exhibiting prolonged antioestrogenic and antifertility activity in mice and rats, *J. Reprod. Fertil.*, 26, 175, 1971.
55. **Jordan, V. C.,** Prolonged antioestrogenic activity of ICI 46, 474 in the ovariectomized mouse, *J. Reprod. Fertil.*, 42, 251, 1975.
56. **Rochefort, H. and Capong, F.,** Binding properties of an antioestrogen to the estradiol receptor of uterine cytosol, *FEBS. Lett.*, 20, 11, 1972,
57. **Ruh, T. S. and Ruh, M. F.,** The effect of antiestrogens on the nuclear binding of the estrogen receptor, *Steroids*, 24, 209, 1974.
58. **Katzenellenbogen, B. S. and Ferguson, E. R.,** Antiestrogen action in the uterus: biological ineffectiveness of nuclear bound estradiol after antiestrogen, *Endocrinology*, 97, 1, 1975.
59. **Baudendistel, L. J. and Ruh, T. S.,** Antiestrogen action: differential nuclear retention and extractability of the estrogen receptor, *Steroids*, 28, 223, 1976.
60. **Clark, J. H., Peck, E. J., and Anderson, J. N.,** Oestrogen receptors and antagonism of steroid hormone action, *Nature (London)*, 251, 446, 1974,
61. **Ferguson, E. R. and Katzenellenbogen, B. S.,** A comparative study of antiestrogen action; temporal patterns of antagonism of estrogen stimulated uterine growth and effects on estrogen receptor levels, *Endocrinology*, 100, 1242, 1977.
62. **Ruh, T. S. and Baudendistel, L. J.,** Different nuclear binding sites for antiestrogen and estrogen receptor complexes, *Endocrinology*, 100, 420, 1977.
63. **Jordan, V. C.,** Antiestrogenic and antitumor properties of tamoxifen in laboratory animals, *Cancer Treat. Rep.*, 60, 1409, 1976.
64. **Clark, J. H. and Peck, E. J., Jr., Eds.,** *Female Sex Steroids: Receptors and Function*, Springer-Verlag, Berlin, 1979, 1.
65. **Markaverich, B. M., Upchurch, S., Glasser, S. R., McCormak, S. A., and Clark, J. H.,** Oestrogen stimulation of uterine growth: effects of steroidal and non-steroidal oestrogen antagonists, in *Non-Steroidal Antioestrogens*, Sutherland, R. L. and Jordan, V. C., Eds., Academic Press, New York, 1981, 113.
66. **Martin, L. and Middleton, E.,** Prolonged oestrogenic and mitogenic activity of tamoxifen in the ovariectomized mouse, *J. Endocrinol.*, 78, 125, 1978.
67. **Joyce, B. G., Nicholson, R. I., Morton, M. S., and Griffiths, K.,** Studies with steroid-fluorescein conjugates on oestrogen target tissues, *Eur. J. Cancer*, 18, 1147, 1982.
68. **Rao, B. R., Patrick, T. B., and Sweet, F.,** Steroid-albumin conjugate interaction with steroid binding proteins, *Endocrinology*, 106, 356, 1980.
69. **Berthois, Y., Laugier, R., Mittre, H., Tubiana, N., Pourreau-Schneider, N., and Martin, P. M.,** Fixation of an estrogen-macromolecular complex on MCF 7 cell plasma membranes can be displaced by anti-estrogens and induces a variation in membrane potential, *J. Steroid Biochem.*, 19, 37S, 1983.
70. **Thomas, T., Leung, B. S., Yu, W. E. Y., and Kiang, D. T.,** Diverse mechanisms of estrogen receptor activation, *Fed. Proc. Fed. Am. Soc. Exp. Biol.*, 42, 1877, 1983.
71. **Wiehle, R. D. and Wittliff, J. L.,** Multiple forms of estrogen receptors during differentiation of the mammary gland of the rat, *Fed. Proc. Fed. Am. Soc. Exp. Biol.*, 42, 1877, 1983.

Chapter 5

ESTROGEN RECEPTORS AND HORMONE RESPONSIVENESS IN SERIALLY TRANSPLANTED MAMMARY TUMORS IN RATS*

Chung Lee, Charmayne Jesik, Marayart Mangkornkanok, Julia Sensibar, and Sin Hang Lee

TABLE OF CONTENTS

I. Introduction 86

II. Serial Transplantation of DMBA-Induced Mammary Tumors in Rats 86

III. Biological Behavior of Serially Transplanted Rat Mammary Tumors 86

IV. Biochemical Determination of Estrogen Receptors 86

V. Histochemical Assessment of Estrogen Binding Sites 87
- A. Primary Tumor Sections 88
- B. First Passage Sections 90
- C. Second Passage Sections 90
- D. Third Passage Sections 91
- E. Fourth and Fifth Passage Sections 91

VI. Conclusion 91

References 93

* Experiments reported in this chapter were supported by NIH research grant CA 14727 and by a grant from AMOCO Foundation, Inc. to Northwestern Memorial Hospital.

I. INTRODUCTION

Rat mammary tumors induced by 7,12-dimethylbenz(a)anthracene (DMBA) are often used as an experimental model for hormone-dependent tumors.[1,2] The hormone dependency of this tumor model has been demonstrated by a regression of the tumor in response to ovariectomy or hypophysectomy,[3-5] and by the presence of estrogen receptors (ER) in the tumor tissue.[6-8]

Results of our recent studies indicated that DMBA-induced, hormone-dependent mammary tumors in rats could be converted to hormone less-dependent or independent tumors by serial transplantation into syngeneic hosts.[9] This transition of tumors from hormone dependency to autonomous is characterized by a progressive replacement of hormone-dependent epithelial cells by hormone-independent spindle-shaped cells with a concomitant loss of total ER content in the tumor tissue. Based on the above observation, we have advanced to the hypothesis that the hormone-dependent epithelial cells are associated with ER, while the hormone-independent spindle-shaped cells are devoid of ER. Utilizing the fluorescent staining technique to localize ER-positive cells recently developed by Lee,[10-12] we report in this chapter our approaches and findings of morphologic localization of ER in serially transplanted, DMBA-induced mammary tumors in rats.

II. SERIAL TRANSPLANTATION OF DMBA-INDUCED MAMMARY TUMORS IN RATS

The mammary tumors used in the present study were induced by feeding 10 mg of dimethylbenzanthracene (DMBA) to 50-day-old Fischer 344 rats. Tumors were palpable within 60 days. The rats were lightly anesthetized and the tumors were examined and measured. The tumors were removed, cut into small pieces, and transplanted into syngeneic female recipients of 90 days old. This serial transplantation was performed repeatedly until we had a series of six tumors: a primary tumor, plus first through fifth passage tumors.

III. BIOLOGICAL BEHAVIOR OF SERIALLY TRANSPLANTED RAT MAMMARY TUMORS

Tumors, serially transplanted into intact recipients, are able to grow to 1.5 to 2.0 cm in diameter within 1 to 2 months. Histologically, tumor tissues of early passages are characterized by mixed populations of epithelial and spindle-shaped cells, while tumors of late passages contain only the spindle-shaped cells. As demonstrated in Figure 1, the primary tumors are considered hormone dependent, because ovariectomy in the host animals would result in complete tumor regression. These regressed primary tumors remain dormant in that treatment of host animals with exogeneous estrogen would reactivate tumor growth. Tumors of the second and third passages are hormone responsive. In these cases, ovariectomy of the host animals would result in a partial regression of the tumor followed by a regrowth of the tumor despite the continuing absence of circulating ovarian hormones. Tumors of the third, fourth, and fifth passages are hormone independent because they would grow in intact hosts as well as in ovariectomized hosts.[9]

IV. BIOCHEMICAL DETERMINATION OF ESTROGEN RECEPTORS

ER in the tumor cytosol was measured by a charcoal absorption method with the use of the principle of Scatchard.[13,14] Levels of estrogen receptors were expressed in femtomoles per milligam protein as described previously.[9,15] Figure 2 shows the levels of cytosol ER in mammary tumors of various passages. There was a significant decrease in ER of the mam-

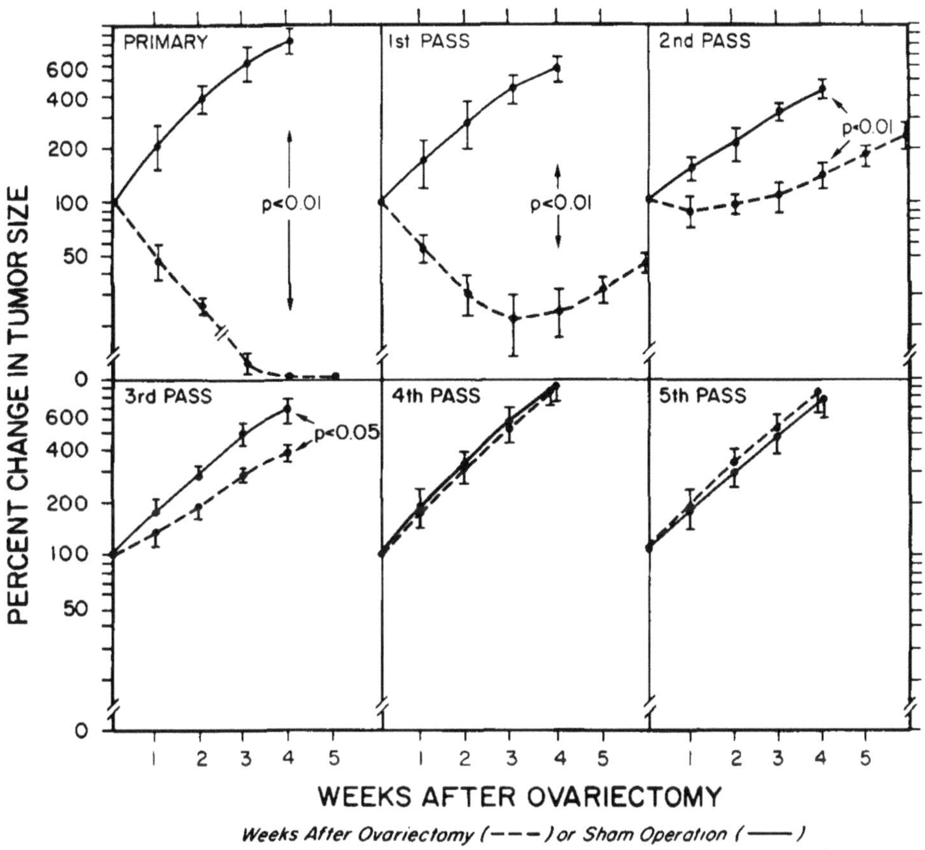

FIGURE 1. Effect of ovariectomy or sham operation on growth (or regression) of established mammary tumors at diffenent generations of transplantation. Each point represents the average of 6 to 10 individual tumors removed serially from animals bearing multiple transplanted tumors. The vertical line denotes a standard error. *P* value is the statistical significance between the two groups compared at the time indicated by the arrows. (Reprinted with permission from *European Journal Cancer Clinical Oncology*, Volume 17, Lee, C., Lapin, V., Oyasu, R., and Battifora, H., Effect of ovariectomy on serially transplanted rat mammary tumors induced by 7,12-dimethylbenzanthracene, Copyright 1981, Pergamon Press, Ltd., New York.)

mary tumors as the generation of tumor transplantation advanced. Contents of ER were high in the primary tumors averaging 45 ± 9.0 fmol/mg protein, and were low in the late passages, being 5.2 fmol/mg.[9]

V. HISTOCHEMICAL ASSESSMENT OF ESTROGEN BINDING SITES

Localization of ER in the serial tumor passages was by the histochemical technique of Lee.[10,12] Serial sections of tumor tissue were cut from tumor pieces mounted in Cytostat mounting in a cryostat kept at −25°C and were placed on uncoated but cleaned, acid-washed glass slides. One section of each series was stained with H and E and mounted in Permamount for histological evaluation.

The experimental sections were air dried for 1 hr in a refrigerator kept at 5°C after they were cut. They were then rehydrated with 2% bovine serum albumin (BSA) for 2 min, then carefully drained. The fluorescent conjugate, 17-beta-estradiol-6-(*0*)-carboxymenthyl)oxime-BSA-fluorescein isothiocyanate (E_2-BSA-FITC), was added to each section so that they were completely covered, and the slides were placed in a darkened humidified chamber for 2 hr.

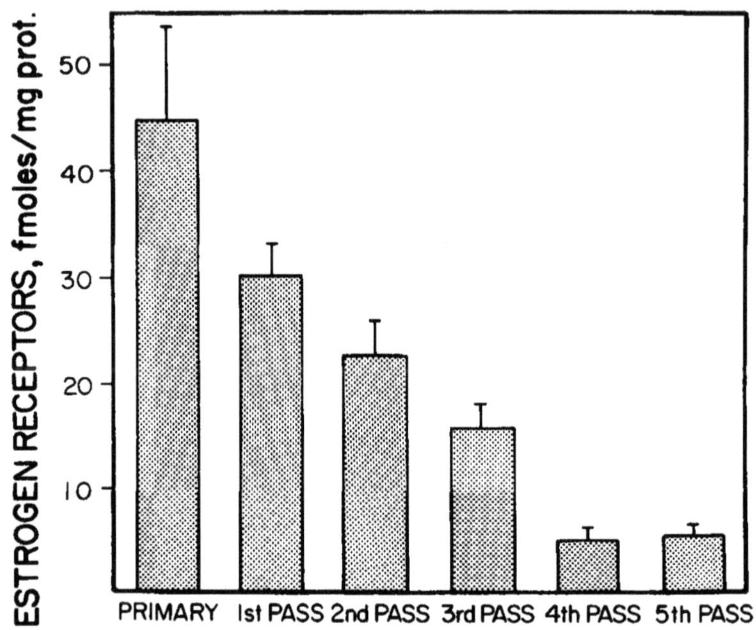

FIGURE 2. Estrogen receptors in rat mammary tumors at different passages of transplantation. Results expressed as mean ± SE. The level of ER in the primary tumors is significantly greater than those of other passages $p<0.01$). The level of ER in tumors of the first passage is significantly greater than those of third ($p<0.05$), fourth, and fifth passages ($p<0.01$). The level of ER in tumors of the second passage is significantly greater than those of fourth and fifth passages ($p<0.01$).

The slides were removed and the sections were drained of excess conjugate; they were washed by inverting the slides in phosphate-buffered saline (PBS) twice for 30 min each time. The sections were then mounted in buffered glycerol and examined with a Leitz fluorescent microscope. The microscope was equipped with an HB-250 mercury vapor lamp, BG-12 exciter filter, and 525 barrier filter.

Controls for nonspecific fluorescence were run simultaneously with the experimental sections. The sections to be tested for nonspecific fluorescence were cut and mounted as described, but aftei rehydration, a conjugate consisting of BSA and FITC alone was added to the sections; they were incubated, washed, mounted, and examined.

To test for specificity of the estradiol conjugate, the serial sections were cut and mounted and rehydrated, but were then incubated with either estradiol or with diethylstilbestrol (DES), at concentrations ranging from 10^{-12} to 10^{-6} *M* for from 3 to 8 hr. The most effective concentration for blocking the activity of the conjugate was 10^{-6} *M* lower concentrations gave inconsistent results. This suggests that both primary and secondary receptor sites were occupied by the estradiol or DES, in agreement with Joyce et al.[16] The tissue sections were then rinsed with PBS for 10 min before they were covered with the estradiol conjugate and incubated. They were washed, mounted, and examined along with the experimental tissues.

A. Primary Tumor Sections

The results of these studies show that primary tumor tissue sections stained with H and E are well differentiated, consisting of epithelial cells and some supporting stromal tissue. The epithelial cells form ducts and tubules; the cells are mostly small and uniform, with small, normal appearing nuclei. Within the stromal matrix are some spindle-shaped cells (Figure 3). When the primary tumor is examined after being stained with E_2-BSA-FITC,

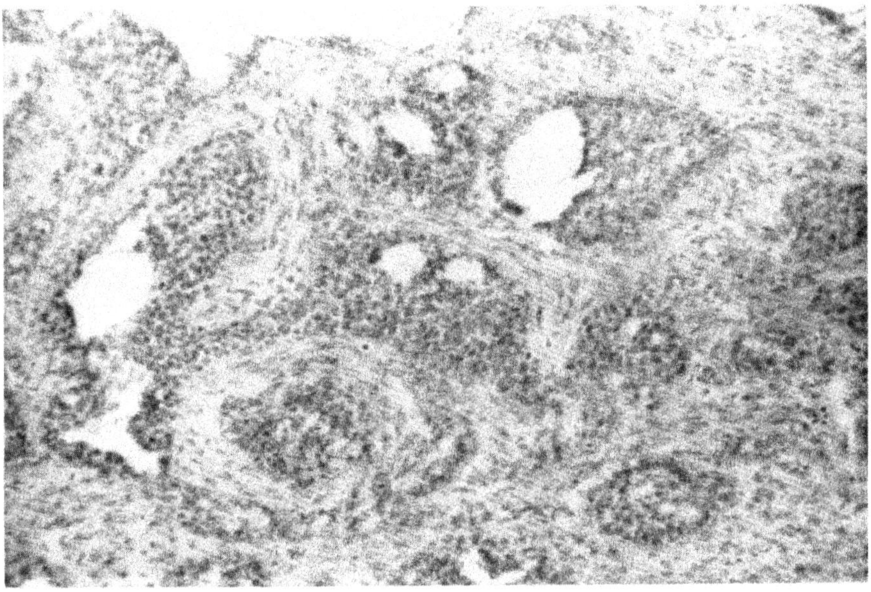

FIGURE 3. H and E sections taken from primary mammary tumor tissue from rats which were fed DMBA. The epithelial and connective tissue form a pattern of highly differentiated ducts and tubules which resemble normal tissue. (Magnification × 160.)

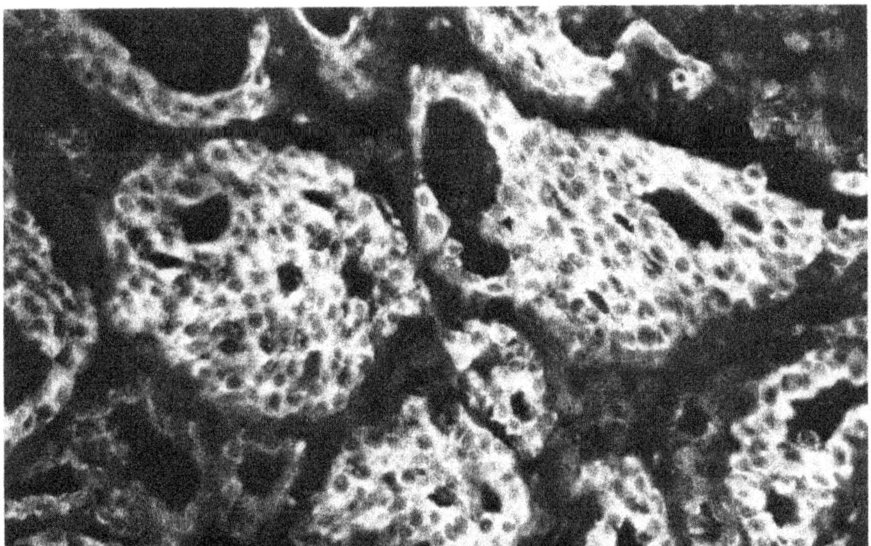

FIGURE 4. Serial sections of the same primary tumor tissue seen in Figure 3, but stained with 17-beta-estradiol-bovine serum albumin-fluorescein isothiocyanate conjugate (E_2-BSA-FITC). The epithelial cells brightly fluoresce; the stromal tissue fluoresces weakly or not at all. (Magnification × 160.)

the majority, but not all of the epithelial cells fluoresce very strongly (Figure 4). The fluorescence is mainly cytoplasmic, although some cells also demonstrate nuclear staining. Those epithelial cells which did not show a positive fluorescence reaction probably had their

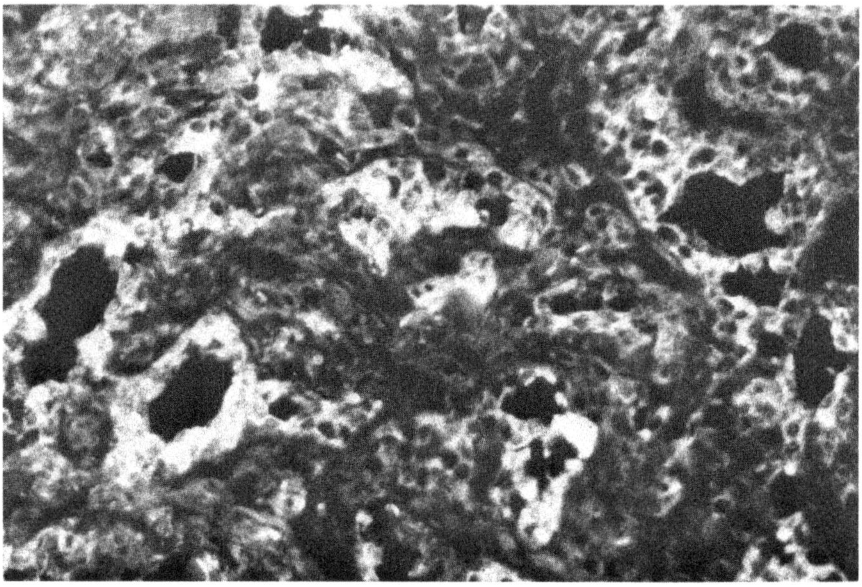

FIGURE 5. First passage tumor tissue serial section stained with E_2-BSA-FITC conjugate. Most epithelial cells still stain intensely with the conjugate, but some fluoresce weakly. (Magnification × 160.)

receptors occupied by endogeneous estrogen. The spindle-shaped cells fluoresce weakly or not at all. When the slides which were used as controls for nonspecific staining were examined, the fluorescence in the spindle-shaped cells was confirmed to be nonspecific. Other cells such as eosinophils also fluoresced, but this was due to autofluorescence.

B. First Passage Sections

When first passage sections were examined, the H and E sections showed a pattern of differentiation which was similar to that seen in primary tumors. However, there appeared to be more spindle-shaped cells in the tumor than there were in primary tumors. The epithelial cells still formed tubules and ducts, and there were still mainly normal epithelial cells in the tumor. Examination of the tumor sections which had been stained with the E_2-BSA-FITC conjugate revealed that the overall pattern of fluorescence was the same as that of the primary tumor, with the majority of epithelial cells staining a bright apple green, and a small number of epithelial cells which fluoresced weakly or not at all. The spindle-shaped cells stained either weakly or not at all due to nonspecific fluorescence (Figure 5).

C. Second Passage Sections

When tissues of the second passage tumors were examined, the H and E stained sections showed that while the epithelial cells still comprised a sizable portion of the tumor, there was less definitive duct and tubule formation; the structures which were formed appeared smaller. Fewer epithelial cells appeared to be normal; many were larger than normal, and had abnormal pleomorphic vesicular nuclei and other unusual morphologies.

It was evident that there was also a larger amount of spindle-shaped cells present in the tumors. When the tissues which had been serially cut from the tumor and stained with the estradiol conjugate were examined, it was apparent that far fewer of the epithelial cells stained with the dye than in the primary and first passage tissues; the overall fluorescence of the sections was comparatively diminished.

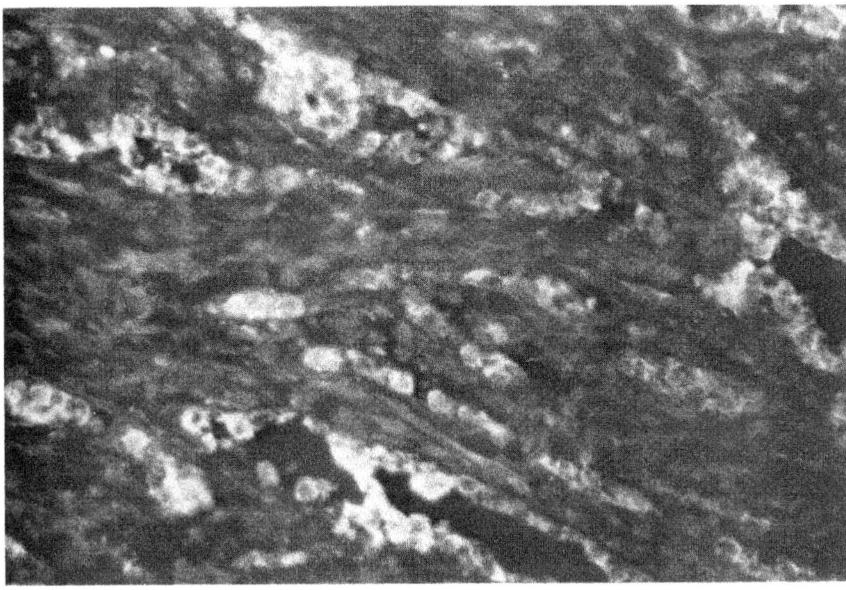

FIGURE 6. Third passage tumor tissue which has been stained with E_2-BSA-FITC conjugate shows an overall loss of fluorescence when compared with primary and first passage tumor tissue. Some areas of epithelial tissue still fluoresce, but many more cells do so weakly or not at all. (Magnification × 250.)

D. Third Passage Sections

Examination of third passage of tissues stained with H and E showed that the epithelial duct and tubule formation was disorganized, with an even larger population of abnormal-appearing epithelial cells present in the sections. Spindle-shaped cells made up an even larger proportion of the malignancy than it did in tumors of the previous passages. Staining the sections of third passage tumor tissues with the fluorescent conjugate showed that, while some epithelial cells still fluoresced, many more did not or fluoresced weakly (Figure 6).

E. Fourth and Fifth Passage Sections

By the fourth passage, the tissues, as shown in sections stained with H and E, consisted almost entirely of spindle-shaped cells, although occasionally epithelial cells were still present. When fourth passage tumor tissue was stained with E_2-BSA-FITC conjugate, the sections showed only a nonspecific staining, aside from an occasional fluorescing epithelial cell. There were few discernible ducts, although blood vessels were still present.

Further passages beyond the fourth passage tumor tissue continued to demonstrate the same staining pattern which we saw in the fourth passage tumors (Figure 7).

We had, then, a series of tumors which have gone from hormone dependent to hormone independent according to a biochemical assay of their estrogen receptor content. Serial sections from these groups of tumors showed a corresponding decrease in estrogen affinity when they were stained with E_2-BSA-FITC conjugate. The staining was, furthermore, specific for the epithelial cells and not for the spindle-shaped cells.

VI. CONCLUSION

DMBA-induced mammary tumors in rats can gradually lose their hormone dependency and finally become hormone independent by serial transplantation into syngeneic hosts.

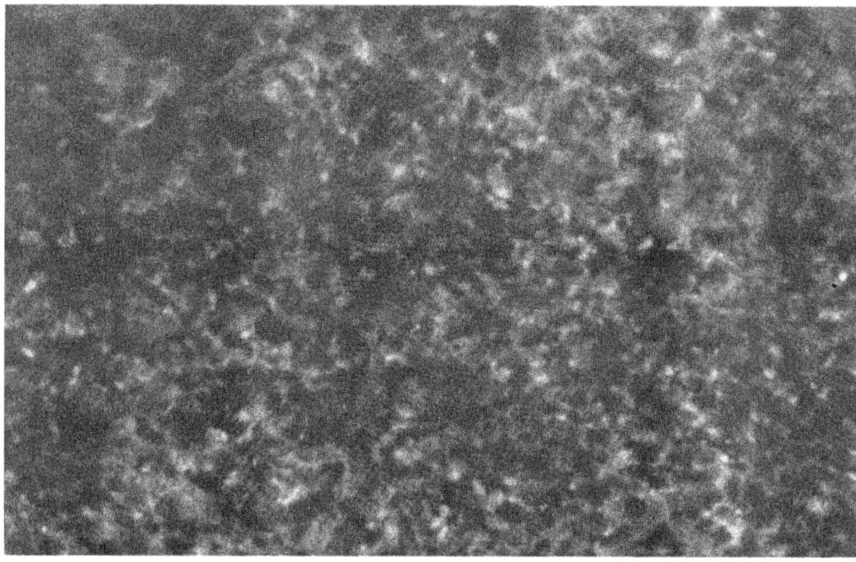

FIGURE 7. Fifth passage tumor tissue which has been stained with E_2-BSA-FITC conjugate shows occasional epithelial cells which fluoresce weakly. The overall brightness of the tissue sections is greatly diminished. Most fluorescence is due to the nonspecific staining of the stromal tissue. (Magnification × 250.)

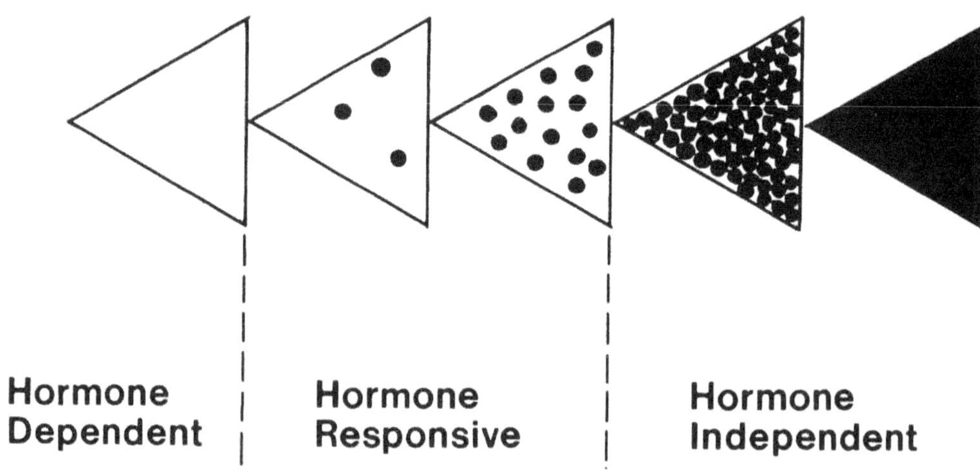

FIGURE 8. Diagrammatic representation of the two-cell type in DMBA-induced mammary tumors in rats upon serial transplantation. Tumors of the primary site and of the first passage consist of predominant ER-positive cells and are, therefore, hormone dependent. As the tumors advance in passage, the number of ER-negative cells increase and the tumors are composed of a mixture of two cell types. These tumors are hormone responsive because ovariectomy in the host would result in partial regression. Tumors of fourth and fifth passages are autonomous because they have mainly ER-negative cells.

Concomitant with the change in hormone dependency, the cell types in the tumor also change from predominant epithelial cells to predominant spindle-shaped cells. Results of the present study have demonstrated that the epithelial cells contain ER and the spindle-shaped cells are devoid of ER.

Based on the present results and on our previous observations, we can further conclude that ER-positive cells are hormone dependent and ER-negative cells are autonomous. Figure 8 is a diagrammatic presentation for the natural history of progression of the serially transplanted mammary tumors. As the primary tumors progress through serial transplantation, the predominant ER-positive cells are subjected to a selection process which favorably permits the outgrowth of ER-negative cells. As a result, the tumor gradually loses its hormone responsiveness. Eventually, as shown during the fourth and the fifth passages, the majority of the tumor consists of ER-negative cells and it becomes autonomous.

Therefore, the procedure of serial transplantation has revealed that DMBA-induced mammary tumors in rats are composed of two malignant cell types: the ER-positive epithelial cells which are present in abundance in primary tumor and the ER-negative spindle-shaped cells which become apparent upon serial transplantation. In addition to the experimental model for studying hormone dependency, the presence of two morphologically and functionally distinct cell types in DMBA-induced mammary tumors offers an excellent experimental system for studying cell selection.

REFERENCES

1. **Huggins, C.,** Two principles in endocrine therapy of cancers: hormone deprival and hormone interference, *Cancer Res.*, 25, 819, 1965.
2. **Bogden, A. E.,** Therapy in experimental breast cancer models, in *Breast Cancer, Advances in Research and Treatment*, Vol. 2, McGuire, W. L., Ed., Plenum Press, New York, 1978, 283.
3. **Huggins, C., Grand, L. G., and Brillantes, F. P.,** Mammary cancer induced by a single feeding of polynuclear hydrocarbons and its suppression, *Nature (London)*, 189, 204, 1961.
4. **Sterental, A., Dominquez, J. M., Weissman, C., and Pearson, O. H.,** Pituitary role in the estrogen dependency of experimental mammary cancer, *Cancer Res.*, 23, 481, 1963.
5. **Daniel, P. M. and Pritchard, M. M. L.,** The response of experimentally induced mammary tumors in rats to ovariectomy, *Br. J. Cancer*, 17, 687, 1964.
6. **King, R. J. B., Cowan, D. M., and Inman, D. R.,** The uptake of (6,7-^{3}H)-oestradiol by dimethylbenzanthracene-induced rat mammary tumors, *J. Endocrinol.*, 32, 83, 1965.
7. **Mobbs, B. G.,** The uptake of tritiated oestradiol by dimethylbenzanthracene induced mammary tumors of the rat, *J. Endocrinol.*, 36, 409, 1966.
8. **McGuire, W. L. and Julian, J. A.,** Comparison of macromolecular binding of estradiol in hormone-dependent and hormone-independent rat mammary carcinoma, *Cancer Res.*, 31, 1440, 1971.
9. **Lee, C., Lapin, V., Oyasu, R., and Battifora, H.,** Effect of ovariectomy on serially transplanted rat mammary tumors induced by 7,12-dimethylbenzanthracene, *Eur. J. Cancer Clin. Oncol.*, 17, 801, 1981.
10. **Lee, S. H.,** Cytochemical study of estrogen receptor in human mammary cancer, *Am. J. Clin. Pathol.*, 70, 197, 1978.
11. **Lee, S. H.,** Cancer cell estrogen receptor of human mammary carcinoma, *Cancer*, 44, 1, 1979.
12. **Lee, S. H.,** Hydrophilic macromolecules of steroid derivatives for the detection of cancer cell receptors, *Cancer*, 46, 2825, 1980.
13. **Scatchard, G.,** The attraction of proteins for small molecules and ions, *Ann. N.Y. Acad. Sci.*, 51, 660, 1949.
14. **Chamness, B. C. and McGuire, W. L.,** Scatchard plots: common errors in correction and interpretation, *Steroids*, 26, 538, 1975.
15. **Shih, A. and Lee, C.,** Fluctuations in levels of cytosol and nuclear estrogen receptors in rat mammary tumors during the estrous cycle, *Endocrinology*, 102, 420, 1978.
16. **Joyce, B. C., Nicholson, R., Morton, M. S., and Griffiths, K.,** Studies with steroid-fluorescein conjugates on oestrogen target tissues, *Eur. J. Cancer Clin. Oncol.*, 11, 1147, 1982.

Chapter 6

FLOW CYTOMETRIC ANALYSIS OF FLUORESCENT ESTROGEN BINDING IN CANCER CELL SUSPENSIONS

Chris Benz, Israel Wiznitzer, and Sin Hang Lee

TABLE OF CONTENTS

I. Introduction 96

II. Flow Cytometry 96

III. Tumor Cells 96

IV. Measuring Fluorescent Estradiol Binding 97

V. Binding Intensity and Light Scatter 100

VI. E-BSA-FITC Binding Specificity 102

VII. Applications of E-BSA-FITC Binding 104

VIII. Summary 109

Acknowledgments 109

References 109

I. INTRODUCTION

Hormone-responsive human tumors include breast, prostate, endometrial, and ovarian carcinomas.[1,2] Surprisingly, estrogen binding proteins (ER) have been found in a wide variety of other human tumors including pancreatic carcinomas,[3,4] but endocrine therapy of these tumors has either never been attempted or has not been adequately correlated with ER studies.[5,6] As is well known, ER analysis can identify a major group of patients with advanced breast cancer who will have a 60% chance for objective response to endocrine therapy. Our failure to extend this therapeutic advance to the nearly 30% of all cancers in this country which are sex hormone related[1] may be attributed, in part, to the difficulties involved in biochemically characterizing ER contained in human tumors. The laborious biochemical extraction and analysis of ER is complicated by the existent heterogeneity of estrogen binding sites, including lower affinity type II and type III binding proteins.[9,10] These difficulties are compounded by potential error in the graphical analysis of Scatchard plots and the resultant quantitation of ER capacity and affinity.[7,8] More often than not, determination of receptor affinity (K_d) is lacking from those clinical studies attempting to correlate tumor ER with endocrine responsiveness. In addition, it has been difficult to prove that the absolute number of ER binding sites is any better predictor of treatment response than simply indicating whether a given tumor is ER positive or negative.[11]

In contrast to biochemical analysis, the histochemical detection of ER in frozen tissue sections offers the potential for rapidly estimating the number of ER-positive and -negative tumor cells within a very small tissue specimen. With recent evidence for a positive correlation between ER-fluorescing breast cancer cells, biochemically measurable ER, and the endocrine responsiveness of these human tumors,[12,13] it is reasonable to suggest that histochemical analysis may soon be available to study a much greater number and variety of potentially endocrine-sensitive human tumors. Unfortunately, the lack of a quantitative measure of cell-bound fluorescence has precluded precise characterization of ER binding by fluorescent ligands. Thus, we have adapted flow cytometric methods to begin quantitating fluorescent estrogen binding in human tumor cell suspensions.

II. FLOW CYTOMETRY

Flow cytometry (FCM) began over 25 years ago as a tool to automatically count and size individual cells from a flowing cell suspension. By the mid 1950s, this early tool was commercially available as the Coulter Counter®. Ten years later, with the aid of more sophisticated spectrophotometric detection and computer coupling, multiparameter analysis of DNA or protein staining cellular constituents could be performed on cells flowing at the rate of 500/sec. These developments resulted in the later availability of the Ortho Cytofluorograph®. Fluorescent dye emission rather than absorption was utilized to increase the signal-to-noise ratio, and the demonstration of stoichiometric binding between fluorochromes and DNA expanded the application of FCM to producing DNA histograms and analyzing the growth of tumor cell suspensions for percent G_1, S, and G_2/M phase cells. Most recently, the Becton-Dickinson Fluorescence Activated Cell Sorter (FACS)® introduced electrostatic sorting of individual cells based on the analysis of light scatter (cell size) or fluorescence intensity. The application of this new and expanding field to medical and biologic problems has been extensively reviewed by Melamed et al.[14] It is now clear that FCM can be successfully used to study the binding of a wide variety of fluorescently labeled ligands to their membrane, cytoplasmic, or nuclear receptors in intact cells.[15]

III. TUMOR CELLS

The human mammary carcinomas MCF-7 and 47-DN (a subline of T-47D) are two well-

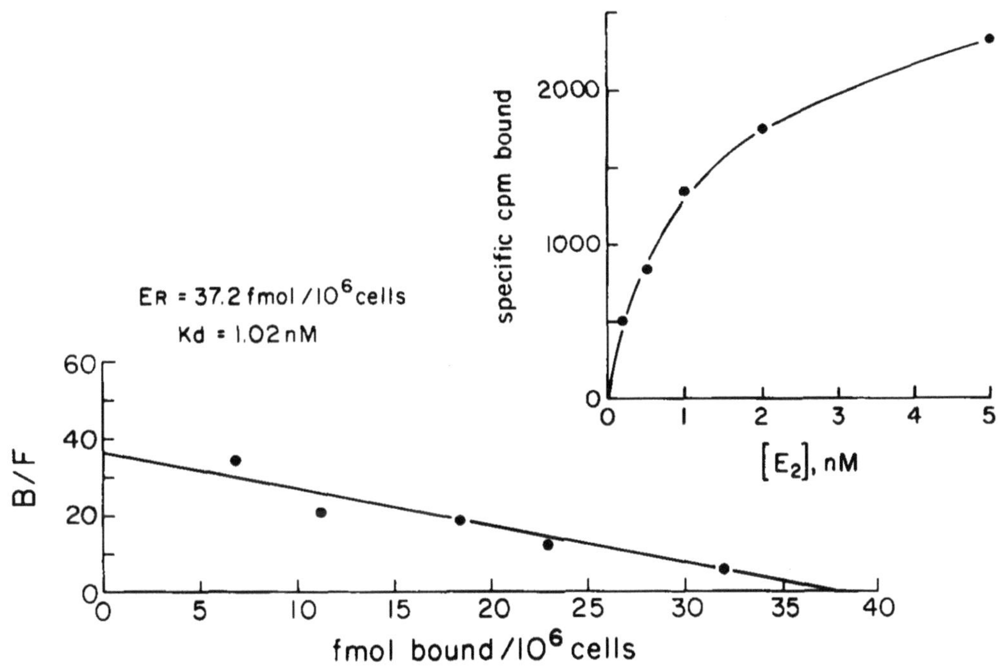

FIGURE 1. Specific binding of estradiol (E_2) in 47-DN cells and the resultant Scatchard determination of ER capacity and affinity (K_d). Using the whole cell assay of Shafie and Brooks,[17] 2 × 10^6 adherent cells per 35 mm culture were well exposed to the indicated concentrations of [^{3}H]E_2 (138 Ci/mmol) ± 100-fold excess DES for 1 hr at 37°C to determine specific binding (cpm or fmol/10^6 cells).

characterized, continuously growing monolayer cell lines that produce cytoplasmic and nuclear ER in culture.[16] COLO-357 and RWP-2 are two other monolayer cell lines derived from human pancreatic adenocarcinomas* which also show specific [^{3}H]-estradiol (E_2) binding as measured by the whole cell assay of Shafie and Brooks.[17] Stock cultures of these four cell lines are maintained in 5% CO_2 incubators at 37°C using RPMI 1640 media (Gibco Labs, N.Y.) supplemented with 10% fetal calf serum. The breast carcinoma cultures are also supplemented with insulin (0.2 IU/mℓ) for maximum growth. Prior to ER determination by either FCM or the whole cell assay, monolayers are washed with phosphate-buffered saline (PBS) and incubated in serum-free media for 1 hr. Single-cell suspensions are prepared by agitating cultures in the presence of trypsin (0.05%) — EDTA (0.02%) and subsequent rinsing with PBS. The whole cell assay for ER is performed on monolayer cells exposed to 0.2 to 10.0 n*M* [^{3}H]E_2 ± 100-fold excess DES (diethylstilbestrol) for 1 hr at 37°C. Figure 1 illustrates the specific, saturable binding of [^{3}H]E_2 in 47-DN and the resultant Scatchard plot quantitating both ER capacity (fmol/10^6 cells) and ER affinity (K_d). Table 1 compares these ER values for each of the four tumor cell lines. Of significance is the relatively major differences noted in ER affinity, and the minor differences noted in ER capacity.

IV. MEASURING FLUORESCENT ESTRADIOL BINDING

The hydrophilic fluorescent estradiol conjugate, 17β-estradiol-6-CMO-BSA-FITC (E-BSA-FITC),[12,18-20] is used to compare the histochemical localization of ER in these four cell lines, as observed under fluorescence microscopy, with the intensity of fluorescence binding as

* Colo-357 cells were provided by Dr. George Moore (University of Colorado, Denver) and RWP-2 cells were provided by Dr. Michael Turner (Brown University, Providence).

Table 1
ESTROGEN RECEPTORS IN FOUR HUMAN CARCINOMA CELL LINES

	$[^3H]E_2$ binding	
Cell lines	(fmol/10^6 cells)	K_d (nM)
MCF-7	35	0.2
47-DN	37	1.0
COLO-357	12	5.0
RWP-2	15	9.2

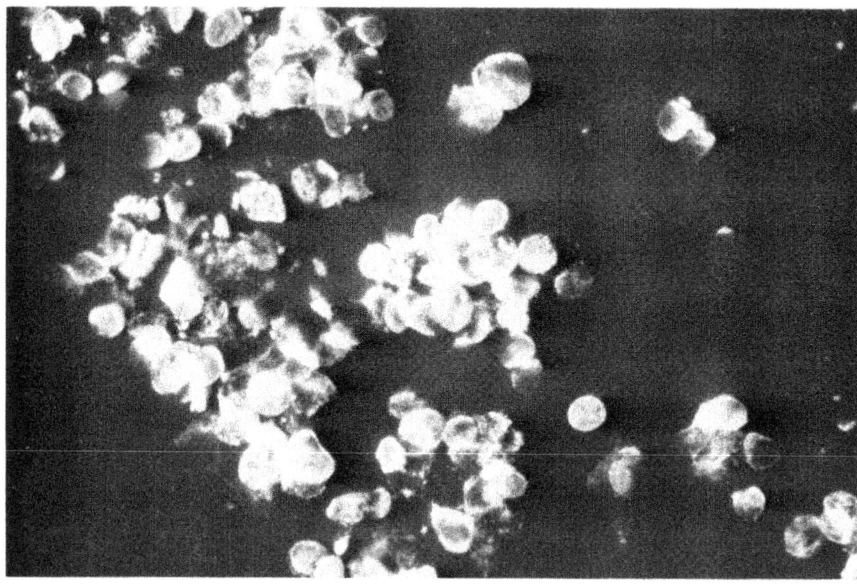

A

FIGURE 2. E-BSA-FITC stained MCF-7 (A), 47-DN (B), COLO-357 (C), and RWP-2 (D) cells. Cell pellets (5×10^6) were frozen, thawed, and suspended in E-BSA-FITC containing 50 μM bound E_2. After washing, each cell suspension was mounted between a microscope slide and a cover slip. Photomicrographs were taken through a Zeiss® fluorescence microscope using a mercury vapor lamp light source.

measured by FCM. Cell pellets (5×10^6 cells) are frozen at -70°C and thawed at room temperature once, then resuspended in 3 mℓ of E-BSA-FITC containing 50 μM bound estradiol derivative diluted in D-G-P saline (0.5 mM dithiothreitol and 10% glycerin in phosphate-buffered saline, pH 7.4).[19] Each mole of BSA in the fluorescent estradiol reagent carries approximately 26 mol of E_2 and 4 mol of FITC. After incubation at 4°C for 2 hr, the cells are washed twice in D-G-P saline, and a drop of the heavy cell suspension is mounted directly between a microscopic slide and a cover slip for fluorescence microscopy. Figure 2 shows the variable binding patterns and staining intensities exhibited by these tumor cells. The heterogeneous, high-intensity cellular staining of MCF-7 and 47-DN is contrasted with the lower-intensity, more homogeneous staining of COLO-357. In addition, the very low-intensity staining of RWP-2 might well be interpreted as background fluorescence and designated ER negative.[12,19] Comparing Figure 2 with the results of Table 1 suggests that

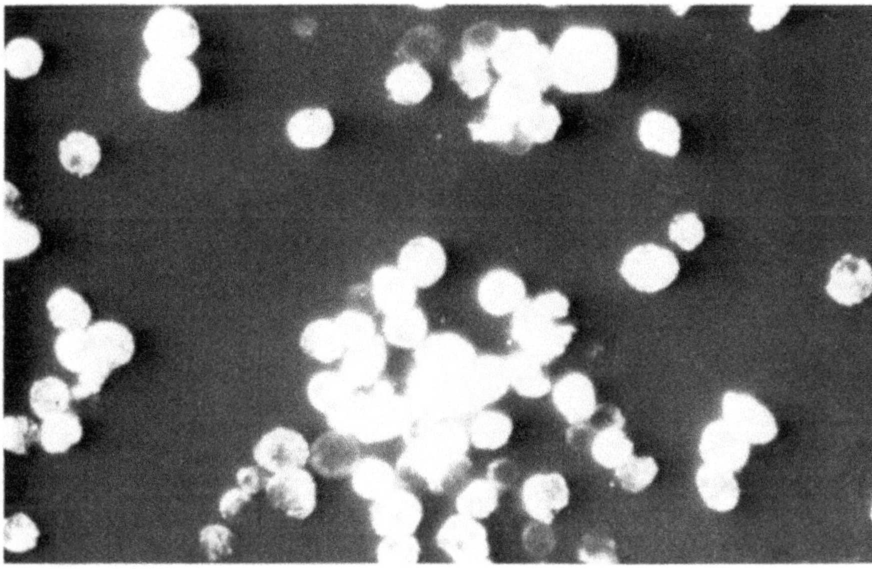

FIGURE 2B.

FIGURE 2C

E-BSA-FITC positivity correlates with specific [^{3}H]E_2 binding affinity rather than binding capacity, with a sensitivity limit <10 n*M* (K_d). Preliminary studies suggest that overall endocrine (tamoxifen, progesterone, and DES) sensitivity of these cell lines in vitro also correlates with E-BSA-FITC positivity and specific [^{3}H]E_2 binding affinity, rather than total capacity.

Washed cell pellets are also prepared for FCM by first freezing at −70°C. Thawed pellets are exposed to varying concentrations of E-BSA-FITC (± competing steroids) in D-G-P

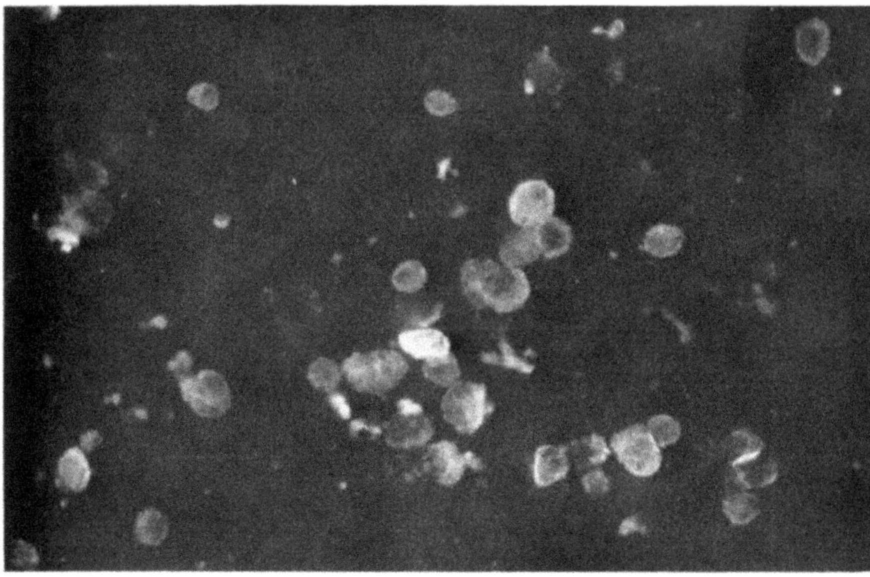

FIGURE 2D

saline at 4°C for 2 to 24 hr, subsequently rinsed twice in 0.5 m*M* dithiothreitol-10% glycerol-PBS, and then rinsed again in cold PBS. Stained and rinsed cells are diluted in PBS to a final concentration of 10^6 cells per milliliter and filtered through a 37-μm nylon mesh just prior to analysis on a Becton-Dickinson FACS IV instrument. Unstained cells are similarly prepared to measure autofluorescence, which primarily reflects the intracellular red-ox state and NAD(P)H level.[21,22] A 488-nm argon laser line (310 mW with 650-V photomultiplier tube) is used to excite fluorescence, and filters are used to permit only emitted wavelength ≥535 nm to be analyzed for pulse-height distribution over 255 channels. Flow rates are maintained between 500 to 1000 cells per second and calibration of the instrument's log-amp fluorescence intensity scale is performed using 1.72-μm fluorescent beads (Polysciences; Warrington, Pa.) such that 72 channel units = 1 log-amp unit. Simultaneous three-parameter measurement (cell number, light scatter, and fluorescence intensity) is recorded and stored for future analysis (peak subtraction) on floppy disks. Channel gating permits analysis of selected cells based on size (e.g., elimination of cell debris) or fluorescence intensity, as well as the rapid determination of a stained population's median channel intensity. Staining heterogeneity could also be quantitated using the histogram's coefficient of variation (CV). Fluorescence intensity and light scatter histograms are recorded on an X-Y recorder.

V. BINDING INTENSITY AND LIGHT SCATTER

Figure 3 compares the fluorescence of fresh and frozen-thawed cells stained with E-BSA-FITC. Light scatter histograms as well as phase microscopy verify that frozen-thawed cells are physically similar to fresh cells. Synthetically inactive and no longer viable in culture, these frozen-thawed cells bind substantially more E-BSA-FITC. It is likely that freezing alters cellular membranes enough to increase permeability of this hydrophilic macromolecule. This is confirmed by the observation that brief exposure of fresh cells to the detergent sodium dodecyl sulfate (SDS) results in binding fluorescence equal to that of frozen-thawed cells. Cells frozen at −70°C for up to 1 year retain their capacity to bind E-BSA-FITC.

Channel gating can be utilized to check for potential staining heterogeneity within a cell

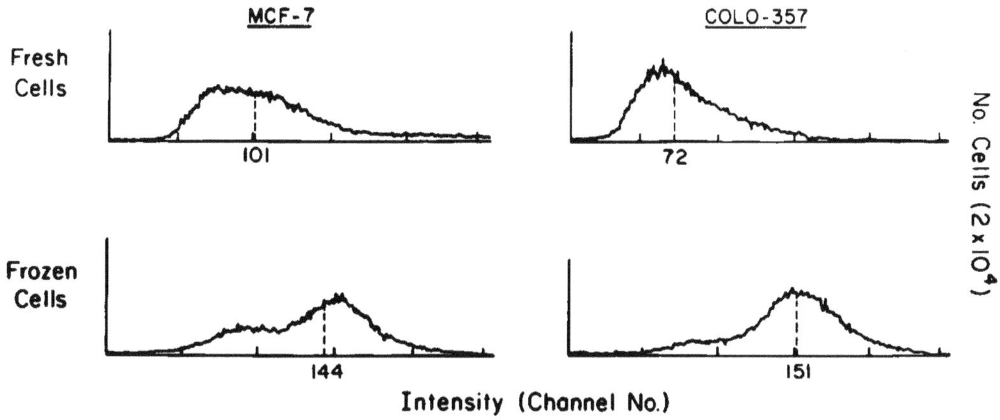

FIGURE 3. E-BSA-FITC binding intensities of fresh (unfrozen) and frozen-thawed MCF-7 and COLO-357 cells. Washed cell pellets (5 × 10^6) were incubated with E-BSA-FITC containing 5 μM bound E_2 for 2 hr at 4°C as described in the text. For each histogram determination, 2 × 10^4 cells were analyzed for fluorescence emission (≥535 nm) over 255 channels, calibrated on a log-amp intensity scale. The median channel number is given as a numerical indication of fluorescence intensity.

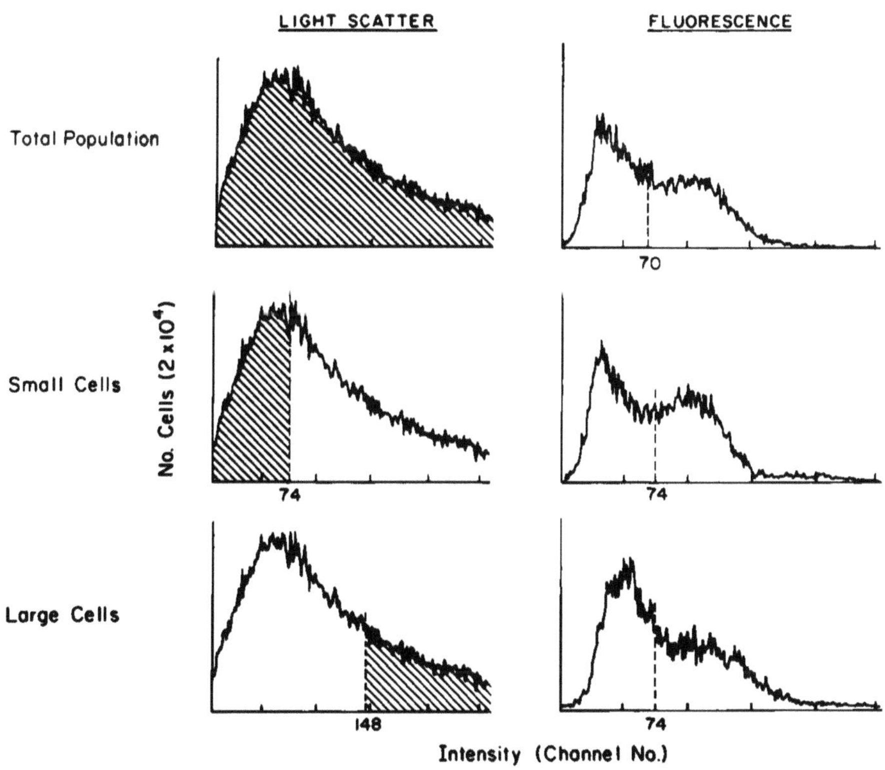

FIGURE 4. Light scatter and fluorescence intensity of MCF-7 cells stained 2 hr with E-BSA-FITC containing 1 μM bound E_2. Histograms were obtained sequentially on the same cell sample, and each represents an analysis of 2 × 10^4 cells. The staining intensities of selected small and large cells were analyzed by gating the light scatter histograms to include only channels 1-74 and 148-255, respectively. The median channel number is given for each fluorescence histogram.

population. Figure 4 shows the light scatter and fluorescence intensity of MCF-7 cells. When small MCF-7 cells (light scatter ≤ channel 74) are compared to large MCF-7 cells (light scatter ⩾ channel 148), only slight differences are observed in their respective E-BSA-FITC histograms.

The E-BSA-FITC binding intensities of different cells must take into account varying contributions from autofluorescence.[21] Figure 5 shows histograms for light scatter, autofluorescence, and E-BSA-FITC intensity; notably, there is considerable overlap in the histograms of autofluorescence and E-BSA-FITC intensity for each of the different carcinoma cell lines. Channel-by-channel subtraction of each autofluorescence histogram from its respective E-BSA-FITC histogram yields another histogram, shown in the far right-hand column of Figure 5, and referred to as corrected intensity. This derived histogram reflects the distribution of net fluorescence enhancement resulting from the cellular binding of E-BSA-FITC. The difference in median channel values between corrected and autofluorescence histograms is a closer numerical approximation of total E-BSA-FITC binding than would be obtained using the uncorrected E-BSA-FITC histogram. Even so, the median corrected intensity values shown in Figure 5 bear little correlation with either the ER affinity or capacity values for these cells as shown in Table 1. Further consideration of specific and nonspecific binding effects is necessary, but it is also important to demonstrate the dose dependence of E-BSA-FITC binding.

When MCF-7 cells are exposed to increasing concentrations of E-BSA-FITC for either 2 or 24 hr, histograms of progressively greater corrected intensity are obtained (Figure 6). Although a 24-hr exposure to 0.5 μM E-BSA-FITC results in as much or greater staining intensity as a 2-hr exposure to 5.0 μM E-BSA-FITC, the 2-hr incubation period is preferred. Incubation of the frozen-thawed cells for 24 hr at 4°C results in a much greater proportion of cellular fragmentation, which must be gated out of the analyzed cell population. A plot of the 2-hr median channel values would show a binding curve that approaches saturation between 1 and 5 μM E-BSA-FITC; and a Scatchard analysis of this data would reveal the presence of a nonspecific binding component.

VI. E-BSA-FITC BINDING SPECIFICITY

In the biochemical analysis of ER it is useful to measure the amount of bound $[^3H]E_2$ in the presence and absence of saturating concentrations of DES. Since type I ER usually have a $K_d \leq 5$ nM, this means that the necessary concentration of added DES is 0.5 μM. The difference in bound $[^3H]E_2 \pm 0.5$ μM DES is, thus, a measure of specific E_2 binding. In an analogous fashion, 0.5 μM DES may be used to define specific E-BSA-FITC binding. This concentration of DES will not completely saturate receptors with $K_d > 0.05$ μM. Thus, a corrected E-BSA-FITC histogram showing a shift to lower intensity in the presence of 0.5 μM DES is presumptive evidence for fluorescent ligand binding to type I or type II ER.

It is reasonable to speculate that the macromolecular nature of E-BSA-FITC reduces its affinity for type I ER. Furthermore, of the 26 molecules of E_2 covalently attached to each molecule of BSA, perhaps only a few can engage in receptor binding at any one time, reducing the known concentration of bound E_2. These two uncertainties provide some justification for believing that 0.5 to 5.0 μM E-BSA-FITC may be necessary to saturate a receptor whose K_d for E_2 (and DES) is 5 nM. Figure 7 shows that all four of the human tumor cell lines show a corrected intensity shift in the presence of 0.5 μM DES. It is interesting that the poorly staining RWP-2 cells (Figure 2), with the lowest affinity for $[^3H]E_2$ (Table 1), also show the smallest specific intensity by FCM (Figure 7).

Specificity is also tested by the ability of other steroids to compete with E-BSA-FITC binding in these cells. Figure 8 shows that DES and E_2 are nearly equal in their competitive effectiveness. Neither fluoxymesterone nor hydrocortisone are capable of reducing the median channel number of the corrected E-BSA-FITC histograms, suggesting that binding of this fluorochrome is, indeed, steroid specific. Results comparing the competitive effective-

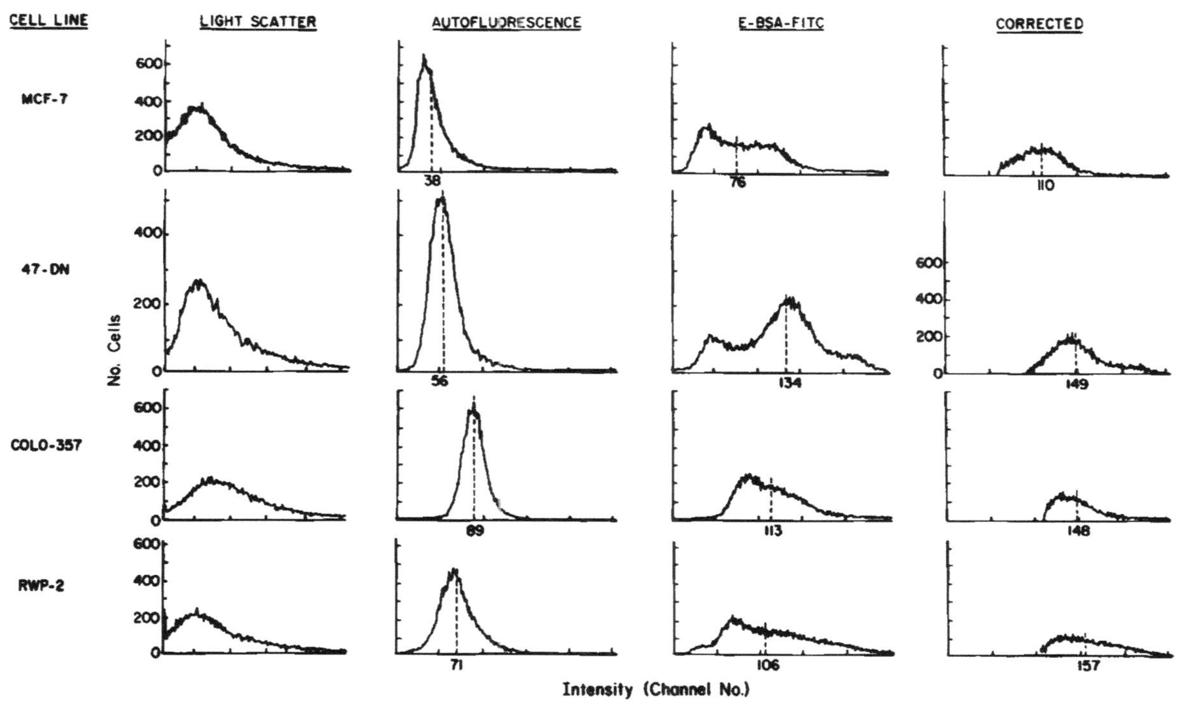

FIGURE 5. Light scatter and fluorescence histograms for each of the human carcinoma cell lines. Stained cells were exposed to E-BSA-FITC containing 0.5 μ*M* bound E_2 for 2 hr at 4°C; autofluorescence was measured in parallel samples incubated in buffer without E-BSA-FITC. For each sample 2×10^4 cells were analyzed with the number of cells/channel indicated on the vertical axis and the median channel number given on the horizontal axis. The distribution of cells with E-BSA-FITC staining intensity greater than autofluorescence (corrected intensity, as shown in the far right column) was derived by histogram subtraction, as described in the text.

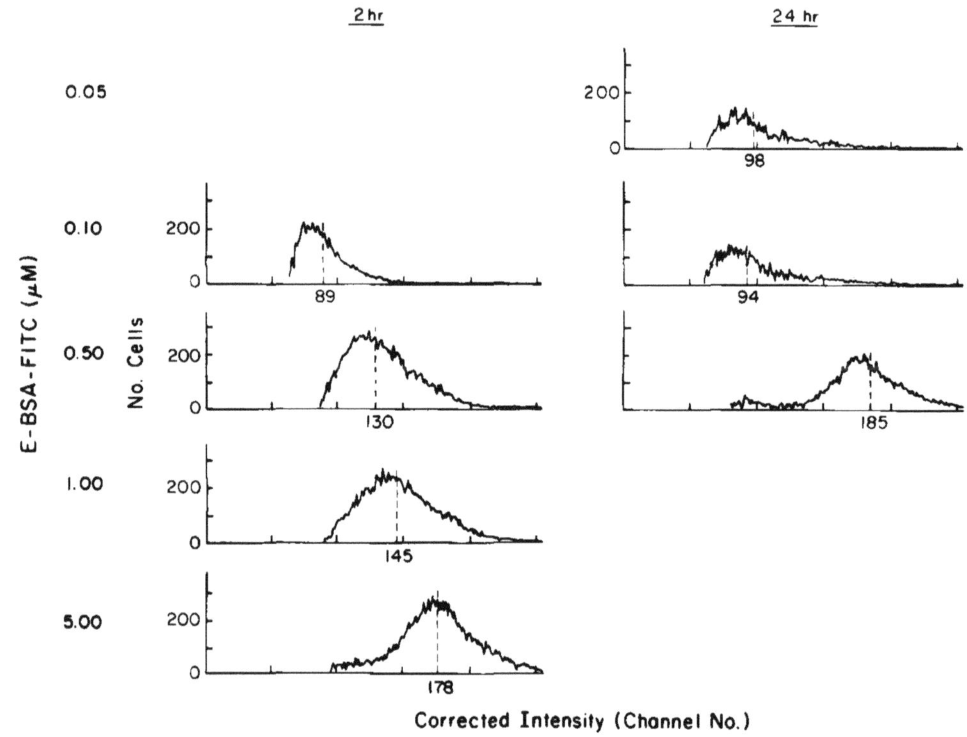

FIGURE 6. Dose dependence of the E-BSA-FITC corrected binding intensity. MCF-7 cells were incubated with varying concentrations of E-BSA-FITC for either 2 or 24 hr at 4°C. The indicated concentrations represent the amount of fluorochrome-bound E_2 present in the incubation solution. Autofluorescence histograms were subtracted from E-BSA-FITC binding histograms to produce the corrected intensity histograms shown here. The median channel numbers are indicated.

ness of progesterone and R5020 are, however, discrepant. Progesterone, which is known to cross react with other steroid receptors, appears to compete with E-BSA-FITC in this experiment. In contrast, the less cross-reactive progestin R5020 appears to enhance rather than inhibit E-BSA-FITC intensity. It is possible that this R5020-induced enhancement is an example of resonance energy transfer or light scattering effect, a phenomenon described in more detail below.

VII. APPLICATIONS OF E-BSA-FITC BINDING

The specific binding of E-BSA-FITC could be used to measure the K_i for DES and, thus, address some of the remaining uncertainties about this fluorochrome's availability for binding to type I or type II ER. In addition, it is possible that fresh human tumors could also be analyzed for ER by FCM using fluorescent ligands after enzymatically disaggregating the solid specimens into single cell suspensions. Of much recent interest, however, is the relationship between tumor ER and growth states, as well as the effects of drug/endocrine treatment on ER content. These latter concerns are easily answered using FCM methodology on human tumor cell lines. Figure 9 shows the difference in E-BSA-FITC specific binding between control and tamoxifen-treated 47-DN cells. In control cells the shift in median channel corrected intensity is 18 units, while in tamoxifen-treated cells this shift is only 8 units (44%). When similarly treated cells are analyzed biochemically, it is found that tamoxifen reduces nuclear ER to 57% of its control value.[16] Agreement between FCM and radioligand receptor methodology is also apparent in a study comparing plateau (confluent)

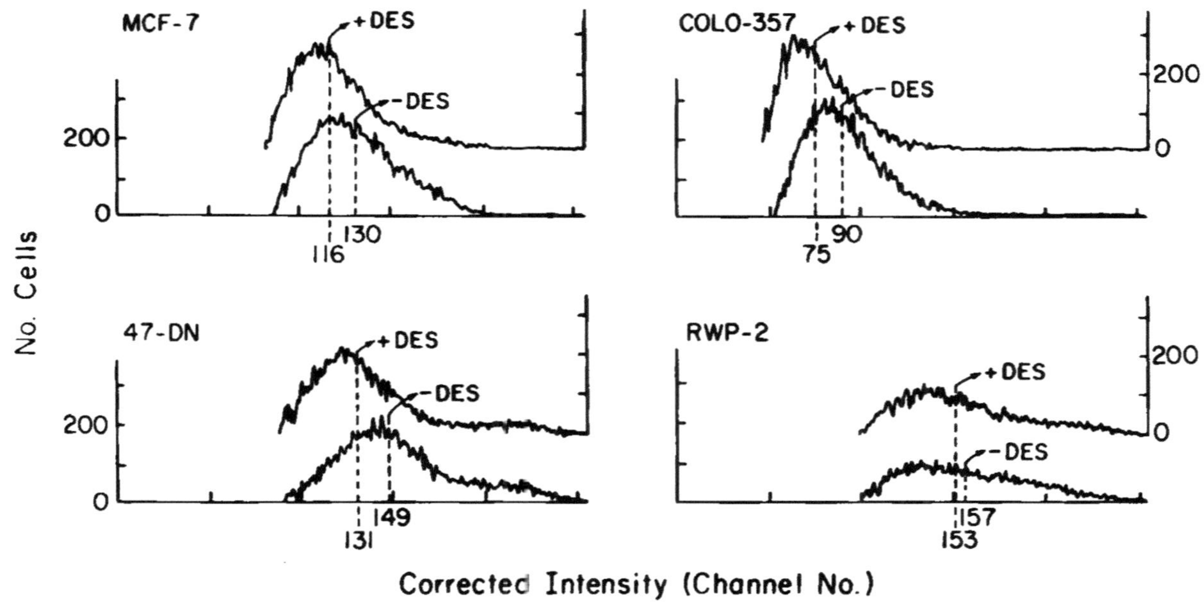

FIGURE 7. E-BSA-FITC specific binding for each of the human carcinoma cell lines. Cell samples were exposed to 0.5 μ*M* E-BSA-FITC ± 0.5 μ*M* DES for 2 hr at 4°C. Autofluorescence was subtracted from the stained histograms resulting in the corrected intensity histograms shown here. The shift in the (−DES) histogram to the lower intensity (+DES) histograms is being used to define specific E-BSA-FITC binding, as discussed in the text. The difference in median channel numbers provides a numerical value for this specific binding.

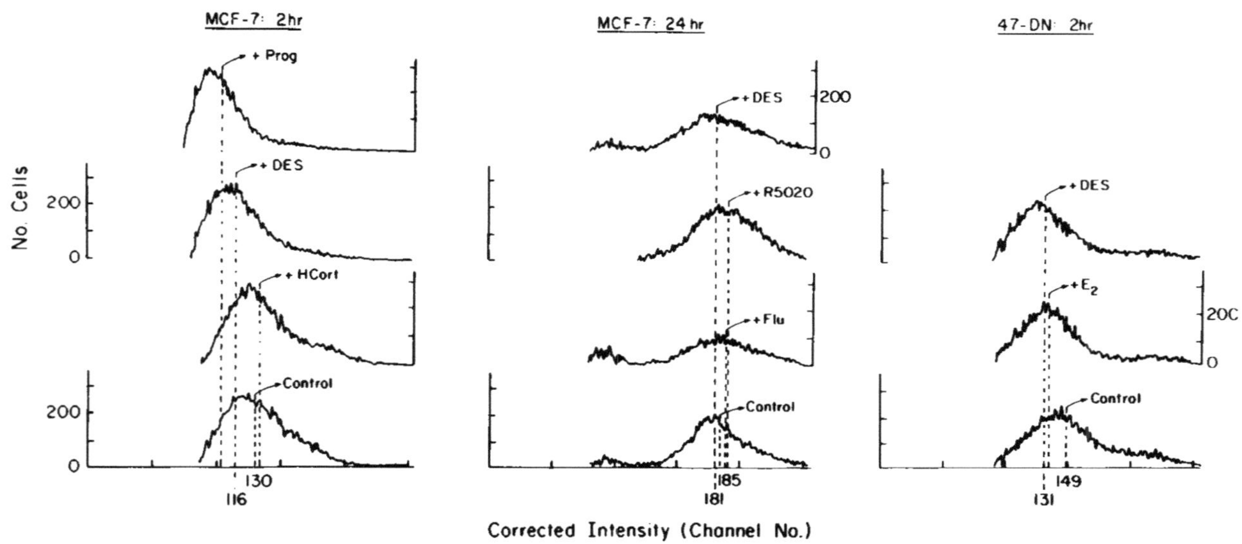

FIGURE 8. Effects of free steroids on E-BSA-FITC binding. MCF-7 and 47-DN cells were exposed for either 2 or 24 hr to 0.5 μM E-BSA-FITC at 4°C. Corrected intensity histograms were obtained in the absence (control) and presence of 0.5 μM free steroid — E_2, DES, R5020, Prog (progesterone), Flu (fluoxymesterone), and HCort (hydrocortisone). The median channel number for each histogram is marked by the broken vertical line.

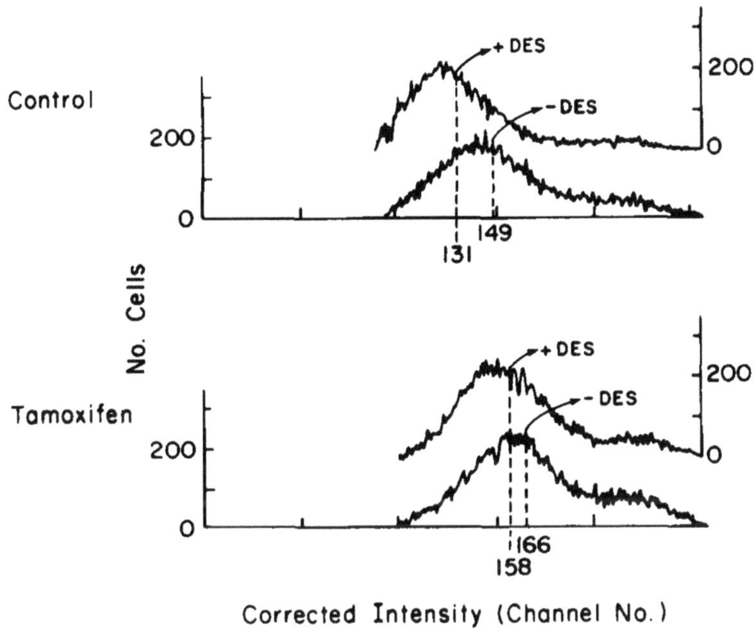

FIGURE 9. E-BSA-FITC specific binding in control and tamoxifen-treated 47-DN. Cells were exposed to 10 μM tamoxifen for 72 hr during early log-phase growth and then washed free of loosely associated drug by a 1-hr incubation in drug and serum-free media. Corrected intensity histograms were obtained on control and treated cells stained with 0.5 μM E-BSA-FITC (2 hr, 4°C) ± 0.5 μM DES. The difference in median channel numbers for (−DES) and (+DES) histograms is a measure of the specific E-BSA-FITC binding for each condition.

and early log-phase (preconfluent) cell cultures. Scatchard analysis using $[^3H]E_2$ reveals that confluent MCF-7 cells have an ER capacity of 4.1 fmol/10^6 cells and K_d = 0.8 nM, while preconfluent cells have an ER capacity of 35 fmol/10^6 cells and K_d = 0.2 nM. In Figure 10, E-BSA-FITC specific binding is given under comparable culture conditions. With 0.5 μM E-BSA-FITC ± 0.5 μM DES, confluent cultures show specific binding intensity of 10 units, while preconfluent cultures show a specific binding intensity of 31 units. By either methodology, there appears to be a greater amount of specific estrogen binding in cells during early log-phase growth. Double staining these cells with E-BSA-FITC (±DES) and a DNA binding fluorochrome such as propidium iodide (which has a noninterfering emission peak at 625 nm) would permit simultaneous FCM analysis of both ER and DNA content, and, thus, relate cell cycle phase to ER production.

Resonance energy transfer (RET) and the use of Stryer's "spectroscopic ruler" offers another potentially powerful application of the FCM methodology.[23,24] By staining cells with a second fluorochrome whose excitation or emission spectrum overlaps with the emission or excitation spectrum of E-BSA-FITC, and measuring the enhancement or quenching of the acceptor's fluorescence emission, it would be possible to study subcellular interactions occurring within 20 to 100 Å from the ER. Selective probes for DNA or RNA could be used in conjunction with E-BSA-FITC to investigate many aspects of ER activation and processing. Fluorescein isothiocyanate (FITC) and tetramethylrhodamine isothiocyanate (TMRITC) have been used as the donor and acceptor fluorochromes in RET studies of Concanavalin-A receptors on fibroblast membranes.[25] Similarly, it would be possible to measure the physical proximity of E-BSA-FITC bound ER to progesterone receptor, bound by a TMRITC conjugate.

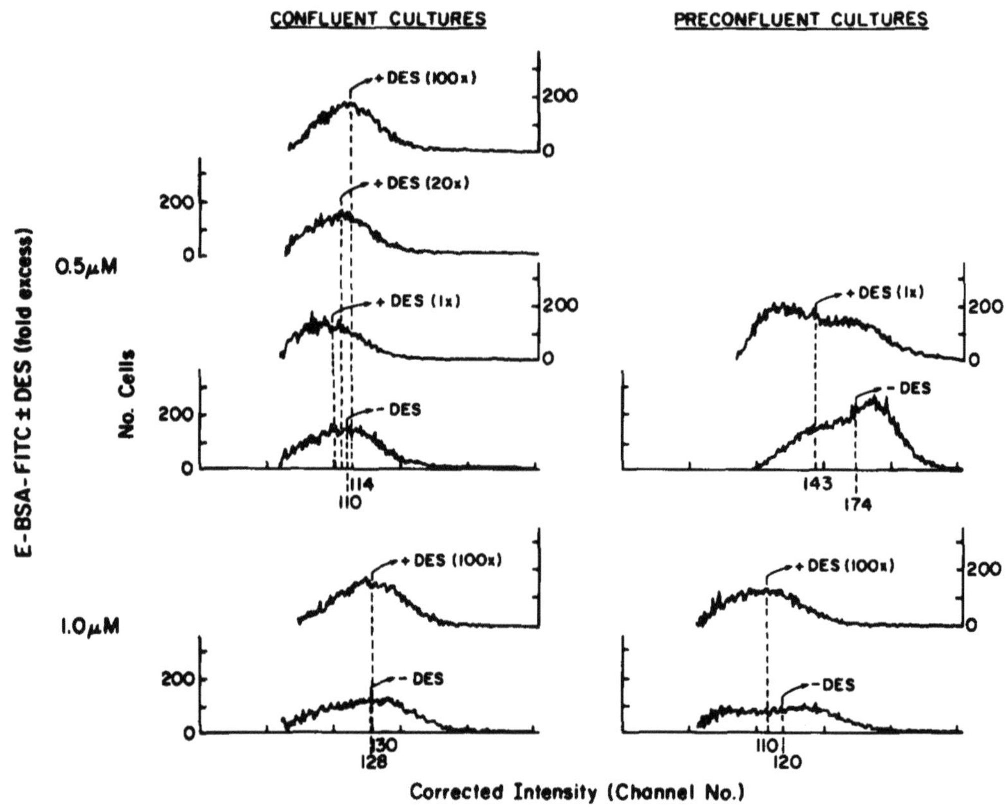

FIGURE 10. Effects of DES on E-BSA-FITC binding in confluent and preconfluent cultures of MCF-7. Confluent cells were harvested during the plateau-phase of culture growth while preconfluent cells were harvested during the early log-phase of growth. Corrected intensity histograms were obtained in the absence (−DES) and presence of varying concentrations of DES, indicated by the degree of excess DES present in relationship to the staining concentration of E-BSA-FITC. All incubations occurred over 2 hr at 4°C. For a given sample, the difference in median channel numbers (broken vertical lines) between histograms marked −DES and +DES (1×) represents specific E-BSA-FITC binding. Higher DES concentrations were used to demonstrate the possibility of fluorescence enhancement as discussed in the text.

Light scattering effects or RET phenomena may potentially interfere with the study of steroids competing with E-BSA-FITC for binding to ER. The covalently conjugated E-BSA-FITC molecule can nonspecifically bind to DES, R5020, or other steroids which might have light scattering effects at high intracellular concentrations; alternatively, E-BSA-FITC can bind to receptors located in close proximity to the specific intracellular binding sites of these other hydrophobic steroids. Resonance energy may be transferred to E-BSA-FITC from these nonspecifically or specifically bound hydrophobic molecules or scattered excitation energy may be deflected to the bound FITC acceptor, producing unintentional enhancement of E-BSA-FITC fluorescence. Enhanced fluorescence is occasionally observed when R5020 is added to E-BSA-FITC stained MCF-7 (Figure 8). Figure 10 also demonstrates that use of DES at concentrations $> 10\mu M$ (in the presence of 0.5 μM E-BSA-FITC) results in fluorescence enhancement rather than reduction in corrected intensity, as observed with 0.5 μM DES. Thus, this interfering effect may potentially underestimate or even negate the measurement of specific E-BSA-FITC binding. The results of Figure 10 suggest that use of 0.5 μM DES is both sufficient and optimal for determining the specific binding of E-BSA-FITC to ER.

Enhancement of E-BSA-FITC fluorescence by competitors (such as DES or E_2) is a phenomenon that can also be observed by fluorescence microscopy with frozen tissue sections, especially when the concentration of competitor approximates that required for complete blocking of E-BSA-FITC fluorescent staining. In an attempt to elucidate the mechanism for this unintentional fluorescence enhancement, samples of E-BSA-FITC solution were mixed with DES or E_2 in various proportions, and the fluorescence intensity of each of the resultant mixtures was measured in a Perkin-Elmer fluorescence spectrophotometer. No increase in fluorescence was found with completely dissolved competitors using an excitation wavelength set at 488 nm and emission wavelengths varying between 488 and 600 nm. However, when concentrations of DES or E_2 approached insolubility in the presence of E-BSA-FITC, the emission intensity for wavelengths near 490 nm was markedly enhanced. This observation may be related to light scattering and the measurement of deflected excitation radiation. Whether a similar effect can occur with E-BSA-FITC stained cells, wherein the intracellular concentration of competing steroid approaches insolubility, requires further investigation.

VIII. SUMMARY

Adapting FCM to the quantitative measure of E-BSA-FITC binding in ER-positive tumor cell suspensions produces further evidence for the receptor specificity of this fluorescent ligand. Changes in ER capacity or affinity which cannot be appreciated under the fluorescence microscope can be quantitated using FCM; and these FCM measurements correlate with ER determination by other methodologies. It is not yet known whether E-BSA-FITC binding is specific for type I, type II, or a combination of type I and II receptors. It is also not known how precisely FCM methodology correlates with the biochemical definition of ER capacity and affinity. Both sets of questions are currently under investigation. FCM analysis of fluorescent estrogen binding may never become a diagnostic tool, but it may help define the clinical role of E-BSA-FITC in the histochemical detection of ER.

ACKNOWLEDGMENTS

We wish to thank Anita Stepanowski and Owen Blair for their FCM assistance (Grant CA 16359), and Arlene Cashmore for her preparation of figures. These studies were supported by grants CH-235 from the American Cancer Society, CA-36769 and CA-36773 from the National Cancer Institute, a Pharmaceutical Manufacturers Association Foundation Research Starter Grant, a Swebilius Cancer Research Award from the Yale Comprehensive Cancer Center, and B.R.S.G. Grant RR 05358 from the National Institute of Health.

REFERENCES

1. **Henderson, B., Ross, R., Pike, M., and Casagrande, J.,** Endogenous hormones as a major factor in human cancer, *Cancer Res.*, 42, 3232, 1982.
2. **Stoll, B. A.,** Endocrine therapy in cancer, *Practitioner*, 222, 211, 1979.
3. **Stedman, K., Moore, G., and Morgan, R.,** Estrogen receptor proteins in diverse human tumors, *Arch. Surg.*, 115, 244, 1980.
4. **Greenway, B., Igbal, M., Johnson, P., and Williams, R.,** Oestrogen receptor proteins in malignant and fetal pancreas, *Br. Med. J.*, 283, 751, 1981.
5. **Patterson, J. and Battersley, L.,** Tamoxifen: an overview of recent studies in the field of oncology, *Cancer Treat. Rep.*, 64, 775, 1980.

6. **Leake, R., Lain, L., Calman, K., and Macbeth, F.,** Estrogen receptors and antiestrogen therapy in selected human solid tumors, *Cancer Treat. Rep.*, 64, 797, 1980.
7. **Chammness, G. C. and McGuire, W. L.,** Scatchard plots: common errors in correction and interpretation, *Steroids*, 26, 538, 1975.
8. **Rodbard, D., Munson, P. J., and Thakur, A. K.,** Quantitative characterization of hormone receptors, *Cancer*, 46, 2907, 1980.
9. **Clark, J. H., Markaberich, B., Upchurch, S., Eriksson, H., Hardin, J., and Peck, E.,** Heterogeneity of estrogen binding sites: relationship to estrogen receptors and estrogen responses, *Recent Prog. Horm. Res.*, 36, 89, 1980.
10. **Panko, W., Watson, C., and Clark, J.,** The presence of a second specific estrogen binding site in human breast cancer, *J. Steroid Biochem.*, 14, 1311, 1981.
11. **McGuire, W. L., Carbone, P.P., Sears, M. E., and Escher, G. C.,** Estrogen receptors in human breast cancer: an overview, in *Estrogen Receptors in Human Breast Cancer*, McGuire, W. L., Carbone, P. P., and Volmer, E. P., Eds., Raven Press, New York, 1975, 1.
12. **Meijer, C. J. L. M., Van Marle, J., Persijn, J. P., Van Niewenhuizen, W., Baak, J. P. A., Boon, M. E., and Lindeman, J.,** Estrogen receptors in human breast cancer: correlation between the histochemical method and biochemical assay, *Virchows Arch. (Cell Pathol.)*, 40, 27, 1982.
13. **Pertschuk, L. P., Tobin, E. H., Carter, A. C., Eisenberg, K. B., Leo, V. C., Gaetjens, E., and Bloom, N. D.,** Immunohistologic and histochemical methods for detection of steroid binding in breast cancer: a reappraisal, *Breast Cancer Res. Treat.*, 1, 297, 1981.
14. **Melamed, M. R., Mullaney, P. F., and Mendelsohn, M. L., Eds.,** *Flow Cytometry and Sorting*, John Wiley & Sons, New York, 1979.
15. **Sklar, L. A. and Finney, D. A.,** Analysis of ligand-receptor interactions with the fluorescence activated cell sorter, *Cytometry*, 3, 161, 1982.
16. **Benz, C., Cadman, E., Gwin, J., Wu, T., Amara, J., Eisenfeld, A., and Dannies, P.,** Tamoxifen and 5-fluorouracil in breast cancer: cytotoxic synergism in vitro, *Cancer Res.*, 43, 5298, 1983.
17. **Shafie, S. and Brooks, S.,** Effect of prolactin on growth and the estrogen receptor level of breast cancer cells (MCF-7), *Cancer Res.*, 37, 792, 1977.
18. **Van Marle, J., Lindeman, J., Ariens, A., Labruyere, W., and Van Weeren-Kramer, J.,** Estrogen receptors in human breast cancer: specificity of the histochemical localization of estrogen receptors using an estrogen-albumin FITC complex, *Virchows Arch. (Cell Pathol.)*, 40, 17, 1982.
19. **Lee, S. H.,** The histochemistry of estrogen receptors, *Histochemistry*, 71, 491, 1981.
20. **Lee, S. H.,** Estrogen-primed immature rat uterus — a tissue control for histochemical estrogen receptor assay , *Am. J. Clin. Pathol.*, 79, 484, 1983.
21. **Thorell, B.,** Flow cytometric analysis of cellular endogenous fluorescence simultaneously with emission from exogenous fluorochromes, light scatter and absorption, *Cytometry*, 2, 39, 1981.
22. **Thorell, B.,** Intracellular red-ox steady states as basis for cell characterization by flow cytofluorometry, *Blood Cells*, 6, 745, 1980.
23. **Stryer, L.,** Fluorescence energy transfer as a spectroscopic ruler, *Annu. Rev. Biochem.*, 47, 819, 1978.
24. **Jovin, T. M.,** Fluorescence polarization and energy transfer: theory and application, in *Flow Cytometry and Sorting*, Melamed, M. R., Mullaney, P. F., and Mendelsohn, M. L., Eds., John Wiley & Sons, New York, 1979, 137.
25. **Dale, R. E., Novros, J., Roth, S., Edidin, M., and Brand, L.,** Application of Forster long-range excitation energy transfer to the determination of distributions of fluorescently-labelled Concanavalin A-receptor complexes at the surfaces of yeast and of normal and malignant fibroblasts, in *Fluorescent Probes*, Beddard, G. S. and West, M. A., Eds., Academic Press, New York, 1981, 159.

Chapter 7

SEX STEROID ACTION MECHANISM BY CYTOCHEMISTRY IN NORMAL AND NEOPLASTIC TARGET-TISSUES*

Elisabetta Marchetti, Andrea Marzola, Alberto Bagni, Patrizia Querzoli, Guidalberto Fabris, and Italo Nenci

TABLE OF CONTENTS

I. Introduction 112

II. Steroid-Plasma Membrane Interactions 113

III. Intracellular Steroid-Cell Interactions 115

IV. Cytoplasm to Nucleus Translocation Process 117

V. Breast Cancer Biology 118

VI. Prostate Proliferative Diseases 120

VII. Perspectives 121

References 121

* The experimental work was supported in part by Grant No. 83.00810.96 from Special Project "Control of Neoplastic Growth" of the Italian National Research Council.

I. INTRODUCTION

Several methodological approaches to the morphological demonstration of steroid binding sites in target tissues have been progressively introduced in recent years.

The parent technique, which first offered a visual tracing of bound steroids, is histoautoradiography. The poor morphological details offered by this technique and the long procedures inherent to it have subsequently favored its replacement by handier methodologies. Thus, a cytochemical approach to the study of steroid binding in target tissues was pursued and quickly evolved through several generations.

The first generation is hormone immunocytochemistry. This technique is based on the detection of steroid hormones bound to their receptors by specific steroid antibodies, thus coupling the sensitivity of radiometric methods to a good morphological detail. Steroid antibodies are then traced by a second antibody tagged either with fluorescein or with peroxidase, or by PAP technique, for ultraviolet, light, and electron microscopy, respectively.[1-6] The method, originally applied to the demonstration of estrogen receptors in vital cells from human breast cancer, was afterwards applied to tissue slices and to the study of steroid receptors other than the estrogen ones. Actually, this technique was envisaged to detect the sequential steps of steroid action at the cellular level, and soon proved to be ideally suited to the dynamic monitoring of estradiol uptake and intracellular redistribution of estrogen-receptor complexes, exploiting the temperature dependence of such a compartmentalization in vitro. Vital cells were exposed to the hormone at different temperatures and for different lengths of time, washed, and left dry in the cold. Receptor-bound estradiol was then traced by a double immunocytochemical technique. This technique, using peroxidase as a tracer, also allowed the study of hormone uptake and compartmentalization at the subcellular level by electron microscopy.[1,2,5,6]

Hormone receptor immunocytochemistry has also been applied to the demonstration of estrogen receptors in frozen sections from estrogen target tissues. An approach involves the immune tracing of estradiol to which frozen sections have been exposed in vitro. To enhance the visualization of specific receptors, a variation of the double antibody method has been introduced, in which 17β-estradiol was substituted by the polymeric derivative polyestradiol phosphate (PEP), which acts as a larger antigen for the antibody.[7,8] Moreover, frozen sections were exposed to a soluble immunocomplex made of PEP and anti-17β estradiol antibody, which was then displayed by a second fluorescein- or peroxidase-labeled antibody. The availability of estradiol-reactive groups to estradiol antibody of the immunocomplexed polyestradiol was tested by a double immunodiffusion technique.[9]

The second generation technique, that is, affinity cytochemistry, exploits steroid hormones bound to a macromolecular protein carrier, bovine serum albumin (BSA), highly substituted with fluorescein isothiocyanate or horseradish peroxidase.[9-13] Some of these steroid derivatives retain enough affinity for specific receptors as shown by affinity chromatography and are, for this reason, widely utilized to purify receptors.[14] These macromolecular steroid derivatives are only suited to the static evaluation of intracellular steroid binders on tissue sections, since their size interferes with the uptake by intact target cells.[15]

The third generation technique is molecular cytochemistry, which makes use of steroid analogs made of fluorescein moiety directly coupled to the hormone. These fluorescent hormonal probes have been shown to interact with receptors with high affinity and specificity.[16] Moreover, their inherent properties allow their uptake by intact cells both in vivo and in vitro.[17-20] These fluorescent steroid derivatives are, thus, successfully applied to the static evaluation of intracellular binding sites in tissue sections and, especially, to the dynamic monitoring of hormone-cell interactions in an intact cell system.

The last generation technique is receptor immunocytochemistry. The availability of polyclonal and monoclonal antibodies to steroid receptors offers a closer view of receptor physiology and pathophysiology. Glucocorticoid receptor antibodies have been applied to the dynamic tracing of the intracellular compartmentalization of the receptor in vital target cells.[21] Polyclonal and monoclonal antibodies have been applied to trace the receptor distribution in progesterone target tissues by a double immunoperoxidase technique.[22,23] Polyclonal and monoclonal antibodies also against estrogen receptors have been synthesized independently by several laboratories in recent years.[24-27]

Monoclonal antibodies have been produced against receptor from calf uterus, either as the native form[27] or as the nuclear form of the estradiol-receptor complex,[26] against the native receptor from the MCF-7 cell line,[25] and from human myometrium.[28] Many of these monoclonal antibodies have been shown to recognize receptor determinants of mammalian species other than the one used as antigenic source. Though promising tools for elucidating some yet unexplained events of the mechanism of action of steroid hormones, some discrepancies are apparent as to the receptor distribution traced by different antibodies. The monospecificity of monoclonal antibodies, the different molecular state of receptor used as antigenic source, and, lastly, the different experimental conditions adopted in immunocytochemical demonstration could account for the different results. Immune localization of receptors has been carried out on fixed frozen sections and on fixed paraffin-embedded tissue sections.[29,62] Also, different displaying systems have been used: immunoperoxidase, immunofluorescence, peroxidase-antiperoxidase, and the avidin-biotin system.

II. STEROID-PLASMA MEMBRANE INTERACTIONS

Cytochemical techniques have been exploited in appropriate experimental conditions to investigate some events of the mechanism of interaction of steroid hormones with target cells; resorting to cytochemistry is issued from the need to overcome the disadvantages inherent in the biochemical approach of the study of hormone action, that is, the use of disrupting procedures.

The evidence for the cytoplasm-to-nucleus receptor system of steroid action and the use of fractionated, not intact cell systems to investigate the complex interaction of steroid hormones with target cells has contributed to ignore a possible involvement of the cell plasma membrane. Steroid hormones have always been assumed to enter cells by a simple, passive diffusion mechanism, at variance with polypeptide hormones which primarily interact with the cell surface. Such a rigid distinction between the mechanism of action of steroid and polypeptide hormones has, nevertheless, been repeatedly argued by more recent evidence. It is well known that many polypeptide hormones enter target cells and interact with subcellular components.[30] On the other hand, steroid hormones have been reported to interact with some constituents of the plasma membrane and to enter target cells by a membrane-mediated process.[31,32]

The possibility that the plasma membrane plays a role other than a passive barrier to the intracellular diffusion of the hormone has, thus, been tested. Actually, some evidence for the interaction of steroids with the plasma membrane of target cells had been obtained by previous ultrastructural studies showing that antiestradiol antibody heavily decorates the plasma membrane of estradiol preexposed breast cancer cells.[6]

A cytochemical approach to an intact cell system has been set up in appropriate experimental conditions, which could in some way prevent the hormone from entering into the cell, while allowing the detection of a specific interaction, if any, of the hormone with the plasma membrane.[33-35]

Isolated viable cells in suspension obtained by mechanical means from freshly removed breast cancer specimens were used as target cells. On occasion, the breast cancer continuous cell lines MCF-7 and ZR-75-1 have been used in this study. Cells were reacted in suspension with estrogen macromolecular fluorescent derivatives. A diffuse surface staining was apparent in cells maintained at 4°C through the labeling, washing, and observing steps. Occasional raising of the temperature as well as the postincubation at 37°C of cells labeled at 4°C, brought about the formation of clusters and patches. Such a topographical rearrangement eventually resulted in the appearance of a discrete fluorescent area with a location apparently dependent on the cell size, shape, and reciprocal organization which, on isolated cells, looked like a fluorescent cap structure with a polar orientation.

The temperature-dependent redistribution of the ligand on the cell surface is obviously artifactual and results from the binding of a multivalent ligand, like the exploited macromolecular hormonal probes, and cannot be expected to occur in the physiological interplay between monovalent steroids and the plasma membrane of target cells. As the two-dimensional translational diffusion is a well-known property of membrane-intrinsic constituents according to the fluid mosaic membrane model,[36,37] it can be concluded that the observed surface binding sites for estradiol fulfill the properties of a structural nonlipid component fully integrated in the plasma membrane.

The binding of estradiol to the cell surface is restricted to target cells. Actually, no membrane positivity was ever observed in nonepithelial, nontarget cells, such as blood cells, which contaminated some breast cancer cell preparations, or several cell lines (HEp-2, BHK, mouse fibroblasts) which are not estrogen responsive.

A similar surface labeling was obtained when cells were reacted first in suspension with an immunocomplex made of PEP and specific antiestradiol antibody and, lastly, on slides, with a second fluorescein- or peroxidase-tagged antiantibody.[33] In particular, such a modified, double immunoperoxidase technique was exploited to observe the binding with scanning and transmission electron microscopy. With scanning electron microscopy and electron probe X-ray microanalysis, a random distribution of the bound macromolecule was evident on the surface of cells incubated at 4°C, while, again, in cells incubated at 37°C the surface-bound immunocomplex coalesced in large aggregates and polar caps.

Several sets of control experiments have reliably excluded the possibility that the observed binding to the cell surface of the macromolecular estradiol fluorescent derivatives was dependent on the interaction of the protein carrier BSA with some membrane components.[35] Unlabeled BSA did not affect the binding of the fluorescent estradiol derivative; fluoresceinated BSA did not appreciably bind to the plasma membrane and, lastly, no surface positivity was observed if estradiol determinants of the fluorescent analog had been obliterated by specific antibodies. The steroid specificity of the observed binding was strengthened by some competition experiments with the same or other classes of unlabeled steroid hormones. The labeling was best prevented by unlabeled protein-linked estrogens. Native "cold" estradiol inhibited the binding only if added in excess to the fluorescent analog solution. This weaker inhibition could be due to a very ephemeral interaction of native estradiol with the plasma membrane or to a better recognition by membrane sites of the hormone in the protein-carried state. The surface positivity was not affected by other steroid hormones such as testosterone and progesterone, both as native and protein-linked forms.

At variance was the binding competition studies for cytoplasmic receptor with diethylstilbestrol and the antiestrogen tamoxifen. No inhibition of membrane positivity was obtained with the latter compounds.

The specificity of the interaction of the estradiol moiety of the fluorescent analog with the plasma membrane was again ascertained by preexposing cells to "cold" macromolecular estradiol at 37°C. Cells treated in this way failed to bind the fluorescent estrogen, suggesting a polar concentration of the cold ligand into caps, the surface binding sites being no longer available for the fluorescent analog.

The presence of specific membrane sites for estrogen was later investigated and confirmed by an affinity binding approach in the reverse direction. Breast cancer cells were reacted in suspension with an insoluble support derivatized with estrogen.[38,39] Such a support was made of agarose beads coated with a multichain polymer of lysine and alanine to which either diethylstilbestrol or estradiol were covalently attached. These agarose beads have been exploited to purify estrogen receptors by affinity chromatography.[40] A tight adhesion of cells to estradiol-derivatized beads could be observed when both were allowed to react in suspension. Such an interaction displayed the same characteristics as the binding observed by fluorescent analogs. Actually, it was prevented by previous blocking of estradiol determinants on beads by specific antibody, by preincubating cells with protein-linked estrogen analogs, and did not occur if cells were reacted with beads derivatized with diethylstilbestrol.

The meaning of this additional level of interaction of estradiol with target cells is not as plain as the evidence for the binding. Yet, some hypotheses can be drawn.

A membrane incorporation of the classic cytoplasmic receptor cannot be discarded. Binding characteristics, however, seem to militate against such a possibility. Nevertheless, the different subcellular location could account for the different binding affinity of diethylstilbestrol and tamoxifen. The lack of competition by testosterone makes the involvement of steroid binding globulins (SBG) quite unlikely.

The possibility must be taken into account that surface binding sites go along and mirror specific stages of differentiation, by analogy with recently reported similar observation on peripheral lymphocytes.[41]

A direct modulation of peculiar membrane function by steroid hormones, such as transmembrane ion exchange, has already been demonstrated and more recently confirmed.[42,43]

Lastly, recognition sites for the hormone on the plasma membrane could subserve a mechanism of local concentration at the level of target cells, thus, enabling its diffusion through the plasma membrane.

Alternatively, a role in the intracellular transport of the hormone could be suggested. In this respect it is worth stressing that a micromolecular estrogen derivative, made of a fluorescein moiety coupled to estradiol, behaves as native estradiol and is taken up by estrogen target cells, but not by nontarget cells. If estrogen sensitivity of a cell were dependent exclusively on the presence of cytoplasmic receptors responsible for the retention of the freely diffused hormone, then the fluorescent hormone would be also expected to enter estrogen nontarget cells. The failure of staining of nontarget cells when reacted with fluorescent estradiol suggests an active involvement of membrane binding sites in the transmembrane crossing of the hormone. This possibility has been investigated by exploiting the different color of emission of the fluorescein-tagged estrogen analogs, which fluoresce green, and of coumestrol, a natural phytoestrogen which fluoresces blue.[35] Actually, cells in which surface binding sites have been saturated by the macromolecular fluorescent estrogen analog do not take up coumestrol to the same extent as untreated cells.

III. INTRACELLULAR STEROID-CELL INTERACTIONS

Besides the tracing of the specific interaction of estrogen with the cell plasma membrane, cytochemical techniques have allowed the visualization of the sequential steps following the entry of the hormone into the cell.

For this study the target tissues of choice have been breast cancer cells in suspension. On occasion, the breast cancer continuous cell lines ZR-75-1 and MCF-7 were used. These intact cell systems appeared particularly suited to the dynamic monitoring of the intracellular action of the hormone. Tissue sections of estrogen target tissues were additionally exploited to integrate the tracing with the structural information obtainable from a tissue as a whole.

The exploited cytochemical techniques have been hormone immunocytochemistry, molecular cytochemistry with fluorescent micromolecular estrogen, and affinity cytochemistry by macromolecular fluorescent estrogen analogs.[1,2,9,18] The latter were utilized on tissue sections only because, as above specified, the presence of the carrier protein prevents these ligands from entering intact cells.

Different experimental conditions were adopted to selectively visualize the steps of the intracellular pathway of steroid-receptor complexes.[1,2,18] Incubation of cells with estradiol at 4°C allowed the visualization of the hormone bound to cytoplasmic receptors, the hormone-receptor complexes being immobilized in the cytoplasmic compartment by the low temperature. The major contribution to this cytoplasmic positivity by electron microscopy appeared to issue from a specific staining of ribosomes, often arranged in microsomal outlines.[5,6] The association with particulate cytoplasmic components is quite at variance with the assumption that cytoplasmic receptor is mainly localized in the cytosol fraction of cellular homogenates. It must be stressed, however, that receptors have repeatedly been identified in association with membranes, cytoskeleton, and microsomes.[44,45] Both the synthesis of receptors and modulation of the translational process by hormone-receptor complexes could account for the microsomal localization. The prevalent cytoplasmic localization of receptors in the absence of the ligand was also confirmed in tissue sections by affinity cytochemistry,[9,18] and more recently by receptor immunocytochemistry.[63]

The step following the initial hormone-receptor binding, that is, the movement of the complex from the cytoplasm to the nucleus, is, perhaps, the key event for the action of the hormone to take place. Actually, nuclear translocation of the complexes correlates with the acquired ability of interaction with nuclear structures and is the end result of a series of phenomena which receptor and hormone-receptor complexes undergo, such as activation and transformation.

The selective visualization of the translocation process has been pursued taking advantage of its well-known temperature dependence in vitro. Thus, in target cells first reacted with estradiol at 4°C and postincubated at 37°C, a nuclear positivity was observed; moreover, only the nuclear localization was apparent in cells incubated with estradiol at 37°C from the beginning. The nuclear translocation of the hormone-receptor complexes was reproducible by all the cytochemical techniques in vitro in intact cell systems and was confirmed, also, on tissue sections in in vivo systems. For instance, a nuclear localization of receptors has been demonstrated by both hormone immunocytochemistry[2] and receptor immunocytochemistry by receptor antibodies,[63] at the endometrial level of spayed rats, after estradiol administration.

The structural counterpart of the transportation process was further investigated by electron microscopy with hormone immunocytochemistry by a double immunoperoxidase technique.[5,6] For this purpose, a slow-down of the translocation process was obtained by modulating the experimental temperature. In particular, postincubation at 20°C of cells previously reacted with the hormone at 4°C resulted in a maximal perinuclear concentration of the bound ligand, as occurs in target tissues of very immature animals.[2] This has been suggested to depend on delayed nuclear translocation. At electron microscopy, such a perinuclear positivity appeared as a heavy, discontinuous staining of the nuclear membrane and as many particles in close contact with the nuclear membrane. These particles, measuring about 50 nm in diameter, were also found in the perinuclear area and in the cytoplasm far from the nucleus. The morphological counterpart of these antibody-stained particles was also looked for by standard electron microscopy in cells reacted with estradiol. Under these experimental conditions it was possible to recognize in cells, incubated with the hormone at 20°C, particles having the same size and the same intracellular distribution as those traced by the antibody. Moreover, both the antibody-traced particles and their morphological counterpart could be observed to adhere and to incorporate to a variable extent with the nuclear membrane and to dissolve in the nuclear environment.

These particles, which look like the sytem of coated vesicles, have been identified as receptosomes,[5,6] by analogy with the dynamic components resulting from receptor-mediated endocytosis of polypeptide hormones and subserving their intracellular transport and delivery to cytoplasmic compartments.[46,47]

By monitoring the temperature-dependent redistribution of the bound ligand, it was possible to distinctly appreciate a nuclear positivity in cells incubated at 37°C. The diffuse nuclear staining was soon replaced by a finite number of long-lasting retention sites on the chromatin network and by a nucleolar localization.[1,2,18] At electron microscopy, with hormone immunocytochemistry, the fine punctuate nuclear positivity appeared to correspond with the decoration of some globoid structures; moreover, the nucleolus also appeared specifically stained.[5,6] These globoid structures have been identified as nuclear bodies, physiological organelles with a DNA coat, which are the morphological expression of estrogen-induced transcriptional activity.[48]

IV. CYTOPLASM TO NUCLEUS TRANSLOCATION PROCESS

Some facets of the translocation process deserve some additional notes, as revealed by the cytochemical assay of receptor distribution in steroid target tissues.

A somewhat puzzling and diverging intracellular distribution of receptor was displayed by affinity cytochemistry, exploiting fluorescent steroid analogs, on cryostatic sections from human prostate cancer. The intracellular distribution was soon realized to be closely dependent on the method of removal of the specimen; that is, surgical prostatectomy or needle biopsy, on one hand, and transurethral diathermic electroresection, on the other hand. The latter is a way of removing tissue exploiting the heat that is produced according to Joule's law by a current flowing through a resistor.

While in prostate specimens obtained by surgical resection the cytoplasmic localization of steroid receptors was prevalent, a major binding at the nuclear level was apparent in specimens obtained by transurethral electrocutting, particularly, in the more peripheral areas of the specimen.[39,49] This dependence on the intracellular distribution of the binding by way of removal of the tissue was also confirmed on sequential specimens from the same patient who underwent first a needle biopsy, and afterwards a transurethral electrocutting. An analogous binding distribution was observed in specimens from human mammary tissue (normal, hyperplastic, and neoplastic) in which a prevalent nuclear binding could be observed in specimens removed by means of a bovie, especially in the peripheral areas of the smaller ones.

These observations have prompted a search for the possibility that the intracellular distribution of steroid receptors could be artifactually affected by the heat produced by electrocutting. In particular, a possible induction of nuclear translocation of receptors in the absence of the hormone was suggested to occur upon heating intact cells over physiological temperatures. An experimental system was, therefore, devised to test the effect of a hyperthermic treatment on the intracellular distribution of steroid receptors in a variety of target tissues, such as human breast and prostate, and rat uterus and prostate.

Small tissue fragments were immersed in buffered saline preheated at 37° or 40°C for increasing periods of time (15, 30, 60 min), then frozen with CO_2. Cytochemical demonstration of steroid binding was then carried out on cryostatic sections by macromolecular fluorescent steroid analogs. While in tissues incubated at 37°C the cytoplasm was the prevalent receptor localization, in the specimens immersed at 40°C the major binding could be appreciated at the nuclear level. Moreover, the longer the incubation at 40°C, the brighter was the nuclear positivity compared with the cytoplasmic staining.

The observed cytochemical binding displayed steroid and tissue specificity. Actually, binding of fluorescent macromolecular steroid analogs was inhibited by the respective "cold" macromolecular steroids and was unaffected by unrelated cold steroids. Moreover, the heat-induced nuclear binding was never observed in steroid nontarget tissues processed as the target ones.

Lastly, it is worth stressing that the nuclear binding was unaffected by precoincubation with dithiotreitol, a specific inhibitor of type II binding sites.[50]

The observed redistribution of binding sites upon heating was further investigated in other experimental conditions. First, a mixed in vivo-in vitro model was exploited, involving the in vivo heating of the uterus of ether-anesthesized spayed rats by filling the peritoneal cavity with warm saline (40° to 41 °C). Cytochemical assay of steroid binding carried out on serially obtained specimens of the uterus also displayed, in these experimental conditions, a major nuclear staining. Lastly, a hyperthermic treatment was carried out directly in vivo, by injecting spayed rats with the pyrogenic drug α-dinitrophenol. Fragments of uterus and mammary glands were then surgically removed 15, 30, and 60 min after the body temperature had reached 40° to 41°C. Cytochemical assay of steroid binding displayed a subcellular redistribution of binding in favor of the nucleus upon hyperthermic treatment in this in vivo system.

It can be concluded that in intact cell systems a nuclear shifting of receptors is induced by a heat shock, in the absence of hormone. It must be stressed that the temperature used in these experiments is higher than the physiological temperature (37°C) which is exploited in vitro to trigger the nuclear translocation of steroid receptor complexes. In this respect, it is also worth noting that in tissue fragments immersed in buffered saline at 37°C in the absence of the hormone, the cytochemical assay has always showed a prevalent cytoplasmic localization of steroid binders. Lastly, though it is known that in cell-free homogenates heating does not produce activation of the receptor in the absence of the hormone, but, on the contrary, brings about its denaturation, it may be that properties and behavior of the receptors are quite different if the cellular microenvironment is preserved, as in the exploited vital cell system.[51]

At present, the mechanism whereby heat shock produces receptor redistribution is unknown. A similar phenomenon has been observed, also, in lymphocytes where the nuclear redistribution of glucocorticoid receptors was induced by stimulation with phytohemagglutinin.[52] It can only be suggested that heat shock, which is certainly not a physiological pathway, could in vitro modulate the activity of some subcellular components, such as enzymatic systems, responsible for the physiological translocation of receptors. Heat treatment is known to activate some phosphokinases with the ability to phosphorylate specific proteins.[53] In this respect it is interesting to note that one of the steps necessary for translocation to take place, that is, receptor activation, has been suggested to depend on the phosphorylation process.[54,55]

V. BREAST CANCER BIOLOGY

Cytochemical techniques are most valuable tools when an integrated functional and structural evaluation of a biological phenomenon is pursued.[9,56] In particular, much insight into endocrine biology has been offered by the possibility of detecting on morphological grounds the presence and function of the cellular components responsible for hormone sensitivity of tissues. A closer view to physiology and pathophysiology of steroid receptors has, thus, been possible with a major contribution to the understanding of steroid responsiveness of some tumors.

When applied to the study of hormone responsiveness of breast cancer, one of the most meaningful results offered by the cytochemical assay of steroid receptors is the recognition of the intratumor diversity of hormone receptivity of tumor cells.

Among all the examined tumors only very few were composed of homogeneous cell populations with respect to cell hormone receptivity, while the majority appeared to be composed of receptor-positive and -negative cell populations in a variable proportion. Moreover, a variable degree of staining intensity was apparent among positive cells, suggesting a different receptor endowment among positive cells. Such variable staining is still unexplained, though it could be tentatively suggested to be inherent to the neoplastic state of the cell or, otherwise, to depend on the phase of the cell cycle.

It is worth noting that the intratumor heterogeneity of cell hormone receptivity has been first recognized by cytochemical techniques which trace receptors through their binding capacity, and more recently observed, also by receptor immunocytochemistry by receptor monoclonal antibodies, which trace receptors as such.[63]

Generally, neither the overall receptor endowment of a given tumor nor the presence of receptors at the singular cell level could be correlated with particular histological and cytological features. Nevertheless, a trend could be observed for some histological types of tumors, such as medullary carcinomas and comedocarcinomas, toward a more frequent receptor negativity. Moreover, at the cytological level, differentiated features appeared more often correlated with the presence of estrogen receptors.

Much information with respect to hormone responsiveness has not only issued from the tracing of the cytoplasmic binding of the hormone in tissue sections and in vital cell suspensions, but also from the dynamic monitoring of the intracellular distribution of the bound hormone in vital cells. The selective visualization of the steps of hormone action subsequent to cytoplasmic binding, enabled by the modulation of the experimental temperature, have focused some defective events.

Actually, it was possible to recognize a failed nuclear translocation of the bound hormone in cells which otherwise exhibited a normal cytoplasmic uptake. The impaired nuclear translocation of the hormone-receptor complexes, appearing as a positive perinuclear ring, bears a strong analogy with a physiological phenomenon observed in target tissues of very young rats, where a perinuclear concentration of the hormone was apparent at a time when the nuclear replenishment has occurred in older animals.[2] These observations, together with the ultrastructural tracing of hormone-receptor complexes at the nuclear membrane level, support a regulatory role of the nuclear membrane in the steroid action mechanism.

An apparent impairment of intranuclear events could also be recognized by cytochemical means. The different pools of receptors usually appreciable at the nuclear level as a bulky, quickly disappearing positivity and as long-lasting, finely punctuate retention sites, respectively, were not appreciable in some cells. Actually, long-lasting retention sites could not be observed in some cells, in spite of an effective nuclear translocation.

It can be concluded that not only the initial hormone binding, but also, further events, such as intranuclear translocation and binding, must be effective for a cell to be hormone responsive. Therefore, predicting hormone responsiveness through the recognition of the first step only can be misleading and inaccurate. For this reason the prediction of hormone responsiveness might be better carried out if the end result of the pathway, such as the presence of nuclear receptors or a phenotypic marker of hormone action, rather than the cytoplasmic receptors were assayed.[57]

Cytochemical assay of tissue sections has displayed, as expected, a main cytoplasmic localization of unoccupied receptors; in addition, a not-infrequent combined cytoplasmic and nuclear positivity of the same cell was apparent. Lastly, though quite rarely, only a nuclear positivity could be appreciated, indicating the presence of unoccupied nuclear receptors even in the absence of cytosolic hormone binders.

Above all, it must be stressed that the cytochemical recognition of intratumor heterogeneity in cell hormone receptivity strongly affects the clinical management of breast cancer. Actually, it has long been recognized that the presence of variable proportions of receptor-positive and -negative populations in a given tumor strongly influences the quantitative biochemical assay of receptors, so that a cytochemical positivity of 20% of cells is enough for a tumor to be considered receptor-positive by biochemical assays.[8,56] Moreover, since less than 10% of the several hundred tumors so far examined are composed of homogeneous cell populations (all cells positive or negative), heterogeneity in hormone receptivity should be considered a trait of breast cancer.

VI. PROSTATE PROLIFERATIVE DISEASES

Cytochemical techniques, first devised to study steroid hormone action and estrogen responsiveness of human breast cancer, have since been applied to study the control of the prostate gland by sex hormones.

A deeper insight into the hormonal modulation of prostatic function appears mandatory in order to understand the mechanism of action of steroid hormones in the proliferative diseases which affect the gland, such as benign hyperplasia (BPH) and cancer (PCA).

The functional integrity of the gland is the result of a close interaction between epithelial and stromal components and is maintained by a complex hormonal environment. In this respect, while the role of androgen is more definitely established, much less information is available on the control by estrogen.

An integrated morphological and functional approach as afforded by affinity cytochemistry seems to be ideally suited, also, to the study of steroid hormone receptivity of proliferative diseases of human prostate.

In a first phase an animal system has been studied, namely, rat ventral prostate from mature and castrated animals, which is a deeply investigated endocrine model with well-established correlations between the endocrine status and steroid receptor endowment. An evident cytoplasmic positivity was observed in epithelial elements with both androgen and estrogen analogs in mature rats. Steroid binding endowment appeared to fluctuate and to be dependent on the time elapsed after castration. The obtained results were confirmed also by ligand competition control experiments, and agree with biochemical assay of steroid receptors in rat ventral prostate in the same endocrine conditions.[58,59]

Cytochemical investigation of steroid receptors was, thus, shifted to the two proliferative processes affecting the human gland: BPH and PCA.

In BPH, a different distribution of the staining by steroid analogs in epithelial and stromal cells, respectively, was observed. In agreement with biochemical and histochemical assays of receptors in BPH,[60,61] while estrogen binding sites were usually seen more in stromal than in epithelial cells, the latter appeared endowed more with androgen than with estrogen binders. Moreover, a nuclear localization of binders was more often displayed by fluorescent estrogen than by androgen.

The different distribution of steroid binders may suggest, therefore, a differential hormonal modulation of growth and function of epithelial and stromal components, respectively. In particular, the preferential stromal distribution of estrogen binders is worthy of note and lends support to a suggested key role of estrogen modulation of stromal cell function in the development of BPH. Estrogen binding at the stromal level has been suggested, not only to trigger a proliferative response of stromal cells, but, also, to enhance the 5α reductase activity.[61] This, in turn, would eventually result in a local increase of dihydrotestosterone, the steroid ultimately responsible for the overstimulation of epithelial growth.

A preliminary study has been carried out on frozen sections of prostate cancer specimens obtained by surgical resection. A cytoplasmic staining of neoplastic cells was prevalent with both fluorescent steroid analogs. As reported, also, by others,[49] no clear-cut correlation could be drawn, in our experience, between hormone receptivity and histologic tumor grade. Steroid receptivity of stromal cells present in cancer specimens often displayed a pattern quite like that of stromal cells in BPH. A more evident estrogen than androgen binding was observed in stromal cells and, again, a nuclear localization was not infrequent.

The different intracellular distribution of steroid binders experienced in surgical specimens, and in specimens obtained by transurethral diathermic electroresection, should be taken into account when assaying steroid binders to predict hormone responsiveness of prostatic carcinoma. As detailed above, levels of binders, nuclear and cytoplasmic, respectively, are deeply affected by the method of tissue removal so that a comparison cannot be made between specimens obtained by different ways. Moreover, a combined evaluation of cytoplasmic and nuclear binding seems to allow the prediction of hormone responsiveness with more reliability if the assay is performed in specimens obtained by transurethral diathermic electroresection.

VII. PERSPECTIVES

Cytochemical techniques for steroid receptors may be expected to allow a deeper insight into some topics relevant to the knowledge of the natural history of human breast cancer and to its treatment.

For instance, the progression toward autonomy is a regular characteristic of hormonal-dependent tumors, breast cancer included; the escape from the need of hormonal support develops during their continuous dependent growth. What is the mechanism of this escape phenomenon? Could it be attributed to the stepwise emergence and selection of mutant cells? Alternatively, could such a phenomenon reflect phenotypic cell differentiation? And if so, must the apparent phenotypic heterogeneity be related to continuous stem cell differentiation, to environmental cell adaptation or to functional fluctuation between interchanging cell status? The answer to the above queries, which have important clinical implications, may be obtained from model systems suited to the study of receptor cell heterogeneity by means of cytochemical techniques and, eventually, may help to achieve its control.

Moreover, it seems conceivable that quantitative and qualitative changes in responsiveness to hormonal stimulation may be a key promotional step in the mammary carcinogenic process. These changes appear to lie centrally in an abnormal response of initiated cells to an essentially normal endocrine environment. Since hormones exert their effect via specific receptors, the abnormal responsiveness could result from variations in the amount of active hormone receptors inducing an unregulated sustained cell proliferation. If so, the histo chemical recognition of cells "turned on" for receptor production could make possible the identification of true preneoplastic changes, with the opportunity to interfere with the accomplishment of the full carcinogenic process.

REFERENCES

1. **Nenci, I., Beccati, M. D., Piffanelli, A., and Lanza, G.,** Detection and dynamic localization of estradiol-receptor complexes in intact target cells by immunofluorescence technique, *J. Steroid Biochem.*, 7, 505, 1976.
2. **Nenci, I., Piffanelli, A., Beccati, M. D., and Lanza, G.,** *In vivo* and *in vitro* immunofluorescent approach to the physiopathology of estradiol kinetics in target cells, *J. Steroid Biochem.*, 7, 883, 1976.
3. **Kurzon, R. M. and Sternberger, L. A.,** Estrogen receptor immunocytochemistry, *J. Histochem. Cytochem.*, 26, 803, 1978.
4. **Shimizu, M., Vajima, O., Miura, M., and Katayama, I.,** PAP immunoperoxidase method demonstrating endogenous estrogen in breast carcinomas, *Cancer*, 52, 486, 1983.
5. **Nenci, I., Fabris, G., Marchetti, E., and Marzola, A.,** Intracellular flow of particulate steroid-receptor complexes in steroid target cells, *Virchows Arch. B, Cell Pathol.*, 32, 139, 1980.
6. **Nenci, I., Fabris, G., Marzola, A., and Marchetti, E.,** Steroid-cell interactions revealed by immunological probes and electron microscopy, in *Pharmacological Modulation of Steroid Action*, Mainwaring, W. I. P., Genazzani, E., and Di Carlo, F., Eds., Raven Press, New York, 1980, 99.

7. **Pertschuk, L. P.**, Detection of estrogen binding in human mammary carcinoma by immunofluorescence: a new technic utilizing the binding hormone in a polymerized state, *Res. Commun. Chem. Pathol. Pharmacol.*, 14, 771, 1976.
8. **Pertschuk, L. P., Tobin, E. H., Brigati, D. J., Kim, D. S., Bloom, N. D., Gaetjens, E., Berman, P. J. , Carter, A. C., and Degenshein, G. A.**, Immunofluorescent detection of estrogen receptors in breast cancer. Comparison with dextran-coated charcoal and sucrose gradient assays, *Cancer*, 41, 907, 1978.
9. **Nenci, I., Fabris, G., Marzola, A., and Marchetti, E.**, Hormone receptor cytochemistry in human breast cancer, in *Hormones and Cancer*, Iacobelli, S., King, R. J. B., Lindner, H. R., and Lippman, M. E., Eds., Raven Press, New York, 1980, 227.
10. **Lee, S. H.**, Cytochemical study of estrogen receptor in human mammary cancer, *Am. J. Clin. Pathol.*, 70, 197, 1978.
11. **Lee, S. H.** , Cancer cell estrogen receptor of human mammary carcinoma, *Cancer*, 44, 1, 1979.
12. **Pertschuk, L. P., Gaetjens, E., Carter, A. C., Brigati, D. J., Kim, D. S., and Fealey, T. E.**, An improved histochemical method for detection of estrogen receptors in mammary cancer, *Am. J. Clin. Pathol.*, 71, 504, 1979.
13. **Walker, R. A., Cove, D. H., and Howell, A.**, Histological detection of estrogen receptor in human breast carcinomas, *Lancet*, 1, 171, 1980.
14. **Parikh, I., Sica, V., Nola, E., Puca, G. A., and Cuatrecasas, P.**, Estrogen receptors, *Methods Enzymol.*, 34, 670, 1974.
15. **Muller, R. E. and Wotiz, H. H.**, Kinetics of estradiol entry into uterine cells, *Endocrinology*, 105, 1107, 1979.
16. **Dandliker, W. B., Brawn, R. J., Hsu, M. L., Brawn, P. N., Levin, J., Meyers, C. Y., and Kolb, V. M.**, Investigation of hormone-receptor interactions by means of fluorescence labeling, *Cancer Res.*, 38, 4212, 1978.
17. **Barrows, G. H., Stroupe, S. B., and Riehm, J. D.**, Nuclear uptake of a 17β estradiol-fluorescein derivative as a marker of estrogen dependence, *Am. J. Clin. Pathol.*, 73, 330, 1980.
18. **Nenci, I., Dandliker, W. B., Meyers, C. Y., Marchetti, E., Marzola, A., and Fabris, G.**, Estrogen receptor cytochemistry by fluorescent estrogen, *J. Histochem. Cytochem.*, 28, 1081, 1980.
19. **Fisher, B., Gunduz, N., Zheng, S., and Saffer, E. A.**, Fluoresceinated estrone binding by human and mouse breast cancer cells, *Cancer Res.*, 42, 540, 1982.
20. **Gunduz, N., Zheng, S., and Fisher, B.**, Fluoresceinated estrone binding by cells from human breast cancers obtained by needle aspiration, *Cancer*, 52, 1251, 1983.
21. **Govindan, M. V.**, Immunofluorescence microscopy of the intracellular translocation of glucocorticoid-receptor complexes in rat hepatoma (HTC) cells, *Exp. Cell Res.*, 127, 293, 1980.
22. **Logeat, F., Hai, M. T. V., and Milgrom, E.**, Antibodies to rabbit progesterone receptor: crossreaction with human receptor, *Proc. Natl. Acad. Sci. U.S.A.*, 78, 1426, 1981.
23. **Gasc, J.-M., Renoir, J.-M., Radanyi, C., Joab, I., and Baulieu, E.-E.**, Etude immunohistologique de l'oviducte de poulet à l'aide d'anticorps anti-recepteur de la progesterone, *C. R. Acad. Sci. Paris*, 295, 707, 1982.
24. **Raam, S., Nemeth, E., Tamura, H., O'Brian, D. S., and Cohen, J. L.**, Immunohistochemical localization of estrogen receptors in human mammary carcinoma using antibodies to the receptor proteins, *Eur. J. Cancer Clin. Oncol.*, 18, 1, 1982.
25. **Greene, G. L., Nolan, D., Engler, J- P., and Jensen, E. V.**, Monoclonal antibodies to human estrogen receptor, *Proc. Natl. Acad. Sci. U.S.A.*, 77, 5115, 1980.
26. **Greene, G. L. and Jensen, E. V.**, Monoclonal antibodies as probes for estrogen receptor detection and characterization, *J. Steroid Biochem.*, 16, 353, 1982.
27. **Moncharmont, B., Su, J. L., and Parikh, I.**, Monoclonal antibodies against estrogen receptor: interaction with different molecular forms and functions of the receptor, *Biochemistry*, 21, 6916, 1982.
28. **King, R. J. B., Gilbert, J., Coffer, A., and Lewis, K.**, Monoclonal antibodies to human estradiol receptor: histochemical studies, *J. Steroid Biochem.*, 19 (Suppl.), 32, 1983.
29. **Greene, G. L. and King, W. J.**, Immunochemical analysis and localization of estrogen receptor with monoclonal antibodies in estrogen-sensitive tissues and tumors, *J. Steroid Biochem.*, 19 (Suppl.), 129, 1983.
30. **Goldfine, I. D.**, Interaction of insulin, polypeptide hormones, and growth factors with intracellular membranes, *Biochim. Biophys. Acta*, 650, 53, 1981.
31. **Szego, C. M. and Pietras, R. J.**, Membrane recognition and effector sites in steroid hormone action, in *Biochemical Actions of Hormones*, Vol. 8, Litwack, G., Ed., Academic Press, New York, 1981, chap. 6.
32. **Rao, G. S.**, Mode of entry of steroid and thyroid hormones into cells, *Mol. Cell. Endocrinol.*, 21, 97, 1981.
33. **Nenci, I., Fabris, G., Marchetti, E., and Marzola, A.**, Cytochemical evidence for steroid binding sites in the plasma membrane of target cells, in *Perspectives in Steroid Receptor Research*, Bresciani, F., Ed., Raven Press, New York, 1980, 61.

34. **Nenci, I., Marchetti, E., Marzola, A., and Fabris, G.,** Affinity cytochemistry visualizes specific estrogen binding sites on the plasma membrane of breast cancer cells, *J. Steroid Biochem.*, 14, 1139, 1981.
35. **Nenci, I., Fabris, G., Marzola, A., and Marchetti, E.,** The plasma membrane as an additional level of steroid-cell interaction, *J. Steroid Biochem.*, 15, 231, 1981.
36. **Singer, S. J.,** The molecular organization of membranes, *Annu. Rev. Biochem.*, 43, 805, 1974.
37. **Poste, G. and Nicolson, G. L., Eds.,** *Dynamic Aspects of Cell Surface Organization*, North-Holland, Amsterdam, 1977.
38. **Nenci, I.,** Specific cell adhesion to estradiol-derivatized agarose beads, *J. Steroid Biochem.*, 19, 109, 1983.
39. **Nenci, I., Fabris, G., Marzola, A., Bagni, A., Poli, G., and Marchetti, E.,** Charting steroid-cell interaction by visual means, in *Regulation of Target Cell Responsiveness*, McKerns, K. W., Aakvaag, A., and Hansson, V., Eds., Plenum Press, New York, 1986, 159.
40. **Bresciani, F., Sica, V., and Weisz, A.,** Properties of estrogen receptor purified to homogeneity, in *Biochemical Actions of Hormones*, Vol. 6, Litwack, G., Ed., Academic Press, New York, 1979, 461.
41. **Tubiana, N., Seignourin, J. M., Lecaer, F., Pincemaille, A., Martin, P. M., and Carcassonne, Y.,** Sex steroid binding to human lymphocytes cell lines membrane, *J. Steroid Biochem.*, 19 (Suppl.), 39, 1983.
42. **Pietras, R. J. and Szego, C. M.,** Endometrial cell calcium and estrogen action, *Nature (London)*, 253, 357, 1975.
43. **Berthois, Y., Laugier, R., Mittre, H., Tubiana, N., Pourreau-Schneider, N., and Martin, P. M.,** Fixation of an estrogen-macromolecular complex on MCF 7 cell plasma membranes can be displaced by anti-estrogens and induces a variation in membrane potential, *J. Steroid Biochem.*, 19 (Suppl.), 37, 1983.
44. **Puca, G. A., Nola, E., Molinari, A. M., Medici, N., De Lucia, D., and Sica, V.,** Biochemistry and biology of estrogen receptor: identification of cytoskeletal binding sites for receptor in a membrane model, in *Steroids and Endometrial Cancer*, Jasonni, V. M., Nenci, I., and Flamigni, C., Eds., Raven Press, New York, 1983, 1.
45. **Jungblut, B. J.,** Sequential extraction of various forms of estradiol receptor, *Acta Endocrinol.*, 87 (Suppl. 215), 137, 1978.
46. **Willingham, M. C. and Pastan, I.,** The receptosome: an intermediate organelle of receptor-mediated endocytosis in cultured fibroblasts, *Cell*, 21, 67, 1980.
47. **Pastan, I. H. and Willingbam, M. C.,** Receptor-mediated endocytosis of hormones in cultured cells, *Ann. Rev. Physiol.*, 43, 239, 1981.
48. **Le Goascogne, C. and Baulieu, E. E.,** Hormonally controlled "nuclear bodies" during the development of the prepuberal rat uterus, *Biol. Cell.*, 30, 195, 1977.
49. **Pertschuk, L. P., Rosenthal, H. E., Macchia, R. J., Eisenberg, K. B., Feldman, J. G., Wax, S. H., Kim, D. S., Whitmore, W. F., Abrahams, J. I., Gaetjens, E., Wise, G. J., Herr, H. W., Karr, J. P., Murphy, G. P., and Sandberg, A. A.,** Correlation of histochemical and biochemical analysis of androgen binding in prostatic cancer: relation to therapeutic response, *Cancer*, 49, 984, 1982.
50. **Markaverich, B. M., Williams, M., Upchurch, S., and Clark, J. H.,** Heterogeneity of nuclear estrogen binding sites in the rat uterus: a simple method for the quantitation of the type I and type II sites by [^{3}H] estradiol exchange, *Endocrinology*, 109, 62, 1981.
51. **Milgrom, E.,** Activation of steroid-receptor complexes, in *Biochemical Actions of Hormones*, Vol. 3, Litwack, G., Ed., Academic Press, New York, 1981, 465.
52. **Papamichail, M., Ioannidis, C., Tsawdaroglou, N., and Sekeris, C. E.,** Translocation of glucocorticoid receptor from the cytoplasm into the nucleus of phytohemagglutinin-stimulated human lymphocytes in the absence of the hormone, *Exp. Cell Res.*, 133, 461, 1981.
53. **Rubin, I., Getz, G., and Swift, H.,** Alteration of protein synthesis and induction of specific protein phosphorylation by hyperthermia, *Cancer Res.*, 42, 1395, 1982.
54. **Fleming, H., Blumenthal, R., and Gurpide, E.,** Effects of cyclic nucleotides on estradiol binding in human endometrium, *Endocrinology*, 111, 1671, 1982.
55. **Auricchio, F., Migliaccio, A., Castoria, G., Castoria, S., and Rotondi, A.,** Regulation of hormone binding activity of estradiol receptor by a phosphorylation-dephosphorylation process, in *Regulation of Target Cell Responsiveness*, McKern, K. W., Aakvaag, A., and Hansson, V., Eds., Plenum Press, New York, 1984, 177.
56. **Nenci, I.,** Estrogen receptor cytochemistry in human breast cancer: status and prospects, *Cancer*, 48, 2674, 1981.
57. **McGuire, W. L., Osborne, C. K., Clark, G. M., and Knight, W. A., III,** Hormone receptors and breast cancer, in *Steroids and Endometrial Cancer*, Jasonni, V. M., Nenci, I., and Flamigni, C., Eds., Raven Press, New York, 1983, 29.
58. **Bruchovsky, N. and Rennie, P. S.,** Steroid receptors and the regulation of DNA synthesis, in *Prostate Cancer*, Vol. 48, Coffey, D. S. and Isaacs, J. T., Eds., UICC Technical Report Series, Geneva, 1979, 134.

59. **Jung-Testas, I., Groyer, M.-T., Bruner-Lorand, J., Hechter, O., Baulieu, E.-E., and Robel, P.,** Androgen and estrogen receptors in rat ventral prostate epithelium and stroma, *Endocrinology*, 109, 1287.
60. **Pertschuk, L. P., Zava, D. T., Tobin, E. H., Brigati, D. J., Gaetjens, E., Macchia, R. J., Wise, G. J., Wax, H. S., and Kim, D. S.,** Histochemical detection of steroid hormone receptors in human prostate, in *Prostate Cancer and Hormone Receptors*, Murphy, G. P. and Sandberg, A. A., Eds., Alan Liss, New York, 1979, 113.
61. **Krieg, M., Klötzl, G., Kaufmann, J., and Voigt, K. D.,** Stroma of human benign prostatic hyperplasia: preferential tissue for androgen metabolism and estrogen binding, *Acta Endocrinol.*, 96, 422, 1981.
62. **Greene, G. L. and King, W. J.,** unpublished results.
63. **Marchetti, E., Marzola, A., Bagni, A., Querzoli, P., Fabris, G., and Nenci, I.,** unpublished results.

Chapter 8

IMMUNOHISTOCHEMICAL STUDIES WITH ANTIBODIES TO THE CHICKEN OVIDUCT PROGESTERONE RECEPTOR

Jean-Marie Gasc and Etienne-Emile Baulieu

TABLE OF CONTENTS

I. Introduction 126

II. Antigens and Antibodies 126

III. Studies on the Chicken Oviduct 127
A. Paraffin Sections 127
B. Frozen Sections 131
C. Hormonal Regulation 132

IV. Studies on the Bursa of Fabricius 134

V. Ovary and Adrenal Glands 135

VI. Pituitary and Brain 136

VII. Discussion — Conclusion 137

Acknowledgments 140

References 140

I. INTRODUCTION

The detection of steroid hormone receptors in target cells has been the matter of numerous studies in recent years. Essentially all techniques have involved detection of hormone that is bound by the receptor and, therefore, only indirectly revealed this receptor in histological sections, smears, or cultured cells. Autoradiography with tritiated hormone has been successful in this respect. However, controversial results have been obtained[1] with fluorescent hormone conjugates and with antibodies raised against hormonal ligands. These techniques and appropriate references are found in other chapters of this book.

Visualizing the hormonal ligand necessarily leaves some degree of uncertainty as to whether it is really the receptor that is revealed. Provided their specificity is well defined, the use of antibodies to the receptor molecule overcomes this ambiguity and does not depend on the function of ligand binding, allowing the study of the receptor whether or not occupied by hormones.

In our laboratory, several forms and components of the progesterone receptor (PR) of the chick oviduct[2-5] have been purified and antibodies have been obtained.[5-8] The chick oviduct system is suitable for detailed studies of steroid hormone action, since hormonal responses are relatively well characterized biochemically and cytologically. Egg white proteins, like ovalbumin, conalbumin, etc. that are synthetized under hormone action, are available in pure forms. The progesterone receptor itself is abundant after estrogen priming[9,10] and the molecular biology of the system is under constant investigation.

In the studies summarized here we have used antibodies against the PR and its constituents: (1) to demonstrate in which cells PR is present in the oviduct; (2) to study its subcellular distribution in the presence and absence of hormone, with reference to recent knowledge of progesterone binding and nonprogesterone binding components of this receptor; (3) to investigate other chicken organs such as the bursa of Fabricius, the thymus, the pituitary, and the brain. We encountered technical difficulties mostly related to the manner of fixation of tissues required to maintain and appropriately preserve the antigenic determinants. The results lead us to propose a new argument as to the concept of receptor translocation and a new way to characterize PR-containing cells. In addition, recent work with frozen sections has allowed, for the first time, the simultaneous visualization of a hormone receptor by immunohistochemistry and autoradiography with a radioactive ligand.

II. ANTIGENS AND ANTIBODIES

Different preparation procedures have yielded several forms and components of PR. Nontransformed 8S-PR forms have been purified, using stabilization by molybdate ions.[2,11] They contain two distinct progesterone-binding subunits, A and B (Mr = 79K and 110K, respectively),[3,12] and a third protein component of Mr $\sim$ 90K which does not bind progesterone.[4] Three antibodies were used in this study. Their specificity was tested by density gradient centrifugation technique and by immunoblots after SDS-PAGE. A goat polyclonal antibody (IgG-G3)[6] reacts with the subunits A, B, and 90K, as well as with the 8S-PR.[2] The rat monoclonal antibody BF4[7] specifically reacts with the 90K protein. IgG-RB[5,8] is a rabbit polyclonal antibody obtained after immunization with purified B subunit. It interacts with both the A and B subunits but not with the 90K. All preparations used were IgG-G fractions from immune sera or culture medium. The three antibodies were used alone, and also in conjunction with purified proteins utilized in presaturation experiments in order to control the specificity of the immunodetection of receptor. Thus, the 8S-PR, B subunit and the 90K protein were purified and systematically used. Other proteins also occasionally used in presaturation control experiments were the oviduct proteins ovalbumin and conalbumin, or serum proteins.

III. STUDIES ON THE CHICKEN OVIDUCT

Young chickens of the Warren strain were received from a farm. In few cases the oviducts of unstimulated immature chickens were studied. Otherwise, the chicks were injected with estrogens. Then, they were sacrificed after primary stimulation (ten daily injections of 1 mg estradiol benzoate per animal), or after secondary stimulation (five additional identical injections after 1 week withdrawal), or, also, in the withdrawn state (4 weeks after the end of the first estrogenic treatment). Initially, the oviduct of an immature chicken is composed of an outside flat epithelium (mesothelium), a thin layer of muscle fibers, and a thick envelope of stroma bordered inwards by the luminal epithelium. There are no glands. After primary or secondary estradiol stimulation, the stroma is invaded by glandular structures containing secretory granules. The glands are more developed after secondary than after primary stimulation, and contain many secretory granules. In withdrawn animals, glands persist, but only a few granules are visible in the glandular cells.

A. Paraffin Sections

The fixation procedure appeared to be critical for the successful immunohistochemistry of PR.

Initially, long fixation periods (6 hr) in Carnoy's fluid or 1% glutaraldehyde were used.[13] Such conditions ensured a good detection of the 90K protein and, also, though less clearly, of the B subunit, in all cells of the luminal epithelium. In the glands, IgG-G3 and BF4 did not demonstrate the presence of PR. As the glandular cells were expected to be the main targets for progesterone, this negative result prompted us to search for more appropriate fixation conditions. Much shorter fixation periods (60 to 90 min) and the use of two other fixatives (Bouin's fluid and acidified alcohol, that is, 1% acetic acid in absolute alcohol) improved results. At the same time IgG-RB became available and its specificity for the progesterone-binding subunit contributed to clarify the interpretation of our results.

Our present fixation conditions are as follows.[14] Oviducts are taken from the animals within minutes after sacrifice and immediately cut in small pieces not larger than 5 mm. Fragments are immersed in ice-cold fixative (Bouin's fluid, acidified alcohol, or 0.5% glutaraldehyde in 0.1 *M* Sorensen buffer at pH 7.4) for 60 to 90 min and then rinsed (in 70% alcohol, 100% alcohol, or 0.1 *M* Sorensen buffer, respectively). The Carnoy's fluid, which we previously recommended, is no longer used since, unlike other fixatives, it prevents any nuclear reaction with antibodies under all hormonal conditions. This is probably because of strong extraction of the nuclear content. Tissues are dehydrated in graded alcohols, (one 20- to 30-min bath with 95%, and two baths in 100%), cleared in 1-butanol (2 × 30-min baths), transferred to 56°C paraffin (2 × 30-min changes), and embedded. The whole procedure from fixation to embedding does not exceed 6 to 7 hr.

Paraffin sections are cut at 7 μm and, after rehydration, processed according to an immunoperoxidase technique using a biotinylated secondary antibody and the avidin-biotin-peroxidase complex (Vectastain Reagents; Vector Laboratories, Burlingame, Calif.).

When prepared under the above-described conditions, sections of estradiol-primed oviduct showed a similar pattern of cellular distribution. However, some differences were observed according to which fixative was used. *Glutaraldehyde* appeared as the most suitable fixative for all tissues: luminal epithelium, stroma, glands, muscle fibers, and mesothelium. The reaction was only nuclear with IgG-RB (Figure 1a), whereas it was both cytoplasmic and nuclear with IgG-G3 (Figure 1b) and BF4 (Figure 1c). Glutaraldehyde, however, was difficult to use since, among simultaneously fixed tissues, some pieces were underfixed and histologically of very poor quality, and others were overfixed, thus, showing little reaction with the antibodies. After *acidified alcohol* fixation, a similar distribution of positive cells was observed with the three antibodies (Figure 2) as after glutaraldehyde fixation. The histology

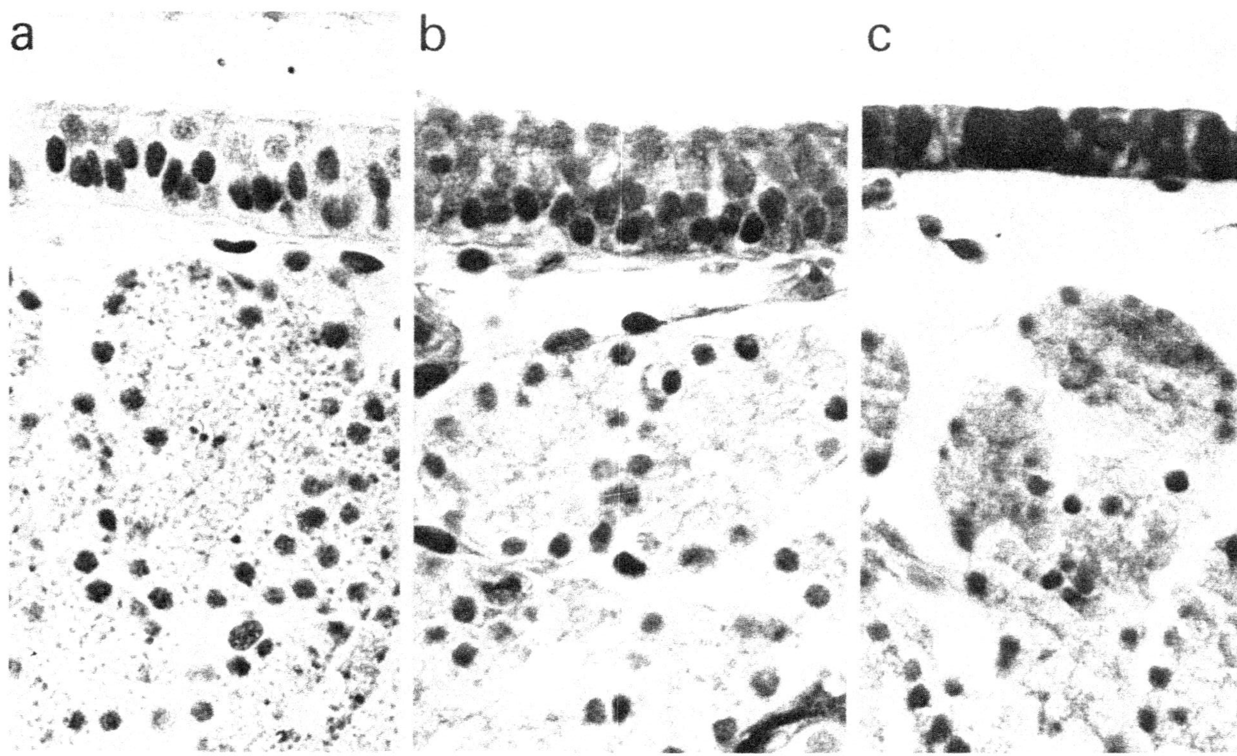

FIGURE 1. Immunohistochemical detection of PR on glutaraldehyde fixed, paraffin-embedded oviduct sections of estradiol-stimulated chickens. (a) IgG-RB, the antibody against the B subunit, reacts in cell nuclei of the luminal epithelium, the glands, and stroma. Note the difference between the two types of nuclei in the luminal epithelium: small, elongated nuclei at the basis of the cells are much more strongly stained than the round nuclei at the top. (b) IgG-G3, the antibody against the nontransformed 8S-PR, which includes the B subunit, reacts in the cell nuclei as IgG-RB and, also, in the round nuclei located at the top of the luminal epithelium. The cytoplasm of epithelial cells is also positive. (c) BF4, the antibody against 90K protein which is part of the nontransformed 8S-PR, reacts in the same cell nuclei as IgG-G3. The cytoplasm of the luminal epithelium cells also displays a very strong reaction. No counterstaining. (Magnification × 650.)

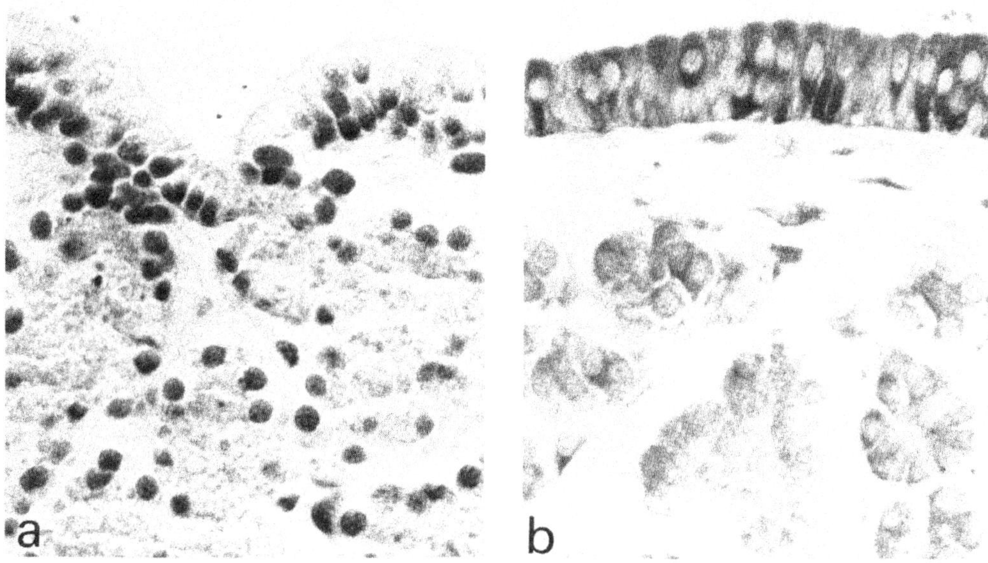

FIGURE 2. Comparative immunodetection of the B subunit of PR (a) and the 90K protein (b) on *acidified-alchohol* fixed and paraffin-embedded oviduct sections of estradiol-stimulated chickens. (a) The B subunit is revealed by IgG-RB in nuclei of the luminal epithelium, glands, and stroma. Like after glutaraldehyde fixation (Figure 1a), the large, round nuclei at the apex of the luminal epithelium are not stained. (b) The 90K protein, which is part of the 8S nontransformed PR, is revealed by BF4: strongly in cytoplasm of some epithelial cells, which appear to correspond to those with round nuclei unstained with IgG-RB, and weakly in the cytoplasm and nuclei of the other epithelial, glandular, and stromal cells. No counterstaining. (Magnification × 650.)

of the sections was poor at the periphery of the tissue. In particular, the luminal epithelium seemed to suffer from a too rapid and drastic fixation. The 90K protein, when detected by its monoclonal antibody BF4, seemed better preserved after glutaraldehyde than after acid ified alcohol fixation. *Bouin's* fixative, though suitable for revealing the presence of PR in the luminal epithelium, stroma, smooth muscle, and mesothelium, did not constantly show a reaction in the glands (Figure 3). Only sporadically were nuclei of glandular cells weakly positive with IgG-RB or IgG-G3. Presumably, the secretion granules which are extremely abundant in glandular cells require special conditions for fixation. Bouin's fixative did not fulfill these conditions, as shown by the poor immunodetection of ovalbumin and conalbumin in cytoplasm, with the antibodies to these secretion proteins. The lack of reaction of IgG-RB and IgG-G3 in the nuclei of the glandular cells, while neighboring nuclei in stromal or epithelial cells were stained, was probably an indirect consequence of the inappropriate fixation of the glands. Under these conditions the detection of a nuclear antigen was impossible, because of loss of a well-maintained cellular structure. Another argument in favor of this interpretation is that, in withdrawn animals and also after primary stimulation, when secretion granules were much less abundant, both cytoplasmic staining with BF4 and nuclear staining with IgG-RB were more easily observed after Bouin's fixation, than in oviduct at the end of the secondary stimulation. The lack of constant and reproducible reaction in the fully differentiated glands fixed in Bouin's fluid with any of the three antibodies leads to false-negative results if Bouin's fixative is not compared with other fixatives. This clearly emphasizes the need for comparison of several fixation conditions.

In summary, the most complete distribution of positive cells was observed after glutar-

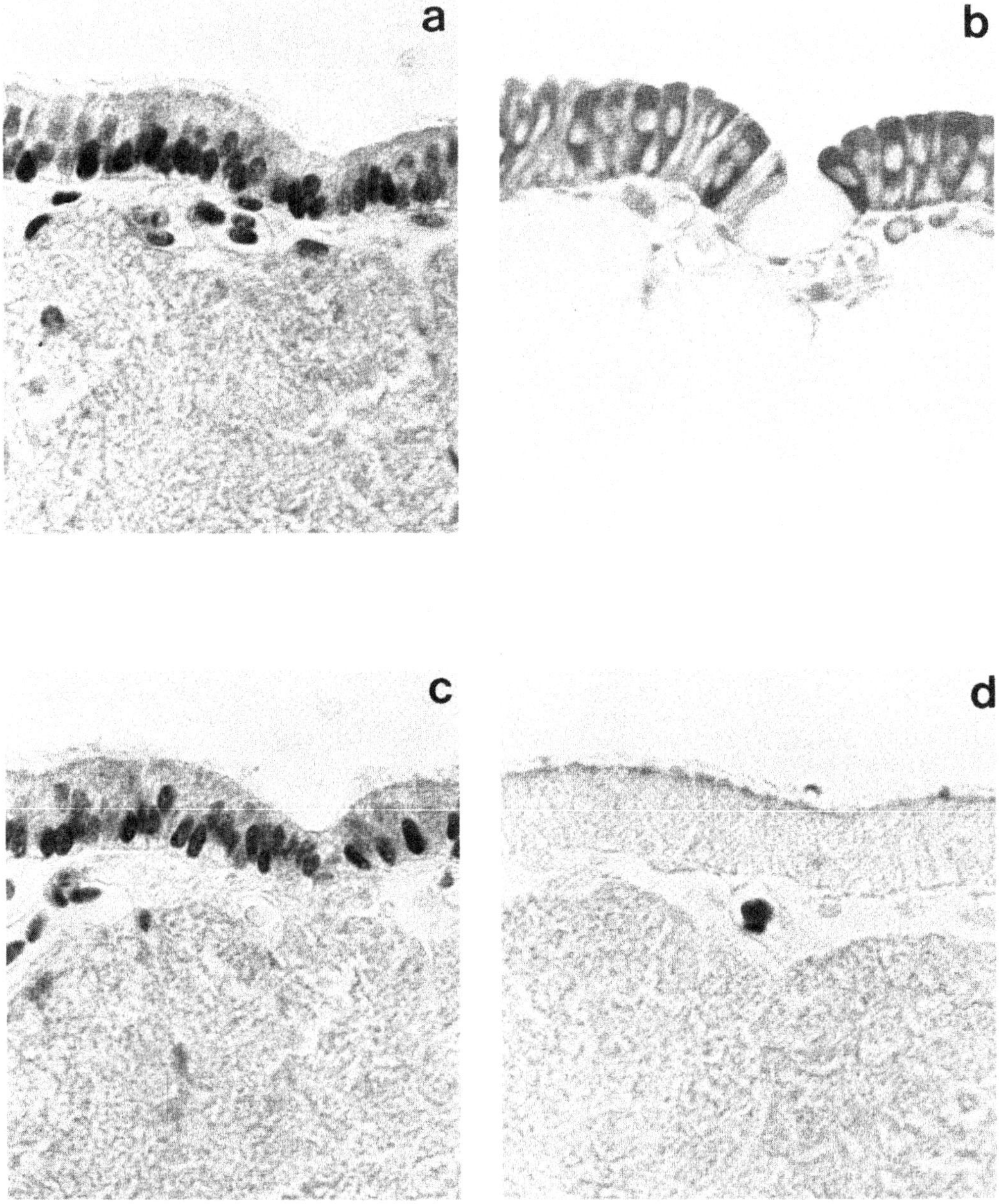

FIGURE 3. Immunodetection of PR on *Bouin* fixed, paraffin-embedded oviduct sections of estradiol-stimulated chickens. (a) IgG-RB reveals the B subunit in nuclei of the luminal epithelium and stromal cells, but not of the glands. (b) BF4 reveals the 90K protein in the cytoplasm of the high columnar cells, probably corresponding to the nuclei which are not, or weakly, stained with IgG-RB. The reaction in nuclei of the luminal epithelium is weak and totally absent in glands. (c) Purified preparation of 90K protein, when used to preabsorb IgG-RB, did not change the reaction of the antibody in cell nuclei. (d) Purified preparation of the B subunit, when used to preabsorb IgG-RB, totally extinguished the reaction of the antibody in cell nuclei, thus, showing the specificity of the reaction for the B subunit. Comparison of Bouin's fixation with glutaraldehyde (Figure 1) or acidified alcohol (Figure 2) shows a more complete reaction in the latter cases than in the former, which fails to show any reaction in glands. Bouin's fixation, however, is easier to control as under- or overfixation is not so critical as with the two other fixatives. No counterstaining. (Magnification × 650.)

aldehyde fixation. However, a comparison of the results obtained with the other fixatives was necessary to avoid false-negative results in some tissues. Altogether, the results gave evidence of nuclear staining with IgG-RB, while BF4 stained both cytoplasm and nuclei of receptor-containing cells. Logically, IgG-G3 also stained both cytoplasm and nuclei. The results were clearly confirmed by presaturation experiments. The nuclear reaction of IgG-RB was extinguished by the B subunit (Figure 3d) and the 8S-PR, but not by the 90K protein (Figure 3c). The nuclear and cytoplasmic reaction of BF4 was abolished by the 90K protein (Figure 3d) and the 8S-PR, but not by the B subunit. The nuclear and cytoplasmic reaction of IgG-G3 was extinguished by 8S-PR, but the B subunit abolished only the nuclear reaction, while the 90K protein mostly abolished the cytoplasmic reaction (though the nuclear reaction was also decreased). The simplest interpretation is that IgG-RB recognized the B (and/or A) subunit in the nucleus of target cells, BF4, the 90k protein present both in the cytoplasm and the nucleus, while IgG-G3 recognized both the B and A subunit as well as the 90K protein. The implications of this interpretation in regard to the function of the different PR components are briefly commented upon in the last part of this chapter.

B. Frozen Sections

Pieces of oviduct from the same animals as those used for paraffin sections were frozen in liquid nitrogen, and cut in a cryostat at $-25°C$. Sections (7 μm) were mounted on a glass slide, fixed, and processed through the same immunoperoxidase technique as the paraffin sections. After several attempts, formaldehyde (3.5% in PBS) appeared as the most suitable fixative which reproduced the results obtained on paraffin sections as regards the detection of the B subunit with IgG-RB and IgG-G3. The detection of the 90K protein with BF4 and IgG-G3 on frozen sections is currently underway. The B (and/or A) subunit was found only in cell nuclei of the luminal epithelium, glands, stroma, muscle fibers, and mesothelium (Figure 4). As in paraffin sections, there was no obvious cytoplasmic staining with IgG-RB, and the weak staining observed with IgG-G3 was almost completely extinguished after presaturation by the 90K protein.

The use of frozen sections does not permit us to completely rule out the possibility of a loss of cytoplasmic PR, even though it seems very unlikely. Conversely, however, the use of frozen sections can almost certainly exclude the possibility of an artifactual translocation of PR from cytoplasm into the nucleus. These results are all compatible with the finding of nuclear PR even in the absence of administered progesterone, and they are against the presence of subunit B (A) in the cytoplasm.

A recent development fully supports the presence of nuclear PR in cells that have not been exposed to progesterone.[15] When autoradiography and immunohistochemistry were carried out on successive frozen sections of the same oviduct explant which had been incubated in the presence of a synthetic tritiated progestin (^{3}H-Org 2058), the two techniques showed the radioactive ligand and the immune reaction in nuclei of the same cells (of luminal epithelium, glands, stroma, muscle fibers, mesothelium). There was, however, a discrepancy in the distribution of the cells on the sections. While nuclei reacting with IgG-RB and IgG-G3 were evenly distributed in the section, nuclei labeled with ^{3}H-Org 2058 displayed a decreasing gradient of intensity from periphery to center of the explant, where cells were not labeled. These cells had not taken up ^{3}H-Org 2058, as shown by autoradiography, nevertheless, PR was immunohistochemically detected in their nuclei. Again, these results indicate that exposure to progestagen is not a prerequisite for PR presence in nuclei.

Double labeling of PR by immunology and radioautography on the same section, after successively applying the two techniques clearly confirms the presence of both the hormonal ligand and the receptor in the nuclei of the same cells.[15] The exact congruence of the two populations of cells by both techniques directly validates these methods for the detection of steroid hormone receptors.

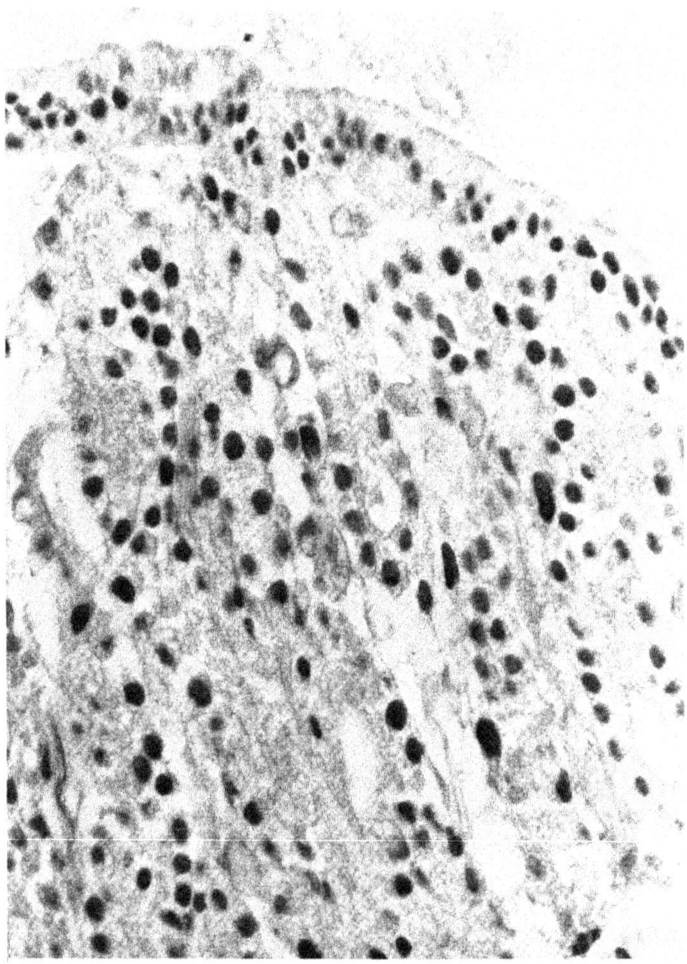

FIGURE 4. Immunodetection of the B subunit of PR on a *frozen formaldehyde fixed* oviduct section of estradiol-stimulated chicken. IgG-RB reveals the B subunit in the cell nuclei of the luminal epithelium, glands, and stroma. This distribution of PR-positive nuclei is the same as after glutaraldehyde (Figure 1a) or acidified alcohol (Figure 2a) fixation on paraffin-embedded sections. No counterstaining. (Magnification × 400.)

C. Hormonal Regulation

The results reported above refer to chickens at the end of the secondary stimulation. In oviduct under primary estrogen stimulation, glands are smaller and cells are filled with fewer secretory granules. The luminal epithelium is also less differentiated. Despite such differences, the distribution of IgG-RB and BF4-positive cells is the same whatever the hormonal treatment, even sometimes easier to demonstrate because of the smaller amount of secretory granules. All technical details being comparable, the intensity of the reaction is somewhat lower, thus, reflecting that oviduct cells under primary stimulation contain fewer PR molecules than after secondary estrogenic stimulation.[9]

In withdrawn chickens, 4 weeks after the end of the estrogenic treatment, the B subunit revealed by IgG-RB (Figure 5a) and IgG-G3 was still present in the nucleus of target cells, and the 90K protein revealed by BF4, mostly or exclusively in the cytoplasm.

In immature chickens, never injected with estradiol, cells reacting with the antibodies

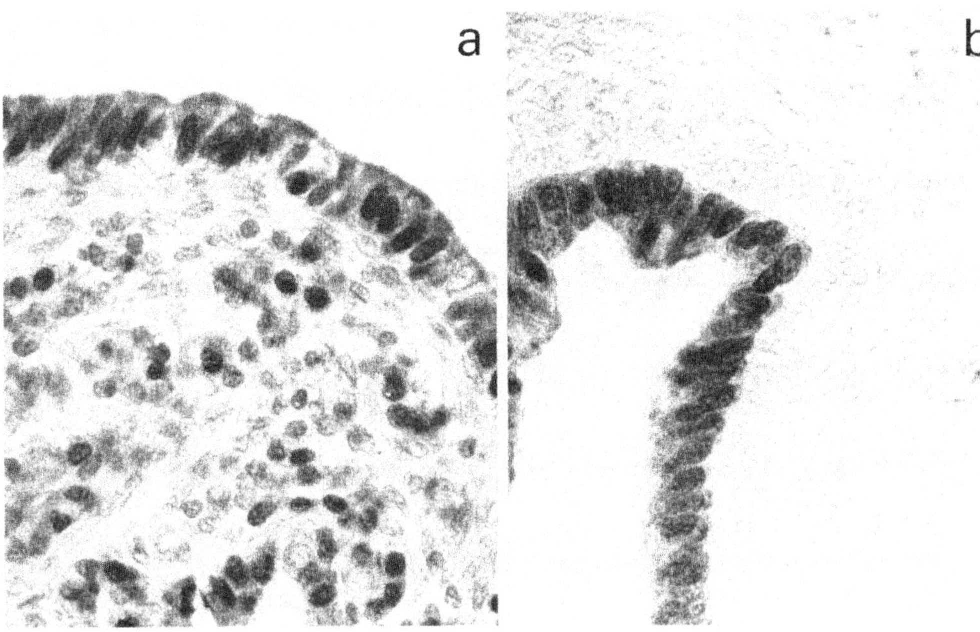

FIGURE 5. Immunodetection of the B subunit of PR on Bouin fixed, paraffin-embedded oviduct sections of chickens after 4 weeks of estrogen withdrawal or never estrogenized. (a) On section of an animal withdrawn from any hormonal treatment for 4 weeks, IgG-RB reveals the B subunit in cell nuclei of the luminal epithelium, glands and stroma. (b) On a section of an immature, unstimulated chicken, the B subunit is revealed by IgG-RB only in the cell nuclei of the luminal epithelium. The stroma, which is devoid of glands, does not show any cell reacting with the antibody. No counterstaining. (Magnification × 650.)

were observed only in the luminal epithelium. The glands were absent, and the other tissues appeared unstained. In the luminal epithelium cells, IgG-RB (Figure 5b) and IgG-G3 reacted with the nuclei, and IgG-G3 and BF4 with the cytoplasm. The failure to detect the BF4-positive antigen in nuclei may be due to too low a level.

Chickens at the end of the secondary stimulation were injected with 2 mg of progesterone and sacrificed after 1 hr. These are conditions which should produce the maximum nuclear receptor according to biochemical studies.[10] No significant differences were observed in the cellular and subcellular distribution of PR, and the three antibodies used showed comparable pictures whether or not the animal was injected with progesterone.

In summary, these immunohistochemical results were in favor of the presence of all elements of PR (A/B and 90K) in nuclei of target cells, whether or not exposed to progesterone. The results obtained in the absence of progesterone are most intriguing. The presence of both progesterone binding and nonprogesterone binding components of PR in the nuclei corroborates the concept of the nontransformed 8S structure that we have proposed,[5] though naturally it does not demonstrate it definitively. This concept indicates the simultaneous presence of either A or B subunit and 90K protein in the 8S structure of the untransformed receptor. It is known that "activated" progestin-receptor complexes have more affinity for polyanions,[16] and particularly for a poorly defined nuclear structure called "acceptor", than the nonactivated 8S-PR. Since immunocytochemistry is topographically conservative, it allows visualization of the components of the nonactivated PR in the nuclei of target cells. However, during the homogenization which is the first step in biochemical studies, it seems very possible that PR components are extracted by the buffer used and is solubilized in the cytosol, i.e., the high speed supernatant of the homogenate. If the receptor has bound hormone and has been activated, increased affinity of the hormone binding units for the nuclear structure make more "nuclear receptors" detectable biochemically when measuring bound radioactive hormone. This gives the illusion of translocation from cytoplasm to nucleus in comparison with the nonactivated receptor. In fact, what occurs is only maintenance of the receptor progesterone binding units in the nucleus because of increased affinity which renders them less extractable biochemically. At the level of precision of immunohistochemistry, we cannot tell whether there is intranuclear translocation of hormone receptor complexes from one site to another.

IV. STUDIES ON THE BURSA OF FABRICIUS

In avian species, the Bursa of Fabricius is a lymphoid organ, located at the extremity of the digestive tract, in which B lymphocytes differentiate.[17] This organ is known to be sensitive to steroid hormones in young animals as well as in embryos.[18] Receptors for androgens,[19,20] estrogens, glucocorticosteroids, and progesterone[21] are present in the bursa of chickens. The role of steroids, in general, and of progesterone, in particular, on the growth and development of the bursa and/or the differentiation of lymphocytes is mostly unknown.

Paraffin sections of bursa of estrogen-treated (secondary stimulated) animals were processed in the same way as for the oviduct. IgG-RB (Figure 6a) reacted only in nuclei of cells located between the follicles.[22] These cells have small, usually elongated and fusiform nuclei, and appear to be undifferentiated stroma cells. Cells of the epithelium and of the follicles did not react with IgG-RB.

Cells stained with IgG-G3 could be classified in three different types. Some interfollicular cells were positive in the cytoplasm but staining was abolished after presaturation with normal chicken serum. It can, therefore, be attributed to contaminant antibodies against immunoglobulins within lymphocytes. Other cells, both inter- and intrafollicular, were positive in the cytoplasm, though generally weaker than the first category. This staining was extinguished after presaturation by the 8S-PR and 90K protein, and revealed the presence of this latter protein in maturating lymphocytes. The third category of cells reacting with IgG-G3 was located only in the interfollicular space and the staining was nuclear. They were morphologically identical to the cells stained by IgG-RB. In addition, with both IgG-RB and IgG-G3 the reaction was abolished after presaturation by the B subunit. With BF4, most cells within the follicles are positive (Figure 6b), the reaction being particularly strong in the cytoplasm. The B subunit, therefore, seems to be present in interfollicular cells, whereas most cells containing the 90K protein (Figure 6b) were located inside follicles and a few others outside. In that respect, the bursa differed from the oviduct where the two components of PR were revealed in the same cells. This difference can be explained if one remembers that the bursa contains other steroid receptors,[21] and that the antigenic determinant recognized by BF4 is also present in other nonactivated steroid receptors.[4,5] Most probably, the cytoplasmic reaction observed with IgG-G3 and BF4, and extinguished after presaturation by 90K protein, was due to receptors other than PR present in intrafollicular cells. The reaction might also be due to 90K protein unrelated to steroid receptors.

Remarkably, the presence of cells containing the B antigen in their nuclei was apparently dependent on pretreatment with estradiol. We did not observe IgG-RB-positive cells in chickens 5 weeks of age not previously stimulated by estradiol. As mentioned for the oviduct, the administration of progesterone did not change the cellular or subcellular distribution of PR.

In other lymphoid organs like thymus and spleen, no cells reacting with IgG-RB were found, while BF4-positive cells were present in thymus, spleen, and lymphatic ganglia in the walls of cloaca.

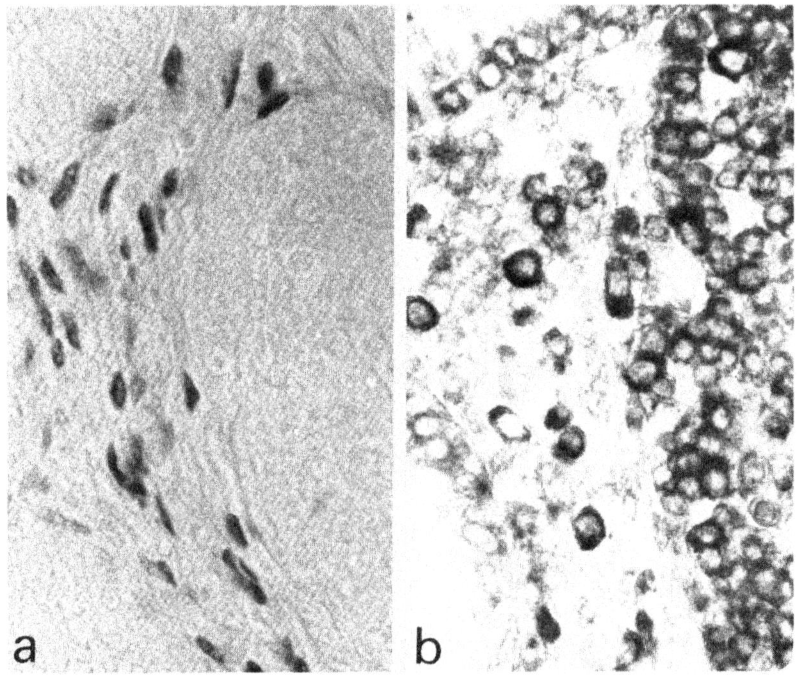

FIGURE 6. Immunodetection of PR on Bouin fixed, paraffin-embedded section of Bursa of Fabricius from an estradiol-stimulated chicken. (a) The B subunit is revealed by IgG-RB in nuclei of interfollicular cells. Cells inside follicle do not show any reaction with the antibody. (b) The 90K protein is revealed by BF4 in cytoplasm and also in nuclei, though less intensely, of intra- and interfollicular cells. Inside the follicles most cells are strongly stained. Outside the follicles, large cells with round nuclei show the strongest reaction while other small cells are weakly or not stained. In the interfollicular space, the cells most strongly stained with BF4 are not of the same type as those in which the B subunit is revealed by IgG-RB (a). No counterstaining. (Magnification × 650.)

V. OVARY AND ADRENAL GLANDS

The use of antibody to hormone receptor is particularly convenient for detecting receptor-containing cells in steroidogenic glands. This is because the technique is not limited by the presence of endogenous hormone occupying the receptor which, thus, may not directly be available for labeling by radioactive or fluorescent ligand.

In ovaries, a nuclear reaction was observed with IgG-RB in the outer layer of the epithelium covering the cortex (Figure 7a). Some cells between the ovarian follicles were also positive, though weakly. A positive reaction with BF4 was also observed in the same cells, in the follicular cells surrounding the ovocytes, and some cells in the interstitial tissue (Figure 7b). Though predominantly cytoplasmic, this reaction was also observed in nuclei after acidified alcohol fixation. As in the case of the oviduct, we are conducting a study on frozen sections in order to confirm that the BF4 antigen is originally present in nuclei and not because of artifactual translocation during tissue preparation. These observations suggested the presence of the B subunit and the 90K protein in the germinal epithelial cells and in some interstitial cells, while the follicle cells contained only the 90K protein, possibly corresponding to another steroid hormone receptor. The presence of PR in nuclei of the germinal epithelium and interfollicular cells perfectly matches the localization of radioactive estradiol demon-

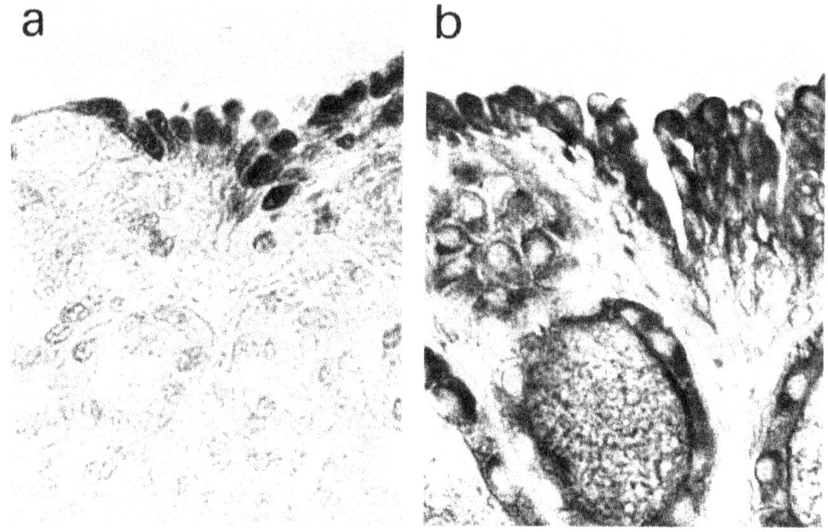

FIGURE 7. Immunodetection of PR in Bouin fixed, paraffin-embedded section of chicken ovary. (a) The B subunit is revealed in cell nuclei of the remnant of the germinal epithelium and also in some stromal cells between the follicles (not shown). No reaction is observed in follicular cells. (b) The 90K protein is observed in cells of the remnant of germinal epithelium like IgG-RB, and also in follicular cells surrounding the ovocyte. Cytoplasm appears more strongly stained than nuclei. No counterstaining. (Magnification × 650.)

strated by autoradiography in the ovaries of chicken embryo,[23] confirming once more the presence of PR in estradiol target cells.[9,16]

In adrenals, no nuclear staining was observed with IgG-RB or IgG-G3. BF4 stained the cytoplasm of strands of cells, presumably cortical, while chromaffin cells were negative (Figure 8a). Neurones in the small ganglia at the periphery of the adrenals were stained with BF4 in cytoplasm (Figure 8b), but were negative with IgG-RB. We concluded, therefore, that PR was absent in adrenal glands, the staining by BF4 possibly being due to other steroid hormone receptors.

VI. PITUITARY AND BRAIN

In the hen, the preovulatory release of luteinizing hormone (LH) is associated with increased concentrations of steroids. Studies under various hormonal conditions, including ovariectomy and "priming" with estradiol, suggested that progesterone induces the preovulatory LH surge.[24] Whether progesterone acts directly at the level of gonadotropic cells in the anterior pituitary, or through LH-RH neurones in the brain, or both, is not known. The use of antibodies against PR offers a simple and novel way of localizing progesterone target cells in the brain and the pituitary either alone or simultaneously with peptide hormones whose synthesis or release is under progesterone control.

As this part of the studies has been undertaken recently, only preliminary results are reported here. As in other tissues, PR in the brain and pituitary is under estrogenic control. For instance, IgG-RB revealed PR in many cells of the pars distalis and pars tuberalis of estradiol-treated male as well as female chickens (Figure 9). It is also possible to combine immunohistochemistry and autoradiography after administration of ^{3}H-Org 2058. This comparison of the two techniques was carried out on the same frozen section of pituitary or

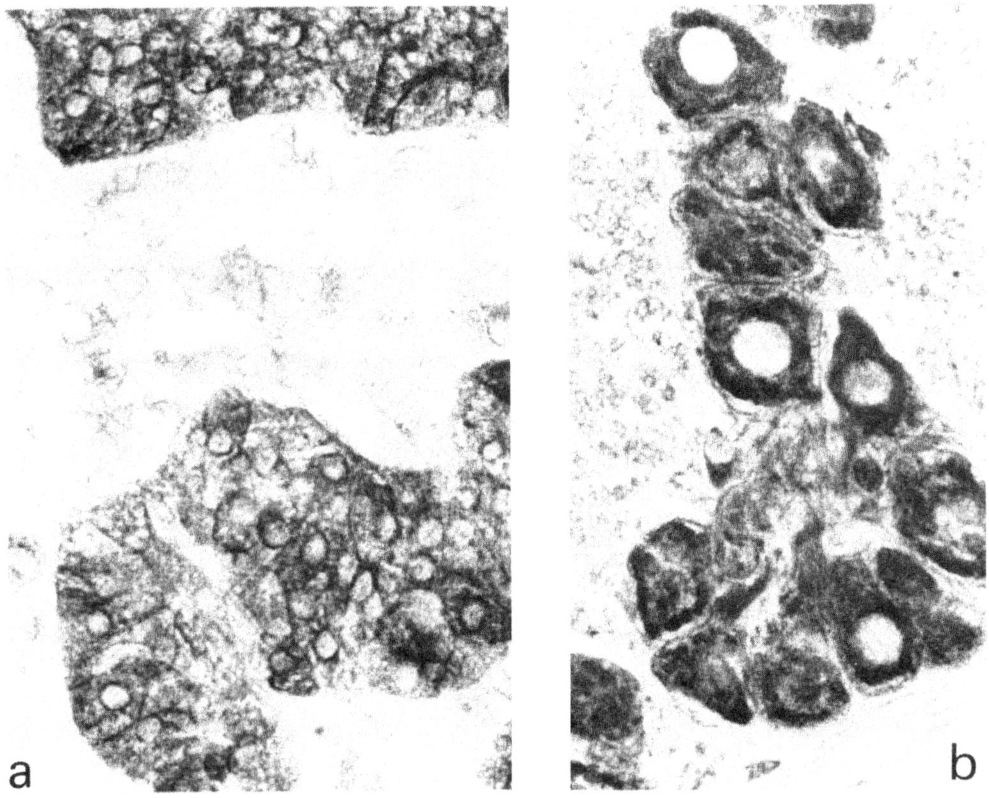

FIGURE 8. Immunodetection of the 90K protein on paraffin sections of chicken adrenal gland after Bouin's fixation. The BF4 antibody reacts in all cells (cytoplasm and nucleus) inside the strands of cortical cells, whereas the chromaffin cells remain unstained (a). Small nerve ganglia at the periphery of the gland display neurons positive with BF4 both in cytoplasm and nucleus (b). The two large nuclei that appear empty on the picture are probably due to artifactual loss of nuclear material; other nuclei are stained. No counterstaining. (Magnification × 650.)

brain.[25] The exact congruence of the two populations of cells, which react with IgG-RB and concentrate ^{3}H-Org 2058 in their nuclei, is a striking validation of the techniques to localize PR in target cells. Further studies are planned to show if PR-positive cells also secrete pituitary hormones, particularly LH. In the brain a complete mapping of PR-positive neurones is currently being performed, together with the dual localization of PR and LH-RH. Preliminary observations show the presence of PR in the preoptic area of estradiol-stimulated chickens (Figure 10).

VII. DISCUSSION — CONCLUSION

The results summarized in this chapter illustrate some applications of the immunolocalization of PR on paraffin and frozen sections. There are limiting factors such as the threshold of detection, i.e., the number of antigenic determinants per cell, and the specificity of the antibodies. Detection does not depend on the presence or absence of progesterone, or on the ability of PR to bind the ligand. Thus, immunohistochemistry for the first time offers a means to detect PR not bound to its ligand such as nonactivated forms, or PR molecules in the process of recycling or metabolism.

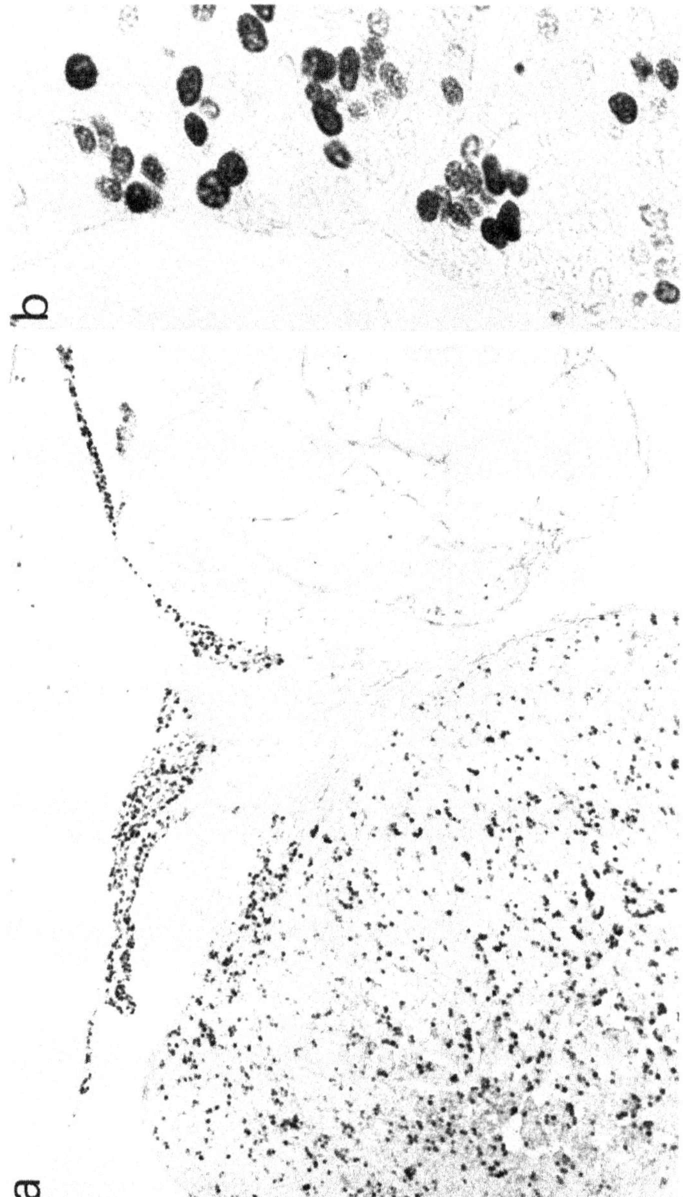

FIGURE 9. Immunodetection of PR on paraffin section of pituitary from an estradiol-stimulated chicken (Bouin's fixation). PR is revealed by IgG-RB throughout the cephalic and caudal lobes of the pars distalis and also in many cells of the pars tuberalis (a). Only nuclei are stained, and the intensity of staining is very different from one cell to another. (b). No counterstaining. (Magnification: a × 100, b × 650.)

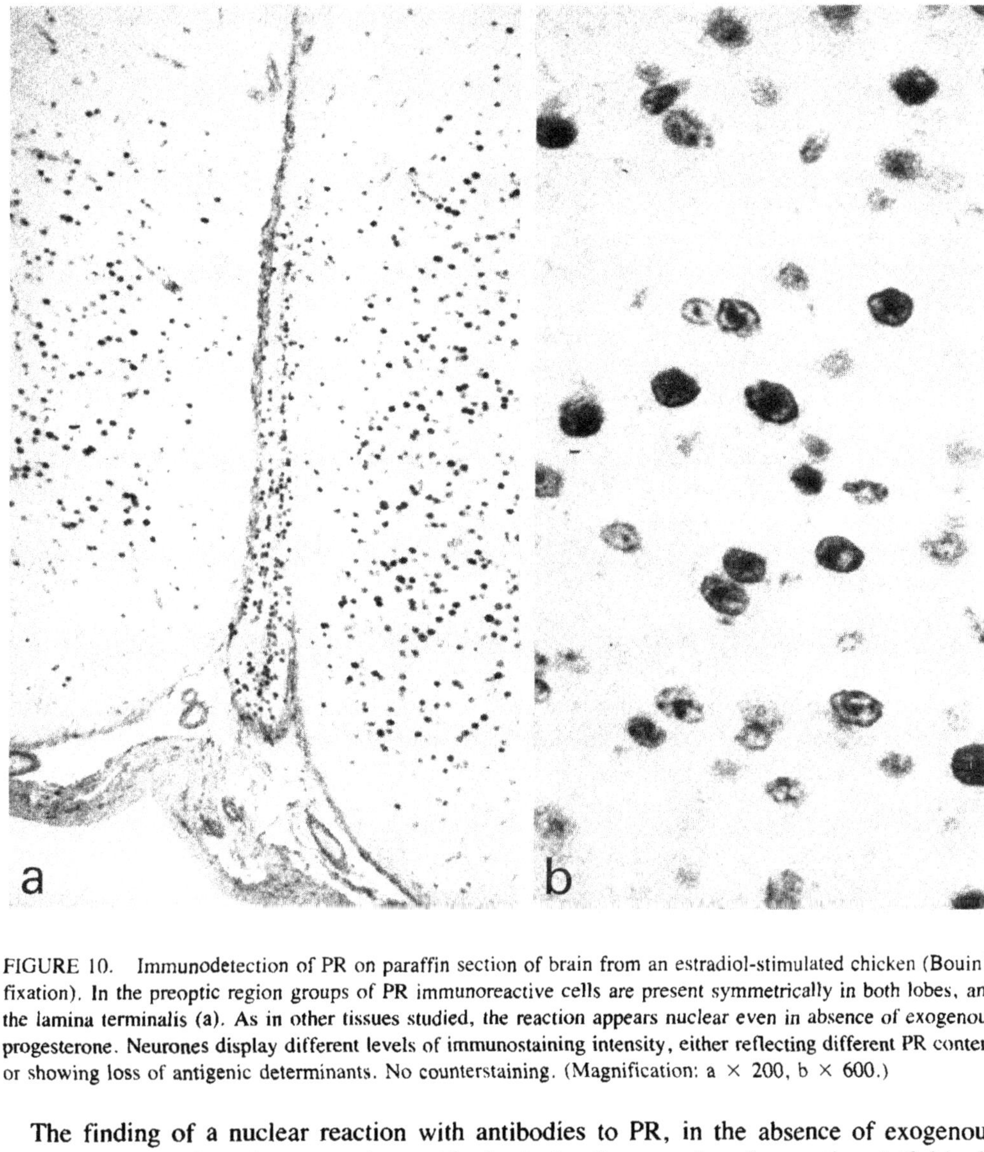

FIGURE 10. Immunodetection of PR on paraffin section of brain from an estradiol-stimulated chicken (Bouin's fixation). In the preoptic region groups of PR immunoreactive cells are present symmetrically in both lobes, and the lamina terminalis (a). As in other tissues studied, the reaction appears nuclear even in absence of exogenous progesterone. Neurones display different levels of immunostaining intensity, either reflecting different PR content or showing loss of antigenic determinants. No counterstaining. (Magnification: a × 200, b × 600.)

The finding of a nuclear reaction with antibodies to PR, in the absence of exogenous progesterone, and at the same time with the lack of a cytoplasmic reaction definitively attributable to a progesterone binding subunit of PR, is in apparent contradiction with biochemical data.[10] However, it only changes the interpretation suggesting that exposure to hormone is necessary, to produce nuclear localization. The presence of nuclear steroid receptor with little or no levels of the corresponding hormone has already been reported for estradiol.[26] The results reported here emphasize a need for reinterpretation of published biochemical data. It does not modify, however, the concept of the receptor reacting with chromatin and/or DNA to initiate events specific to the corresponding hormone. Regarding the lack of receptor in the cytoplasm, this cannot be so definitively established as the presence of PR in the nucleus, since it is always difficult to rely on negative results. In any case, PR as well as estradiol receptor have already been shown in cytoplasm of rat pituitary cells at the electron microscopic level.[25,26] We favor a concept which would distinguish between two states of the receptor, nonactivated in absence of hormone under the intracellular conditions and activated in presence of hormone. The nonactivated receptor is distributed between the cytoplasmic and the nuclear compartment in different proportions according to different target organs, species, and the different steroid receptors. In the case of the chick oviduct and probably in many other cases, it is mostly, or even only, concentrated in nuclei. The hormone-receptor complexes, in the so-called activated (transformed) state, would have a higher affinity for nuclear structures at the level of which response is initiated. The activated hormone-receptor complexes can also bind with more affinity than the nonactivated receptor to nonspecific, nonresponsive nuclear structure(s). In the case of antihormone, the distinction between binding of activated receptor to nonresponsive and to responsive nuclear structures may be envisaged, since there are cases where antihormone-receptor complexes are detected in the nuclear fraction, while no hormonal response can be recorded. This is the situation in the chick oviduct after tamoxifen administration.[29]

In all tissues in which the 90K protein was revealed by either IgG-G3 or BF4, its distribution was more widespread than that of the B subunit of PR. This observation probably reflects the fact that 90K protein is not only common to all steroid hormone "8S" receptors in the chicken[4] and can, therefore, be detected in all steroid target cells, but is also present in excess to the amount corresponding to receptors.[30]

ACKNOWLEDGMENTS

The authors acknowledge the technical help of Francine Delahaye, the secretarial assistance of Françoise Boussac and Martine Rossillon, and the generous gift of antibodies by J. M. Renoir and Thierry Buchou (IgG-G3), Pr P. Tuohimaa (IgG-RB), I. Joab, and C. Radanyi (BF4).

REFERENCES

1. **Chamness, G. C., Mercer, W. D., and McGuire, W. L.,** Are histochemical methods for estrogen receptor valid?, *J. Histochem. Cytochem.*, 28, 792, 1980.
2. **Renoir, J. M., Yang, C. R., Formstecher, P., Lustenberger, P., Wolfson, A., Redeuilh, G., Mester, J., Richard-Foy, H., and Baulieu, E. E.,** Progesterone receptor from chick oviduct: purification of molybdate-stabilized form and primary characterization, *Eur. J. Biochem.*, 127, 71, 1982.
3. **Renoir, J. M., Mester, J., Buchou, T., Catelli, M. G., Tuohimaa, P., Binart, N., Joab, I., Radanyi, C., and Baulieu, E. E.,** Purification by affinity chromatography and immunological characterization of a 110kDa component of the chick oviduct progesterone receptor, *Biochem. J.*, 218, 685, 1984.
4. **Joab, I., Radanyi, C., Renoir, J. M., Buchou, T., Catelli, M. G., Binart, N., Mester, J., and Baulieu, E. E.,** Immunological evidence for a common non hormone-binding component in "non-transformed" chick oviduct receptors of four steroid hormones, *Nature (London)*, 308, 850, 1984.
5. **Baulieu, E. E., Binart, N., Buchou, T., Catelli, M. G., Garcia, T., Gasc, J. M., Groyer, A., Joab, I., Moncharmont, B., Radanyi, C., Renoir, J. M., Tuohimaa, P., and Mester, J.,** Biochemical and immunological studies of the chick oviduct cytosol progesterone receptor, in *Nobel Sym. on Steroid Hormone Receptors: Structure and Function*, Gustafsson, J. A. and Eriksson, H., Eds., Elsevier, Amsterdam, 1983.
6. **Renoir, J. M., Radanyi, C., Yang, C. R., and Baulieu, E. E.,** Antibodies against progesterone receptor from chick oviduct. Cross-reactivity with mammalian progesterone receptors, *Eur. J. Biochem.*, 127, 81, 1982.
7. **Radanyi, C., Joab, I., Renoir, J. M., Richard-Foy, H., and Baulieu, E. E.,** Monoclonal antibody to chicken oviduct progesterone receptor, *Proc. Natl. Acad. Sci. U.S.A.*, 80, 2854, 1983.
8. **Tuohimaa, P., Renoir, J. M., Radanyi, C., Mester, J., Joab, I., Buchou, T., and Baulieu, E. E.,** Antibodies against highly purified B-subunit of the chick oviduct progesterone receptor, *Biochem. Biophys. Res. Commun.*, in press.

9. **Toft, D. O. and O'Malley, B. W.,** Target tissue receptors for progesterone: the influence of estrogen treatment, *Endocrinology*, 90, 1041, 1972.
10. **Mester, J. and Baulieu, E. E.,** Progesterone receptors in the chick oviduct: determination of the total concentration of binding sites in the cytosol and nuclear fraction and effect of progesterone on their distribution, *Eur. J. Biochem.*, 72, 405, 1977.
11. **Dougherty, J. J. and Toft, D.,** Characterization of two 8S forms of chick oviducts progesterone receptors, *J. Biol. Chem.*, 257, 3113, 1982.
12. **Schrader, W. T. and O'Malley, B. W.,** Progesterone-binding components of chick oviduct. Characterization of purified subunits, *J. Biol. Chem.*, 247, 51, 1972.
13. **Gasc, J. M., Renoir, J. M., Radanyi, C., Joab, I., and Baulieu, E. E.,** Etude immunohistologique de l'oviducte de poulet à l'aide d'anticorps anti-récepteur de la progestérone, *C. R. Acad. Sci. Paris*, 295, 707, 1982.
14. **Gasc, J. M., Renoir, J. M., Radanyi, C., Joab, I., Tuohimaa, P., and Baulieu, E. E.,** Progesterone receptor in the chick oviduct: an immunohistochemical study with antibodies to distinct receptor components, *J. Cell Biol.*, 99, 1193, 1984.
15. **Gasc, J. M., Ennis, B. W., Baulieu, E. E., and Stumpf, W. E.,** Récepteur de la progestérone de l'oviducte de poulet: double révélation par immunohistochimie avec des anticorps anti-récepteur et par autoradiographie à l'aide d'un progestagène tritié, *C. R. Acad. Sci. Paris*, 297, 477, 1983.
16. **Milgrom, E., Atger, M., and Baulieu, E. E.,** Progesterone in uterus and plasma. IV. Progesterone receptor(s) in guinea pig uterus cytosol, *Steroids*, 16, 741, 1970.
17. **Glick, B.,** The bursa of Fabricius and immunoglobin synthesis, *Int. Rev. Cytol.*, 48, 345, 1977.
18. **Erickson, A. E. and Pincus, G.,** Modification of embryonic development of reproductive and lymphoid organs in the chick, *J. Embryol Exp. Morphol.*, 16, 211, 1966.
19. **LeDouarin, N. M., Michel, G., and Baulieu, E. E.,** Studies of testosterone induced involution of the bursa of Fabricius, *Dev. Biol.*, 75, 288, 1980.
20. **Gasc, J. M. and Stumpf, W. E.,** The bursa of Fabricius of the chicken embryo: localization and ontogenic evolution of sex steroid target cells, *J. Embryol. Exp. Morphol.*, 63, 225, 1981.
21. **Sullivan, D. A. and Wira, C. R.,** Sex hormones and glucocorticoid receptors in the bursa of Fabricius of immature chicks, *J. Immunol.*, 122, 2617, 1979.
22. **Ylikomi, T., Gasc, J. M., Isola, J., Tuohimaa, P., and Baulieu, E. E.,** Progesterone receptor in the chick bursa of Fabricus: characterization and immunohistochemical detection, *Endocrinology*, in press.
23. **Gasc, J. M.,** Estrogen target cells in gonads of the chicken embryo during sexual differentiation, *J. Embryol. Exp. Morphol.*, 55, 331, 1980.
24. **Williams, J. B. and Sharp, P. J.,** Control of the preovulatory surge of luteinizing hormone in the hen *(Gallus Domesticus)*: the role of progesterone and androgens, *J. Endocrinol.*, 77, 57, 1978.
25. **Stumpf, W. E., Gasc, J. M., and Baulieu, E. E.,** Progesterone receptor in pituitary and brain: combined autoradiography-immunohistochemistry with tritium-labeled ligand and receptor antibodies, *Mikroskopie*, 40, 359, 1983.
26. **Mester, J. and Baulieu, E. E.,** Nuclear estrogen receptor of chick liver, *Biochim. Biophys. Acta*, 261, 236, 1972.
27. **Morel, G., Dubois, P., Gustafsson, J., Radojcic, M., Radanyi, C., Renoir, M., and Baulieu, E. E.,** Ultrastructural evidence of progesterone receptor by immunochemistry, *Exp. Cell Res.*, 155, 283, 1984.
28. **Morel, G., Dubois, P., Benassayag, C., Nunez, E., Radanyi, C., Redeuilh, G., Richard-Roy, H., and Baulieu, E. E.,** Ultrastructural evidence of oestradiol receptor by immunochemistry, *Exp. Cell Res.*, 132, 249, 1981.
29. **Sutherland, R. L., Mester, J., and Baulieu, E. E.,** Tamoxifen is a potent "pure" anti-oestrogen in chick oviduct, *Nature (London)*, 267, 434, 1977.
30. **Moncharmont, B. and Catelli, M. G.,** unpublished.

Chapter 9

METHODS FOR THE QUANTIFICATION OF HISTOCHEMICAL STEROID BINDING ASSAYS*

Louis P. Pertschuk

TABLE OF CONTENTS

I. Introduction 144

II. Semiquantitative Analyses 144
- A. Methods 144
- B. Correlation of Semiquantified Estrogen Binding with Clinical Hormone Responsiveness in Breast Cancer 145
- C. Correlation of Semiquantified Progesterone Binding with Clinical Hormone Responsiveness in Breast Cancer 146
- D. Semiquantitative Androgen Binding Determinations in Prostatic Cancer 146
- E. Comparison of Semiquantified Estrogen Binding Values with Biochemical ER in Breast Cancer 147
- F. Semiquantification of Estrogen Receptor Immunocytochemical Assay... 148

III. Quantification Analysis 148
- A. Equipment Utilized 148
- B. Quantification Technique 149
- C. Results 149
- D. Comparison of Quantified Estrogen Binding with Clinical Response to Endocrine Therapy in Breast Cancer 155
- E. Quantification and Competitive Binding Studies 156

IV. Basic Problems with Methods of Quantification and Semiquantification 156

Acknowledgments 159

References 159

* Supported by USPHS Grants Nos. CA23623 and CA25760 from NCI.

I. INTRODUCTION

Not only the presence, but the amount of estrogen receptor (ER) has been shown to correlate with clinical response to endocrine therapy in breast cancer.[1,2] As a consequence, it may be necessary to develop methods to semiquantify or quantify histochemical techniques for the detection of steroid binding, in order to determine if the amount of ligand which may be bound is of significance. This chapter details our experience with such methodologies.

Any attempt at quantification should take into account a number of different variables. Heterogeneity of composition of malignant neoplasms is a phenomenon familiar to all surgical pathologists. This is particularly true for breast and prostate carcinomas, both of which are under endocrine control. Some breast cancers are composed almost entirely of neoplastic cells with but a small amount of supporting blood vessels and stroma. Medullary carcinomas fit this description as well as lymph nodes which have been replaced by metastatic tumor. On the other hand, tumors of the scirrhous variety and some of the invasive lobular carcinomas consist almost entirely of connective tissue stroma and contain a paucity of maligant cells. The majority of cancers have a composition which may vary anywhere between these two extremes. It is also apparent that in both breast and prostatic carcinomas, tumor cells may be intermingled with benign epithelial components which, also, may be quite capable of binding steroids. To further complicate matters, all workers in the field are in agreement that steroid hormone cellular binding heterogeneity is a ubiquitous occurence and that such mosaicism is the rule rather than the rarity. Although it is not uncommon to examine tumor sections where all the malignant cells are low or negative steroid binders, it is uncommon to see a neoplasm composed entirely of positive cells. Furthermore, among the positive cells, binding of steroid may be low, moderate, or intense and may be distributed in cytoplasm, nucleus, and/or nucleolus.

The only techniques capable of being quantified with any degree of accuracy are those which employ labeled steroids where the ratio of steroid to label remains constant. Methods in which steroid is bound directly to label would lend themselves most readily to quantification. With those ligands which have a backbone of bovine serum albumin (BSA), any change in the number of fluorescein moles bound per mole BSA would prevent any comparison of results. Furthermore, this ratio of fluorescein to BSA must not be allowed to vary from one conjugate to another if it is desirous to compare the level of estrogen bound with, for example, progesterone. To this end, it is necessary to first label the BSA with fluorescein and then to conjugate the steroid thereafter. All of the quantitative data discussed in this chapter were obtained with ligands containing 4.5 mol of fluorescein per mole BSA. Despite some ingenious attempts, quantification of methods which employ antibodies are fraught with difficulties and only semiquantifications are possible at best.

II. SEMIQUANTITATIVE ANALYSES

A. Methods

The simplest method of semiquantification is that of Lee.[3,4] In this technique, binding of ligand-conjugate to benign epithelial elements is used as a standard. Any binding in tumor cells equal to this level is termed positive. Binding of an intensity above this level is called strongly positive while staining below this level is designated as negative. The main problem with this scheme lies in the fact that it is common to see tissue blocks from tumors which do not contain any normal ducts or lobules. In addition, since no allowance is made for specimen composition, it becomes difficult to correlate results with conventional biochemical receptor assays which report the total receptor content, either related to tissue weight, DNA content, or amount of protein. Meijer and colleagues[5] modified Lee's method in a manner similar to that used in our laboratory as well as utilizing limiting dilution techniques.

Our method of semiquantification consists of several steps.[6-8] Hematoxylin and eosin-stained frozen tumor sections immediately adjacent to those employed for histochemistry are carefully studied by light microscopy and the following features are assessed. The proportion of the specimen composed of viable malignant cells, benign epithelium, and stroma are estimated and recorded. The sections processed for steroid binding are then evaluated by ultraviolet microscopy. Estimates are made of the percentages of negative and positive cells. The positively stained cells are then subdivided into groups which exhibit varying degrees of fluorescent intensity using symbols from ± to + + +. Binding in any nonmalignant tissue component is appraised. Finally, an overall number is computed to reflect the amount of ligand bound only by tumor cells.

In our semiquantitative method we assign arbitrary values for each cellular degree of estimated fluorescent intensity and then average these values for the entire observed area. The resulting computed values are then grouped into one of six classifications: no binding = zero. To the fluorescent intensity of ± a value of 30 is assigned, to +1 a value of 100, to +2 a value of 200, while to +3 the value of 300 is assigned. These values are then multiplied by the fractions of cells displaying the various intensity grades, added together, and then multiplied by the estimated fraction of the specimen actually composed of tumor cells. Finally, the calculated numbers are again arbitrarily divided into six groups as follows: zero, trace = <5, low = 5 to 12, intermediate = >12 to 18, high = >18 to 30, and very high = >30.

The following illustrations show how tissue heterogeneity can affect the calculated value of a specimen and change a lesion which has qualitatively been judged to be positive into a lower category. Conversely, a qualitatively poor specimen may emerge into a much higher classification.

Let us first consider a tumor composed of 95% stroma in which nearly all visible tumor cells bind the ligand-conjugate with an intensity rated as +2. The calculated value of the specimen would be 200 × 1 × 0.05 = 10, placing the tumor into the low category though qualitatively a good ligand binder.

Let us now consider a highly cellular neoplasm containing an estimated 10% stroma in which all the tumor cells are showing very weak binding, i.e., ±. The computed value of this specimen would be 30 × 1 × 0.9 = 27, placing this qualitatively poor binding neoplasm into the "high" classification.

Although these two examples are extreme, they serve to illustrate that there may be major differences between qualitative and quantitative evaluations of cellular positivity. However, for the average lesion some correlation does exist between these two determinations. From the point of view of comparison of histochemical and biochemical steroid binding data, it would seem logical to assume that the semiquantified comparison would be the more valid since in both methods, the average steroid binding capacity would have been determined. Only recently have some biochemists begun to insert corrections into their receptor determinations to compensate for some of the variables pointed out to them by the histochemists.[9-11] In point of fact, there is no way for the biochemists to assess steroid binding heterogeneity, or the potential contribution of nonmalignant cells to the total receptor content.

B. Correlation of Semiquantified Estrogen Binding with Clinical Hormone Responsiveness in Breast Cancer

Using the above scheme we have analyzed semiquantitative estrogen binding (EB) determinations in 111 patients utilizing 17 beta-estradiol-17-hemisuccinyl-BSA-fluorescein isothiocyanate (17-FE) in a concentration of 7×10^{-7} *M* as we have previously described.[8] Results were compared to the clinical response to endocrine therapies in a prospective manner (Table 1). There were 50 patients with semiquantified 17-FE binding values in the zero to low range. In this group, 45 (89%) failed therapy while 5 (11%) responded. There were 12

Table 1
CORRELATION OF SEMIQUANTIFIED HISTOCHEMICAL ESTROGEN BINDING ASSAY (17-FE) RESULTS WITH CLINICAL RESPONSE TO HORMONAL THERAPY IN BREAST CANCER

Assay results[a]	Responded	Failed	Total
Cytoplasmic			
Zero—low	5	45	50
Intermediate	7	5	12
High—very high	12	10	22
Nuclear	8	19	27
Total	32	79	111

[a] See text for derivation of categories.

Table 2
CORRRELATION OF SEMIQUANTIFIED HISTOCHEMICAL PROGESTERONE BINDING ASSAY RESULTS WITH CLINICAL RESPONSE TO HORMONAL THERAPY IN BREAST CANCER

Assay results[a]	Responded	Failed	Total
Cytoplasmic			
Zero—low	9	49	58
Intermediate	5	7	12
High—very high	10	8	18
Nuclear	6	14	20
Total	30	78	108

[a] See text for derivation of categories.

patients with cancers in the intermediate range of whom 7 (58%) responded. There were 22 cases that were placed into the high to very high category. Twelve responded while 5 failed. Thus, 56% of cases which were in the intermediate to very high levels of 17-FE binding responded. Cases displaying predominantly nuclear-bound 17-FE were excluded as these patients were known to do relatively poorly from earlier studies.

C. Correlation of Semiquantified Progesterone Binding with Clinical Hormone Responsiveness in Breast Cancer

The results of progesterone binding (PB) were available for semiquantification in 108 cases. In the zero to low category, 49 of 58 patients (85%) failed hormonal treatment, while in the intermediate to very high range, 15 of 30 (50%) responded (Table 2).

These results show that there exists for breast cancer some correlation between the amount of EB and PB per tumor section and hormone response. However, the overall percentage of responders and regressors is quite similar to those of qualitative analyses as detailed in our chapter on breast cancer.

D. Semiquantitative Androgen Binding Determinations in Prostatic Cancer

Semiquantified data on androgen binding and correlation with endocrine response in 72

Table 3
CORRELATION OF SEMIQUANTIFIED HISTOCHEMICAL ANDROGEN BINDING ASSAY RESULTS WITH CLINICAL RESPONSE TO HORMONAL THERAPY IN PROSTATIC CARCINOMA

Assay results[a]	Responded	Stable	Failed	Total
Zero—low	9	7	22	38
Intermediate	7	4	5	16
High—very high	9	6	3	18
Total	25	17	30	72

[a] See text for derivation of categories.

Table 4
COMPARISON OF SEMIQUANTIFIED ESTROGEN BINDING (17-FE) ASSAY RESULTS WITH ER BY DEXTRAN-COATED CHARCOAL ASSAY

	Histochemical (17-FE)[a]			
Biochemical ER	Zero—trace	Low—intermediate	High—very high	Total
Zero—trace (0—15 fmol/mg protein)	24	12	5	41
Low—intermediate (16—124 fmol/ mg protein)	11	21	17	49
High—very high (>124 fmol/mg protein)	6	9	6	21
Total	41	42	28	111

[a] See text for derivation of categories

men with advanced prostate cancer is shown in Table 3. Using the exact same method for semiquantification as described above with our androgen ligand-conjugates, there were 38 patients in the zero to low categories of whom 22 (58%) failed therapy. There were 34 men in the intermediate to very high groups of whom 26 (77%) responded or evidenced stabilization of their disease.

It is apparent that there is a clear association between the amount of androgen ligand bound and clinical response to endocrine therapy with an increasing proportion of men responding as the amount of androgen bound increased. However, because of the discrepancies between qualitative and semiquantitative analyses, the overall semiquantified results are not as good as were those of a qualitative nature. This is particularly true in prostate cancer where many of the specimens possess a high stromal content often with only a few malignant cells. This fact accounts for a good proportion of those patients either responding or becoming stabilized with computed semiquantitative values in the lower categories.

E. Comparison of Semiquantified Estrogen Binding Values with Biochemical ER in Breast Cancer

In our chapter on steroid binding in prostate cancer we show a table wherein semiquantified results of androgen binding are compared with the results of biochemical androgen receptor assay. Table 4 shows a similar analysis of the 111 breast cancer patients where EB has been semiquantified and compared with ER by the dextran-coated charcoal assay (DCC). For this purpose, the semiquantified histochemical results were grouped into three major categories:

Table 5
COMPARISON OF ER BY DEXTRAN-COATED CHARCOAL ASSAY WITH RESULTS OF SEMIQUANTIFIED ER IMMUNOCYTOCHEMICAL ASSAY (ERICA)

	ERICA[a]			
Biochemical ER	Zero—trace	Low—intermediate	High—very high	Total
Zero—trace (0—15 fmol/mg protein)	57	10	4	71
Low—intermediate (16—124 fmol/mg protein)	22	18	14	54
High—very high (>124 fmol/mg protein)	7	18	19	44
Total	86	46	37	169

[a] See text for derivation of categories.

zero to trace, low to intermediate, and high to very high. In order to compare ER by DCC, results of biochemical assay were also divided into three groups: zero to trace = 0 to 15 fmol/mg protein, low to intermediate = 16 to 124 fmol/mg protein, and high to very high = >125 fmol/mg protein. It can be seen that some degree of correlation exists between these assay results although it is far from perfect (46%).

F. Semiquantification of Estrogen Receptor Immunocytochemical Assay

The Estrogen Receptor Immunocytochemical Assay (ERICA) of Greene[12] is a method which employs monoclonal antibodies to ER,[3,14] a bridging antibody, and peroxidase-anti-peroxidase complexes. Semiquantification of this method is possible by estimating the degree of specific antibody bound on a scale of ± to + + +, assigning values for each degree of positivity, multiplying by the fractions of cells exhibiting various staining intensities, and correcting for the proportion of stroma in the tumor section as is done for the estrogen ligand-conjugate. We chose to use the same arbitrarily assigned values for ± to + + +, i.e., 30 to 300 in computing the final results.

Comparison of semiquantified ERICA results and ER by DCC is shown in Table 5. Results of each assay were grouped into zero to trace, low to intermediate, and high to very high categories with the same femto mole range for DCC utilized previously. Results were in concordance in 94 of 169 specimens (56%). When it is considered that DCC measures binding of steroid to unoccupied cytosol ER while ERICA detects occupied or unoccupied receptor-antigen(s), together with the difficulties inherent to the attempted quantification of an immunohistologic assay employing four different antibodies, this degree of agreement is eminently satisfactory, though not nearly as good as in strictly qualitative comparisons (see our chapter on Breast Cancer).

III. QUANTIFICATION ANALYSIS

One of the apparently valid criticisms leveled against the histochemical steroid binding techniques is that, to a considerable extent, interpretation is largely performed in a subjective manner. We have attempted to gauge the accuracy of such a subjective analysis by comparing with objective measurements employing a sophisticated, computer-assisted microfluorometric method.

A. Equipment Utilized

A Farrand microscope spectrum analyzing unit consisting of a minimonochromator assembly and a photomultiplier-photometer was attached to one of the ocular tubes of a Zeiss® Universal Research microscope by means of an internal adapter. An analog to digital converter linked this apparatus to an Apple II® microcomputer with disc drive. A permanent record was provided by an Epson MX-80® printer. The analyzer target opening was 0.5 mm with an exit slit band width of 15 nm. The analyzing monochromator wavelength was set at 530 nm to coincide with the emission wavelength of fluorescein. A thin slice of polished uranium oxide glass, 0.25 nm in diameter, mounted on a glass slide, was used as a standard. A program was written which allowed for computer acceptance of the microfluorometric readings and for their analysis.

B. Quantification Technique

At the start of each session, the apparatus was standardized by measuring the fluorescence of the uranium oxide glass at a magnification × 500 and adjusting the instrument to a set range. All tissue was processed with the same batch of 17-FE containing 4.5 mol of fluorescein per mole BSA. In each case, the background reading from three separate areas was first measured. These readings were fed into the computer where they were averaged. Fluorometric readings were then taken from 100 tumor cells selected from five representative areas of the tissue section. Measurements from cells showing nuclear or cytoplasmic fluorescence were recorded separately and automatically accessioned by the computer which corrected each reading to compensate for the average background fluorescence. At the conclusion of each examination a print-out of all the raw data was obtained and a statistical analysis and graphs were provided. The latter showed the percentages of positive cells per units of relative fluorescent intensity in separate plots for nuclear, cytoplasmic, cytoplasmic and nuclear, and total binding.

C. Results

A total of 130 breast cancers and 26 prostatic cancers were subjected to steroid binding quantification and semiquantification methods to allow for a comparison between techniques. In order to avoid bias, semiquantification and quantification were performed by different examiners on separate days, each being unaware of the results of the other.

Examples of the plots provided by the quantification technique together with representative photomicrographs for comparison are shown in Figures 1 to 3 for breast cancer and Figures 4 and 5 for prostatic carcinoma. It can be seen that, in general, the microfluorometric analyses correlated reasonably well with the subjective examination.

In order to compare semiquantitative and quantitative determinations, the mean value of relative fluorescent intensity of each neoplasm was corrected for the proportion of the specimen estimated to be composed of tumor cells and the resulting values divided into three main groups: zero to trace = 0 to 10, low to intermediate = 11 to 40, and high to very high = >40. Results are shown in Table 6.

There was overall agreement as to the amount of ligand-conjugate bound in 92 specimens of breast cancer (71%). It was apparent that significant disparities were present in many specimens. Three cases designated as zero to trace by the semiquantitative technique fell into the high to very high quantitative range as did 17 cases classified semiquantitatively as low to intermediate. Five cases quantified as zero to trace fell into the low to intermediate semiquantified category and the converse occurred in six cases. Another seven cases fell into the low to intermediate quantified range but into the high to very high semiquantified group. Reasons for this discordance will be discussed a little later.

Comparison of semiquantified and quantified androgen binding results in prostate carcinoma is shown in Table 7. There was agreement as to the level of androgen bound in 21 (81%) with no serious discrepancies.

There are a number of reasons why there may be discordance between specimens assessed

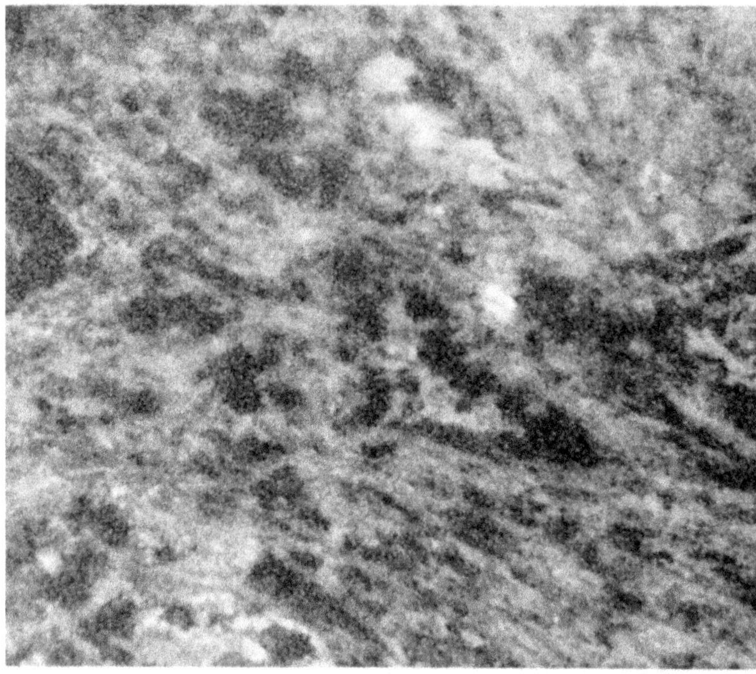

A

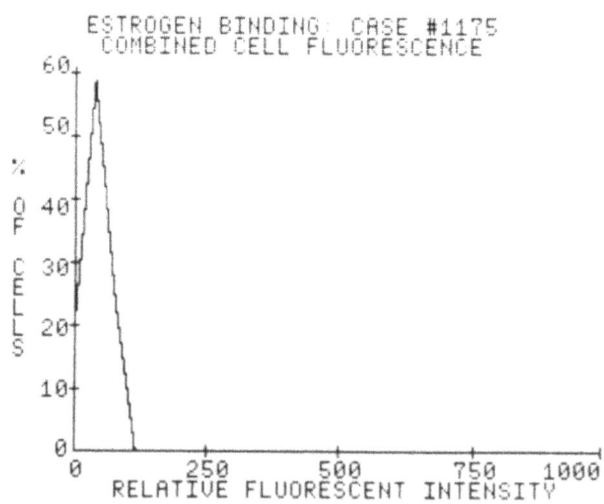

B

FIGURE 1. (A) Infiltrating duct cell carcinoma of breast after incubation with 17-FE exhibiting poor ligand-conjugate uptake. (Magnification × 100.) (B) Computer-derived graph of microfluorometric analysis confirming overall low to negative level of 17-FE binding.

subjectively and objectively. In the subjective approach the observer examines the entire tissue section, usually under low and medium power magnification, and thus makes an estimate after rapidly appraising thousands of cells. In the objective technique, the fluores-

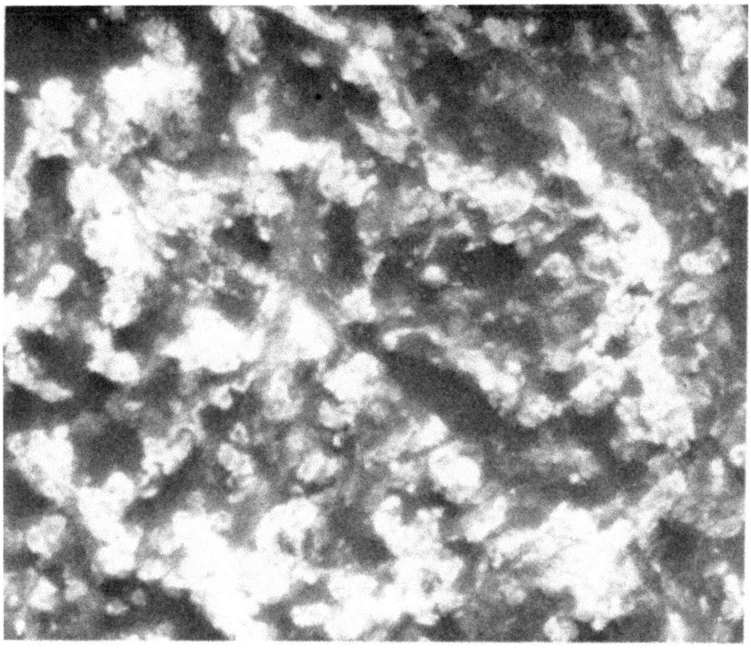

A

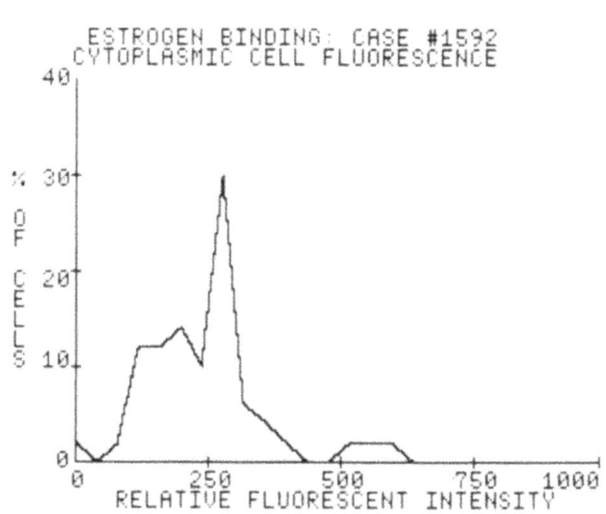

B

FIGURE 2. (A) Infiltrating duct cell carcinoma of breast after exposure to 17-FE. Although unequivocally positive, heterogeneity of ligand binding is present. (Magnification × 100.) (B) Graph of microfluorometer measurements confirms heterogeneity of 17-FE binding sites. One major peak at the +1 level (250 units of relative fluorescent intensity) is apparent.

cence emission of 100 cells is measured and these cells may not be representative of the tumor cell population. However, not only would it be unduly fatiguing to measure a larger number of cells, but the observer would then be faced with the problem of quenching of

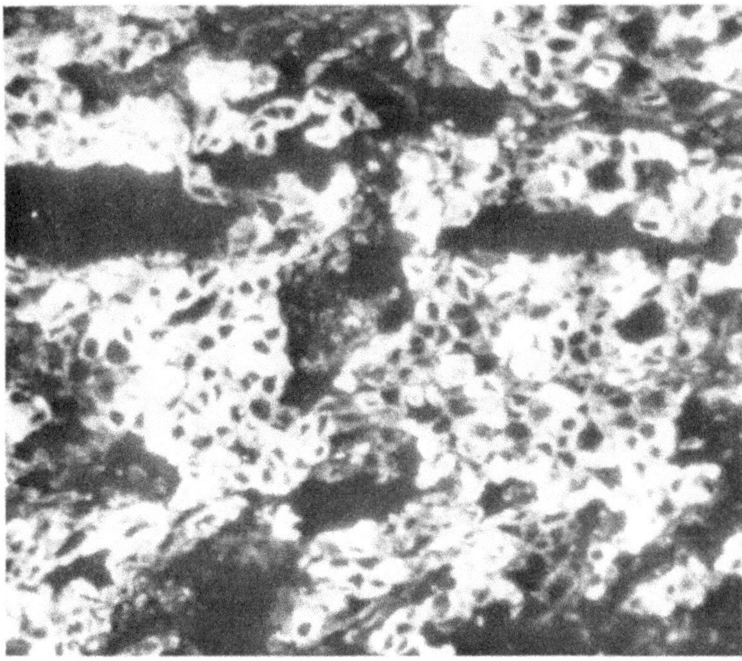

A

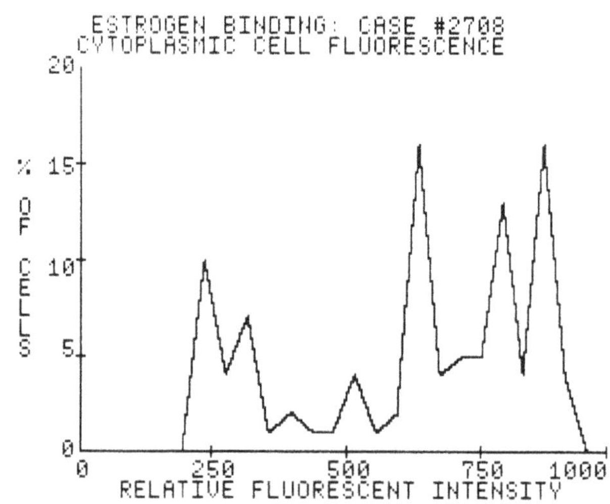

B

FIGURE 3. (A) Infiltrating duct cell carcinoma of breast after reacting with 6-FE. There is bright cytoplasmic fluorescence of the majority of the tumor cells. Heterogeneity of binding is apparent. (Magnification × 100.) (B) Computer-assisted graph of microfluorometric readings showing at least six groups of cells of varying intensities ranging from +1 (250 units of relative fluorescent intensity) to +4 (1000 units of relative fluorescent intensity).

fluorescence emission which would present an almost insurmountable obstacle. We have also noted that upon subjective examination there is a tendency to overestimate the numbers of cells at the extreme ends of the spectrum, i.e., the numbers of cells of zero intensity, or

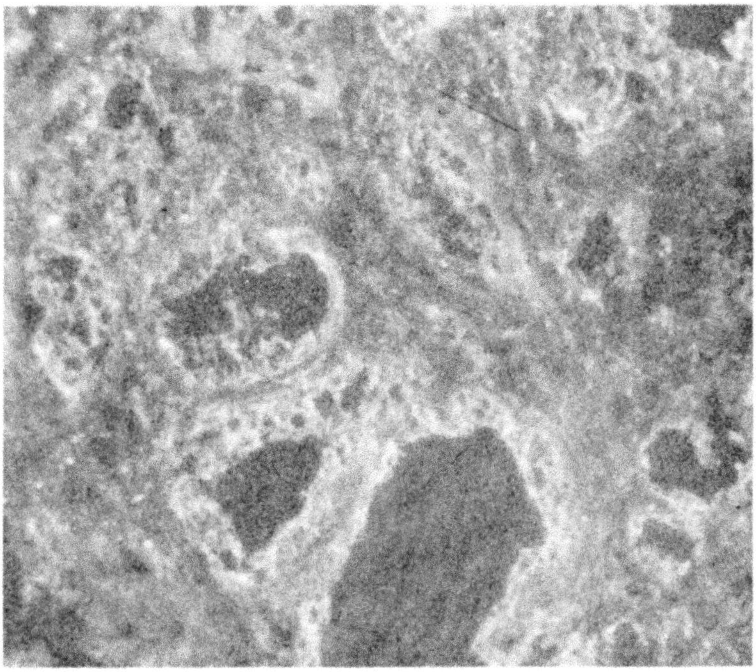

A

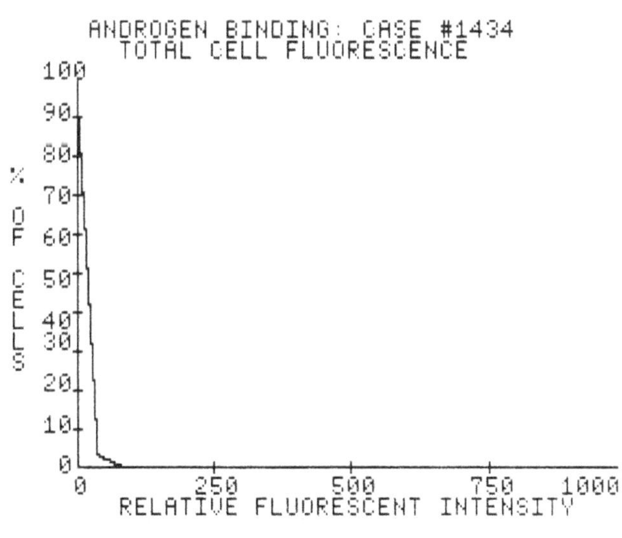

B

FIGURE 4. (A) Prostatic adenocarcinoma (top) adjacent to benign prostatic glands, both exhibiting a very low level of androgen ligand uptake. (Magnification × 100.) (B) Graph of computerized microfluorometric readings confirms the poor androgen binding.

in the highest range. As a consequence, the resultant computed values may place the specimen into the next lower or higher category. The better concordance in prostatic carcinoma (81 vs. 71% in breast cancer) may be due to the fact that an insufficient number of cases were

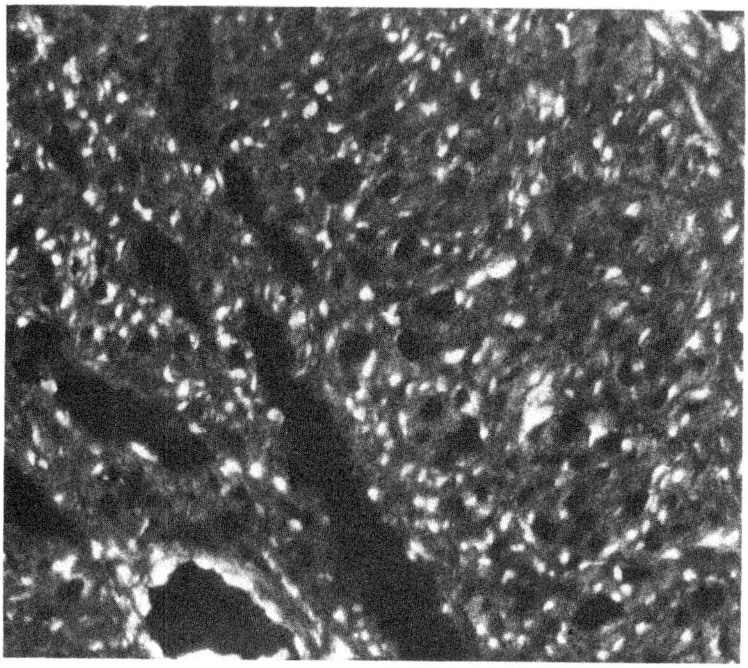

A

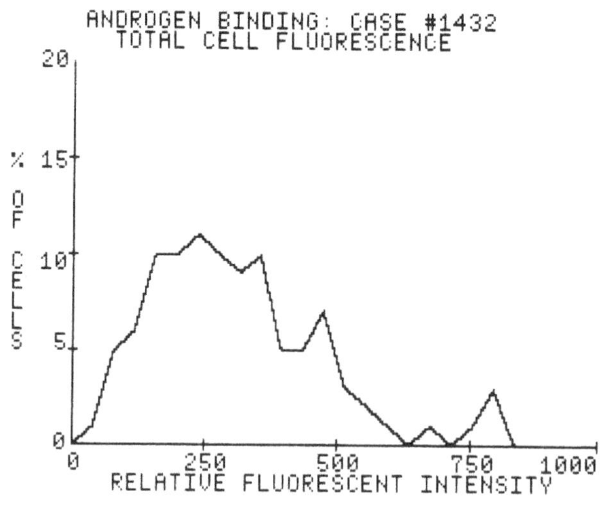

B

FIGURE 5. (A) Infiltrating prostatic adenocarcinoma showing clear nuclear positivity after reacting with androgen ligand. (Magnification × 100.) (B) Graph of microfluorometric measurements confirms good ligand uptake and reveals considerable binding heterogeneity not too apparent on visual inspection.

studied. A more plausible explanation is that nine of the prostate specimens were needle biopsies and, thus, the cells measured were more representative of the specimen content.

Table 6
COMPARISON OF MICROFLUOROMETRIC QUANTIFICATION OF ESTROGEN BINDING (17-FE) WITH SUBJECTIVE SEMIQUANTIFICATION IN BREAST CANCER

	Microfluorometry[a]			
Semiquantification[a]	**Zero—trace**	**Low—intermediate**	**High—very high**	**Total**
Zero—trace	16	6	3	25
Low—intermediate	5	33	17	55
High—very high	0	7	43	50
Total	21	46	63	130

[a] See text for derivation of categories.

Table 7
COMPARISON OF MICROFLUOROMETRIC QUANTIFICATION OF ANDROGEN BINDING WITH SUBJECTIVE SEMIQUANTIFICATION IN PROSTATIC CARCINOMA

	Microfluorometry[a]			
Semiquantification[a]	**Zero—trace**	**Low—intermediate**	**High—very high**	**Total**
Zero—trace	4	1	0	5
Low—intermediate	2	5	0	7
High—very high	0	2	12	14
Total	6	8	12	26

[a] See text for derivation of categories.

Table 8
CORRELATION OF MICROFLUOROMETRIC ESTROGEN BINDING (17-FE) QUANTIFICATION WITH CLINICAL HORMONE RESPONSE IN BREAST CANCER

Quantified assay results[a]	Responded	Stable	Failed	Total
Zero—trace	0	0	3	3
Low—intermediate	4	1	7	12
High—very high	5	1	5	11
Nuclear	1	0	7	8
Total	10	2	22	34

[a] See text for derivation of categories.

D. Comparison of Quantified Estrogen Binding with Clinical Response to Endocrine Therapy in Breast Cancer

In 34 cases of breast cancer we were able, in a retrospective study, to compare clinical endocrine response with quantified EB. Results are shown in Table 8. There were three cases in the zero to trace category, all of whom failed therapy. Out of 12 cases in the low to intermediate group, 5 (42%) responded or were stabilized. There were 11 patients in the high to very high range. Six (55%) either responded or were stabilized. The cases which showed predominantly nuclear binding are shown separately. One with high to very high nuclear 17-FE responded, while the other seven patients in this group failed. Five of the latter were in the low to intermediate group, and one each in the zero to trace and high to very high categories.

It can be seen that the percentage of responding or stable patients in the low to intermediate and high to very high groups is similar to that of the larger series analyzed semiquantitatively (Table 1) and, for that matter, almost identical to the percentage of positive responders as determined qualitatively.

A comparison of semiquantified and quantified EB data utilizing the 6-position labeled estradiol of Lee[3,4] was similar to that for 17-FE. Quantification of the Lee method failed to improve either the predictive value of the assay or the biochemical correlations when compared to either the semiquantified or qualitative results (data not shown.) The biochemical correlations with 17-FE were also not enhanced by quantification.

E. Quantification and Competitive Binding Studies

Several investigators have reported performing competitive binding studies by exposing parallel tissue sections, respectively, to ligand-conjugate and ligand-conjugate plus an excess molar concentration of competitor ligand. A reduction in fluorescent intensity in the section exposed to ligand-conjugate and competitor, compared to that seen with ligand-conjugate alone, is accepted as evidence of successful competition for the binding sites under study. This kind of judgement is, of course, usually made solely by visual inspection and is entirely subjective and, as a consequence, also open to criticism.

One of the virtues of microfluorometry is in assessing competitive binding studies since numerical evidence of inhibition of ligand-conjugate binding can, thus, be obtained. Figures 6A and 6B show the results of a competitive binding study by computer-assisted microfluorometry. Sections of an EB+ breast cancer were exposed in parallel to 7×10^{-7} *M* 17-FE and to the same identical concentration of 17-FE plus a 7×10^{-5} *M* concentration of the antiestrogen CI 628. Binding of ligand was then measured in 100 randomly selected cells in each of the tissue sections. The reduction in fluorescence intensity was subjectively estimated to be approximately 95% and, as can be seen from the graphs, this was confirmed objectively.

IV. BASIC PROBLEMS WITH METHODS OF QUANTIFICATION AND SEMIQUANTIFICATION

Earlier in this chapter we demonstrated how methods designed to semiquantify steroid hormone binding in tumor tissue sections can provide misleading data and actually change the category of a lesion from low to high or from high to low. Unfortunately, this is also true for methods which employ quantitative techniques as the following illustration will show. A scirrhous carcinoma specimen consisting of 99% stroma and containing 100 tumor cells is found to have a high mean fluorescent intensity of 600. Obviously, it is a tumor in which the cells are rich in EB. When corrected for tissue composition, however, it would fall into the zero to trace category ($600 \times 0.01 = 6$). In a similar fashion a medullary carcinoma containing an estimated 10% stroma and lymphocytes and with a low mean fluorescent intensity of 46 would, when corrected for tissue composition, fall into the high to very high group ($46 \times 0.9 = 41$).

Unless the area of the tumor component within a given tissue section were to be actually measured, a complicated and tedious task especially when there are numerous cords and columns and islets of cells, subjective estimations may vary as much as 20% between different observers, while the same observer will usually reevaluate a tumor section within a range

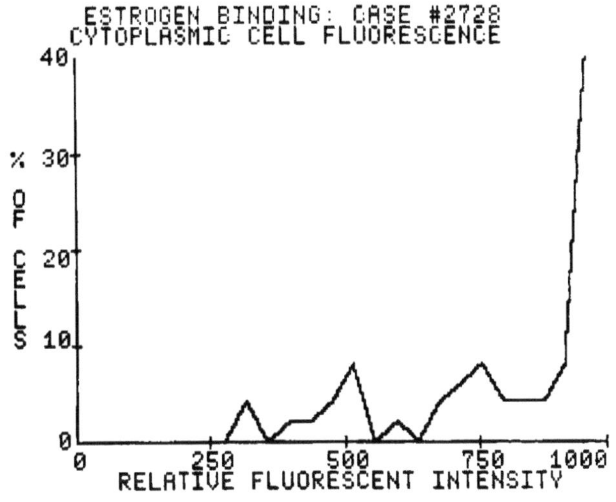

A

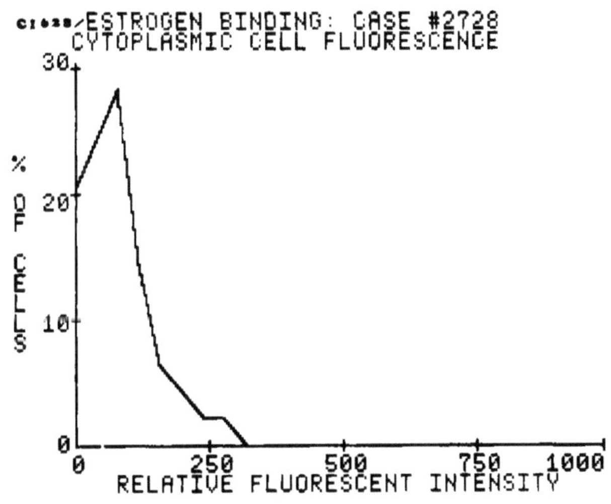

B

FIGURE 6. (A) Computer-derived graph of microfluorometric analysis of 17-FE binding in a concentration of 7×10^{-7} *M* in an infiltrating duct cell breast carcinoma. Some heterogeneity of binding is evident with a high peak at the 4+ level (1000 units of relative fluorescent intensity). (B) Graph of microfluorometric measurements of a parallel tissue section exposed to 7×10^{-7} *M* 17-FE plus 5×10^{-7} *M* of the antiestrogen CI 628. There is clearly diminished binding of 17-FE indicative of a successful competition by CI 628.

of 10% plus or minus. However, even a slight, i.e., 10% variation may be sufficient to change the calculated value of either a quantified or semiquantified analysis significantly as will be shown in the following examples.

A breast cancer processed by ERICA is found to contain 30% + cells. One observer estimates that the neoplasm is comprised of 40% tumor cells, while a second examiner estimates that the percentage occupied by tumor is 50. Calculations based on the first estimate

$(100 \times 0.3) \times 0.4 = 12$ would place the lesion into the low to intermediate range. Computations based upon the second estimation $(100 \times 0.3) \times 0.5 = 15$ would categorize the specimen as high to very high.

Even with the quantitative analysis, a variation of 10% in estimation of specimen composition would result in a significant change in the computed value. Two microscopists using computer-assisted microfluorometry obtain a mean relative fluorescent intensity of 175. One estimates the volume of tumor cells to be 20%, the other 30%. Calculations to correct the mean relative fluorescent intensity for tumor cell volume based on these estimates would result in a final classification of low to intermediate using the first estimate: $(175 \times 0.2) = 35$; and of high to very high with the second: $(175 \times 0.3) = 52.5$.

In a similar manner, incorrect estimations of the percentage of positive cells, or intensity of staining, could alter the final classification of a specimen by the semiquantitative method. Consider a tumor section processed with 17-FE and examined by two observers, both of whom concur that the specimen is comprised of 30% tumor cells. The first observer reports that the observed positively stained cells are divided into 20% ± and 20% +. The computed value would then be $(30 \times 0.2) + (100 \times 0.2) \times 0.3 = 7.8$ placing the lesion in the low binding category. The second observer estimates that 30% of the malignant cells show a + intensity and 20% a + + fluorescent intensity. Calculations based on the latter estimation, i.e., $(100 \times 0.3) + (200 \times 0.2) \times 0.3 = 21$, would place the lesion in the high binding group.

When one considers, therefore, that at best all these techniques are very crude, what is surprising is that in spite of all these very real potential sources for error, there nonetheless exists a significant degree of correlation beyond that which would be expected by chance alone, both with biochemically measured levels of ER as well as with clinical endocrine response in both breast and prostate cancer. These data, therefore, strongly support the assumption that histologic techniques detect binding sites that are significantly associated with biochemical determinations as well as with endocrine response.

Although it might well be concluded from all of the above that there is little value for semiquantitative and quantitative methods other than to confirm competitive binding studies, especially those difficult to judge by the naked eye, there is a sound hypothetical reason to continue the semiquantitative method. It is reasonable to assume that only tumor cells containing steroid binding proteins are responsive to endocrine treatment. Recurrence in a previously responsive patient, presumably, would be due to proliferation of autonomous cells. It follows, therefore, that there should be a direct correlation between the proportion of negative cells and the length of time of hormonal response. To date we have insufficient data to thoroughly test this hypothesis. However, only by continuing to collate data on proportions of weakly positive, moderately positive, strongly positive, and negative cells can this hypothesis be properly examined.

None of the data we have presented should cause the reader to infer that cases with a low volume of tumor cells are more likely to be ER-negative and, thus, less likely to respond to hormonal treatment. In fact, the converse appears to be true. In the past year, among the breast cancers studied were 68 in which the quantity of tumor cells was estimated to occupy 20% or less of the specimen; 48 (71%) were ER-positive and 20 (29%) ER-negative. There were 29 specimens where the tumor volume was thought to be 60% of the specimen, or greater. The latter were almost equally divided into ER-positive and ER-negative (15:14). In 86 cases, where response to endocrine therapy had been determined, there were 69 specimens containing less than 50% tumor cells and 17 estimated to contain more than 50% malignant cells. The response rate was slightly, but not significantly lower for the group with more cellular neoplasms.

For purposes applicable to community hospital use, we believe that any fully validated histological technique can be qualitatively assessed as demonstrating poor, moderate, or high steroid binding capacity. There is seldom any real disparity between different microscopists on this fundamental determination.

ACKNOWLEDGMENTS

We are most grateful to Farrand Optical Co., Inc., Valhalla, N.Y. for the loan of the microscope spectrum analyzer utilized in these studies. Dr. E. Gaetjens supplied the androgen, progesterone, and estrogen (17-FE) ligand-conjugates and Zeus Scientific, Inc., Raritan, N.J. generously donated 6-FE. Warner-Lambert/Parke-Davis donated CI 628. The majority of DCC analyses were performed in the laboratory of Dr. W. L. McGuire, University of Texas Health Sciences Center, San Antonio, Tex. A minority were performed in the Receptor Laboratory, Maimonides Medical Center, Brooklyn, N.Y. The Apple II® computer and associated equipment were loaned by Zeus Scientific, Inc.

REFERENCES

1. **Knight, W. A., Livingston, P. B., Gregory , E. J., and McGuire, W. L.,** Estrogen receptor as an independent prognostic factor for early recurrence in breast cancer, *Cancer Res.*, 37, 4669, 1977.
2. **Jensen, E. V.,** Hormone dependency of breast cancer, *Cancer*, 47, 2319, 1981.
3. **Lee, S. H.,** Cancer cell estrogen receptor of human mammary carcinoma, *Cancer*, 44, 1, 1979.
4. **Lee, S. H.,** Cellular estrogen and progesterone receptors in mammary carcinoma, *Am. J. Clin. Pathol.*, 73, 323, 1980.
5. **Meijer, C. J. L. M., van Marle, J., Persijn, J. P., van Niewenhuizen, W., Baak, J. P. A., Boon, M. E., and Lindeman, J.,** Estrogen receptors in human breast cancer. II. Correlation between the histochemical method and biochemical assay, *Virchows Arch. (Cell Pathol.)*, 40, 27, 1982.
6. **Pertschuk, L. P., Rosenthal, H. E., Macchia, R. J., Eisenberg, K. B., Feldman, J. G., Wax, S. H., Kim, D. S., Whitmore, W. F., Jr., Abrahams, J. I., Gaetjens, E., Wise, G. J., Herr, H. W., Karr, J. P., Murphy, G. P., and Sandberg, A. A.,** Correlation of histochemical and biochemical analyses of androgen binding in prostatic cancer: relation to therapeutic response, *Cancer*, 49, 984, 1982.
7. **Pertschuk, L. P., Tobin, E. H., Carter, A. C., Eisenberg, K. B., Leo, V. C., Gaetjens, E., and Bloom, N. D.,** Immunohistologic and histochemical methods for detection of steroid binding in breast cancer: a reappraisal, *Breast Cancer Res. Treat.*, 1, 297, 1982.
8. **Pertschuk, L. P., Gaetjens, E., Carter, A. C., Brigati, D. J., Kim, D. S., and Fealey, T. E.,** An improved histochemical method for detection of estrogen receptors in mammary cancer, *Am. J. Clin. Pathol.*, 71, 504, 1979.
9. **Mason, R. C., Steele, R. J. C., Hawkins, R. A., Miller, W. R., and Forrest, A. P. M.,** Cellularity and the quantitation of estrogen receptors, *Breast Cancer Res. Treat.*, 2, 239, 1982.
10. **Underwood, J. C. E., Dangerfield, V. J. M., and Parsons, M. A.,** Oestrogen receptor assay of cryostat sections of human breast carcinomas with simultaneous quantitative histology, *J. Clin. Pathol.*, 36, 399, 1983.
11. **Gairard, B., Calderoli, H., Keiling, R., Renaud, R., Bellocq, J. P., and Koehl, C.,** Cancer cell counts and validity of steroid receptor determinations in breast cancer, *Lancet*, 2, 1419, 1981.
12. **Press, M. F., King, W., and Greene, G.,** Immunocytochemical localization of estrogen receptors in the human endometrium using a monoclonal antibody against human estrogen receptor, *Fed. Proc.*, 42, 1178, 1983.
13. **Greene, G. L., Fitch, F. W., and Jensen, E. V.,** Monoclonal antibodies to estrophilin: probes for the study of estrogen receptors, *Proc. Natl. Acad. Sci. U.S.A.*, 77, 1178, 1980.
14. **Greene, G. L., Nolan, C., Engler, J. P., and Jensen, E. V.,** Monoclonal antibodies to human estrogen receptor, *Proc. Natl. Acad. Sci. U.S.A.*, 77, 5115, 1981.

Chapter 10

A TWO-STAGE IMMUNOCYTOCHEMICAL METHOD FOR PUTATIVE ESTROGEN RECEPTOR ANALYSIS

Vincenzo Eusebi and Gianni Bussolati

TABLE OF CONTENTS

I. Introduction 162

II. Methodology 162
  A. First Step: Cytochemical Method 162
  B. Second Step: Immunocytochemical Method 163
    1. Controls 163

III. Results and Discussion 163

Acknowledgment 164

References 164

I. INTRODUCTION

In 1981 we reported an immunocytochemical method for semiquantitative analysis of estrogen receptors (ER).[1]

We felt that an improvement of the existing techniques for assaying estrogen receptors was needed, at the time, in view of conflicting evidence in the literature concerning the lack of correlation between morphological and functional parameters of breast tumors. A correlation between high ER content and a low histological grade of malignancy had in fact been found by some workers,[2,3] but denied by others.[4-6] Several groups of workers found a positive correlation between ER positivity and the lobular type of invasive carcinoma,[7-9] a finding not corroborated by other groups.[2,3,10] Furthermore, while one group reported that five tubular carcinomas were all ER-negative,[6] another group found that four out of six tubular carcinomas were ER-positive.[11] Lastly, while a direct relationship between the degree of tumor elastosis and the presence of ER has been reported,[12] other workers suggested, from a multivariate analysis, that tumor elastosis might be more closely related to patient age rather than to ER status per se.[2]

For a direct demonstration of estrogen receptors on isolated breast carcinoma cells or on fresh tissue sections, several probes have been described (see Reference 21 for a review). Basically, the signal given by the single estrogen binding site was magnified, either making use of a large fluorescent complex or of enzymatic reactions.

Lee[13] introduced a cytochemical fluorescent method in order to visualize the ER-positive cells by means of E-BSA-FITC complexes. Nenci et al.[14] and Pertschuk et al.[15] used immunofluorescent methods employing antiestrogen sera; recently, Walker et al.[16] published a cytochemical method using 17-β-estradiol bound to peroxidase.

It seemed to us that the advantages of the procedures could be increased by using a two-stage method: a fluorescent cytochemical step followed by an immunohistochemical method employing antiestrogen sera.[1] As a second step, to visualize E-BSA complexes we employed antiestradiol serum and an immunoperoxidase procedure with peroxidase antiperoxidase complexes. Recently, we have proposed[19] a novel immuno-galactosidase procedure which we now use to reveal bound estradiol.

II. METHODOLOGY

Two adjacent mirror-image blocks are taken from each tumor and promptly frozen in a cryostat at 32°C. The blocks include the entire lesion in the smaller tumors, while only the edge is sampled from larger carcinomas.

One block is used for a dextran-coated charcoal analysis for ER, the other utilized for the two-step histochemical study.

A. First Step: Cytochemical Method

Four consecutive 5-μm sections are cut from the tissue block. The first and third sections are stained with hematoxylin-eosin while the second and fourth are studied using Lee's[13] method and employing the same fluorescent estrogen-BSA conjugate. Only two modifications are introduced.

We use an adhesive developed in our laboratory,[17] which allows a vigorous washing of the sections. In this way, the reaction time is shortened by 1 hr. In addition, all the reactions are carried out at a temperature of 4°C. This modification is a compromise between obtaining optimal cytochemical results and maintaining good tissue preservation.

The sections are then observed using fluorescent light and, in selected cases, a photomicrograph is taken.

B. Second Step: Immunocytochemical Method

After recording the results of the cytochemical method, the slides are washed in PBS buffer (pH 7.2) and then immersed in Carson's fixative[18] for 1 to 12 hr. The sections are then washed in PBS and treated first with normal swine serum, diluted 1:10, followed by treatment overnight with an antiserum, diluted 1:600, raised in rabbits against 17-β-estradiol-6-CMO-BSA (kindly supplied by Dr. G. F. Bolelli, Bologna). Anti-BSA antibodies are removed by adsorbing the serum with 1% BSA.

The sections are then rinsed in PBS for 30 min and treated with donkey anti-rabbit gamma-globulin β-galactosidase conjugate (Carlo Erba — Farmitalia, Milan). β-Galactosidase activity is then located with 5-bromo-4-chloro-3-indolyl-β-D-galactoside (Serva), according to Bondi et al.[19]

The slides are counterstained with the Van Gieson stain.

1. Controls

Tissues — Rat uterus is used as a postive control and human stomach as a negative control for both the initial cytochemical and the immuno-β-galactosidase method. We usually stain only four cases each time, together with the control slides, in the same batch.

Methods — FITC-conjugated albumin is used in control sections for the cytochemical step. Furthermore, sections are used without prior incubation with fluorescent 17-β-estradiol conjugate. The specific antiserum is substituted with nonimmune antiserum, or with antiserum previously absorbed with 17-β-estradiol or with 17-β-estradiol-6-CHN-BSA-FITC.

III. RESULTS AND DISCUSSION

The positive stain obtained with the immuno-β-galactosidase method (β I-G) is intensely blue colored. We have substituted it for the immunoperoxidase method for several reasons; partly, these are due to the real advantage of the β I-G method over the PAP technique and partly because there are practical advantages in this particular investigation.

The β I-G is faster than the PAP method as no inhibition of endogenous peroxidase is needed and the whole procedure is one step shorter. In addition, no carcinogenic activity has been demonstrated, so far, for the substrate used in the β I-G method nor for the indigo products of the reaction, while the carcinogenic activity of the diaminobenzidine is a matter of concern.[20]

A great advantage of this method resides in its blue positive reaction product which can be counterstained with Van Gieson. The connective tissue is nicely stained in red and the negative epithelial cells in yellow. This makes it easier to appreciate the positive and negative cancer cells and to afford a semiquantitative evaluation. With the β I-G method, like the PAP, it is possible to assess both nuclear and cytoplasmic positivity. To compare the immunogalactosidase and peroxidase procedures we have stained 20 consecutive sections with both methods: similar results have been obtained.

The PAP method was also abandoned because it was realized that, when sections are stained with nonimmune sera, occasionally there is positive staining of the nuclei, especially in the endometrium, and constantly in areas of elastosis. This nonspecific staining of PAP is not found using the β I-G method.

The present method, although more time consuming, has the advantage over the other fluorescent methods[13-15] in that the preparations are permanent. In addition, an enhancement of the signal is obtained.

Recently, doubts on the validity of results using the histochemical methods have been proposed.[21] Admittedly, the estradiol concentration utilized in the first step of our method, as well as in all the other cytochemical procedures, overwhelms the physiological quantities to which the "normal" "in vivo" cell is exposed. Furthermore, we are aware that, using antiestradiol antibodies, it is possible to visualize the presence of the hormone, either in the cytoplasm or in the nucleus, while sometimes it is not possible to assess whether the molecules of the steroid are actually bound to the receptors.

As we have previously pointed out,[1] it is well known that the major fault of the DCC method is the lack of histological control of the specimen analyzed. Furthermore, the insensitivity of the DCC method in the case of poorly cellular tumors has been clearly demonstrated.[7] This is confirmed by the lack of agreement in the literature, among different workers, when estrogen receptor content is considered in invasive lobular carcinomas which by definition are poorly cellular tumors. A further confirmation springs out from our previous study, in which all seven poorly cellular tumors gave low DCC values.[1]

Therefore, we think that at the moment there are no methods which can claim to be the optimal means of assessing the precise receptorial profile in breast tumors. In addition, from our previous work, it is apparent that there is a reasonably good correlation between the immunocytochemical method and the DCC analysis, with the exception of poorly cellular tumors.

At the moment, the biochemical assay and the immunocytochemical methods should, in our opinion, be used in conjunction. This would supply complementary information in the lack of more specific methods.

ACKNOWLEDGMENT

This work was supported by C.N.R. (Rome) Grants "Controllo della Crescita Neoplastica" N. 82.00300.90 and 82.00248.96.

REFERENCES

1. **Eusebi, V., Cerasoli, P. T., Guidelli-Guidi, S., Grilli, S., Bussolati, G., and Azzopardi, J. G.,** A two-stage immunocytochemical method for oestrogen receptor analysis: correlation with morphological parameters of breast carcinomas, *Tumori*, 6, 315, 1981.
2. **Fisher, E. R., Redmond, C. K., Liu, H., Rockette, H., Fisher, B., and collaborating NSAPB investigators,** Correlation of estrogen receptor and pathological characteristics of invasive breast cancer, *Cancer*, 45, 349, 1980.
3. **Maynard, P. V., Davies, C. J., Blamey, R. W., Elston, C. W., Johnson, J., and Griffiths, K.,** Relationship between oestrogen-receptor content and histological grade in human primary breast tumors, *Br. J. Cancer*, 38, 745, 1978.
4. **Johansson, H., Terenius, I., and Thoren, I.,** The binding of estradiol-17 β to human breast cancers and other tissues in vitro, *Cancer Res.*, 30, 692, 1970.
5. **Paszko, Z., Padzik, H., Dabska, M., and Pienkowska, F.,** Estrogen receptor in human breast cancer in relation to tumor morphology and endocrine therapy, *Tumori*, 64, 495, 1978.
6. **Rosen, P. P., Menendez-Botet, C. J., Senie, R. T., Schwartz, M. K., Schottenfeld, D., and Farr, G. H.,** Estrogen receptor protein (ERP) and histopathology of human mammary carcinoma, in *Hormones, Receptors and Breast Cancer*, McGuire, W. L., Ed., Raven Press, New York, 1978, chap. 6.
7. **Antoniades, K. and Spector, H.,** Correlation of estrogen receptor levels with histology and cytomorphology in human mammary cancer, *Am. J. Clin. Pathol.*, 71, 497, 1979.
8. **Eusebi, V., Pich, A., Macchiorlatti, E., and Bussolati, G.,** Morpho-functional differentiation in lobular carcinoma of the breast, *Histopathology*, 1, 301, 1977.
9. **Rosen, P. P., Menendez-Botet, C. J., Nisselbaum, J. S., Urban, J. A., Mike, V., Fracchia, A., and Schwartz, M. K.,** Pathological review of breast lesions analyzed for estrogen receptor protein, *Cancer Res.*, 35, 3187, 1975.
10. **Poulsen, H. S., Ozzello, L., and Andersen, J.,** Oestrogen receptors in human breast cancer, *Virchows Arch. (Anat. Pathol.)*, 397, 103, 1982.

11. **Lagios, M. D., Rose, M. R., and Margolin, F. R.,** Tubular carcinoma of the breast, *Am. J. Clin. Pathol.*, 73, 25, 1980.
12. **Masters, J. R. W., Millis, R. R., King, R. J. B., and Rubens, R. D.,** Elastosis and response to endocrine therapy in human breast cancer, *Br. J. Cancer*, 39, 536, 1979.
13. **Lee, S. H.,** Cytochemical study of estrogen receptor in human mammary cancer, *Am. J. Clin. Pathol.*, 70, 197, 1978.
14. **Nenci, I., Beccati, M. D., Piffanelli, A., and Lanza, G.,** Detection and dynamic localization of estradiol-receptor complexes in intact target cells by immunofluorescence technique, *J. Steroid. Biochem.*, 7, 505, 1976.
15. **Pertschuk, L. P., Tobin, E. H., Brigati, D. J., Kim, D. S., Bloom, N. D., Gaetjens, E., Berman, P. J., Carter, A. C., and Degenshein, G. A.,** Immunofluorescent detection of estrogen receptors in breast cancer, *Cancer*, 41, 907, 1978.
16. **Walker, R. A., Cove, D. H., and Howell, A.,** Histological detection of oestrogen receptor in human breast carcinomas, *Lancet*, 1, 171, 1980.
17. **Bondi, A., Fedeli, F., Eusebi, V., and Bussolati, G.,** Immunoperoxidase method for detection of immunoglobulins, *J. Clin. Pathol.*, 33, 904, 1980.
18. **Carson, F. L., Martin, J. H., and Lynn, J. A.,** Formalin fixation for electron microscopy: a re-evaluation, *Am. J. Clin. Pathol.*, 49, 365, 1973.
19. **Bondi, A., Chieregatti, G., Eusebi, V., Fulcheri, E., and Bussolati, G.,** The use of β-galactosidase as a tracer in immunocytochemistry, *Histochemistry*, 26, 153, 1982.
20. Registry of Toxic Effects of Substances, Vol. 1, Noish, ed., Department of Health and Human Services, Cincinnati, 1979.
21. **Chamness, G. C., Mercer, W. D., and McGuire, W. L.,** Are histochemical methods for estrogen receptors valid?, *J. Histochem. Cytochem.*, 28, 792, 1980.

Chapter 11

ESTRADIOL-BSA CONJUGATES FOR ESTROGEN RECEPTOR LOCALIZATION: THE VIENNA EXPERIENCE

Maximilian Binder, Klaus Czerwenka, Manfred Boehm, Juergen Spona, Raimund Jakesz, Roland Kolb, and Georg Reiner

TABLE OF CONTENTS

I. Introduction 168

II. Current Controversy 168

III. The Stability of Conjugates 169

IV. In Vitro Interaction of Conjugates with Estrogen Receptor Type I 171

V. Conjugate Binding of Human Breast Cancer Cell Lines 176

VI. Histochemical Receptor Determination in Human Breast Tumors: Comparison with the DCC Assay and Histopathology 179

VII. Editors' Comment 180

References 181

I. INTRODUCTION

Estrogen receptor (ER) determination as a means of predicting hormone dependency of human breast cancer is now widely used and generally accepted.[1] The rationale of this method originated in the observation that estrogen responsive reproductive tissue and tissue of certain tumors contain a protein which binds the radiolabeled hormone as a ligand with high affinity and low capacity.[2-4] Based on radiolabeled ligand binding, a variety of biochemical, analytical techniques were devised for ER determination.[5]

In addition, in recent years approaches towards histologically detecting ER were attempted. They comprise antihormone antibody, antireceptor antibody, and ligand-binding methods. Although these histochemical methods initially showed promise for a rapid, convenient, and simple ER assay, they have not yet been proven to be an alternative to standard biochemical techniques. On the contrary, with the exception of autoradiography and antireceptor antibody methods, the histochemical methods were criticized for claiming the localization of ER in tissue sections.[6-10]

II. CURRENT CONTROVERSY

Among the criticized methods, the most popular ones utilize fluorescent estradiol BSA conjugates (17β-estradiol 17-hemisuccinate-bovine serum albumin-fluorescein isothiocyanate = E_2-BSA-FITC, 17β-estradiol 6-*O*-carboxymethyloxime-bovine serum albumin-fluorescein isothiocyanate = E_2-6CMO-FITC) as ligands.[11,12] Objections to the claim of receptor localization originated in discrepancies between the results obtained using these methods, the biochemical assays, and in theoretical considerations.

ER, as measured by the biochemical assay, is readily extracted from homogenates or frozen tissue powder with buffer. Thus, it was criticized that whenever frozen tissue sections are incubated with buffers, which is the case with these methods, ER leaches out from the section into the buffer and is no longer available for the *in situ* interaction with the ligand.[9,13] Lee studied ER binding characteristics in frozen sections using the radiolabeled ligand-binding technique (^3H-E_2 as ligand) and found a binding behavior similar to that obtained with the biochemical assay of cytosolic ER.[14] Lee suggested that ER content measured in cytosol with the biochemical assay represents only a fraction, i.e., the soluble portion, of total receptor content in tissue.

This suggestion is supported by a recent article by Sierralta et al. who found that high speed supernatant contains only about one third of total ER content of tissues.[15] However, the remaining structure-associated ER (bound to nuclei and extranuclear membranes of microsomes) is not readily accessible, since with buffers — as used for cytosol preparation — only minor quantities of ER could be recovered. Satisfactory recoveries were only obtained when certain agents such as detergent in combination with DTT and trypsin were added to the buffers. Lee also suggested that these "cryptic" receptors, in order to be saturated with the ligand, require a higher ligand concentration and more prolonged incubation periods as compared to soluble cytoplasmic receptors.[16] These findings may refute the argument on "receptor-leaching". However, there still remains an issue to be stressed in this context. The ligand used by Sierralta et al. and Lee in their experiments was ^3H-E_2. Conclusions drawn from results of experiments with E_2 cannot be transferred to those with other ligands. In particular, fluorescent BSA conjugates which represent large, bulky ligands compared to E_2, might not reach structure-associated receptors. In the light of the findings of Sierralta, namely, that biochemical assays only measure one third of total receptor content, another frequently expressed criticism against histochemical receptor localization could be rebutted provided that the ligand can interact with all receptors. It was doubted that the number of binding sites in a receptor-positive cell (approximately 20,000 to 40,000) would be sufficient in order to surmount the sensitivity level of conventional fluorescence microscopy.[6,17] If there are three times more binding sites per cell, the chance of being above the sensitivity level would be considerably enhanced. Moreover, the structure-associated receptors are certainly not uniformly distributed throughout the entire cell.

Furthermore, dry- and thaw-mount autoradiography in vivo after injection of ^3H-E_2 and in vitro after tissue slice incubation under supravital conditions show preferential nuclear labeling (little in cytoplasm and none in nucleoli).[9] This is in agreement with the central dogma of steroid hormone action and with biochemical results.[18] In contrast, with fluorescent BSA conjugates preferential cytoplasmic and distinct nucleolar labeling of uterine epithelial cells is obtained, whereas nuclear labeling is only occasionally observed. Two explanations can be offered for this discrepancy. Either staining with conjugates produces an artifact, or the cytoplasmic receptors are not translocated to the nucleus. This would be a reasonable assumption for the particular in vitro conditions. However, autoradiography with ^3H-E_2 under the same conditions does not produce any labeling (except cytoplasm of eosinophils).

Receptor specificity of ligand binding of biochemical and histochemical methods is demonstrated by competitive inhibition of the ligand-receptor interaction by unlabeled ligands with an affinity for the receptor similar to that of E_2. With the biochemical method, competitive inhibition does not cause any problems, whereas with the histochemical techniques, controversial results have been reported.[19] Inhibition of fluorescent staining with E_2, DES, and other antiestrogens was criticized as being incomplete,[6,7,20] even with high concentration of competitor. Theoretically, if one compares the relatively low affinity of fluorescent conjugates with the high affinity of competitors such as E_2, DES, etc., effective inhibition of fluorescent staining should readily occur at lower concentrations than applied. Regarding tissue specificity of steroid hormone binding, it is now evident that more tissues exhibit specific estrogen binding than previously thought.[21-24] However, some of these tissues also served as negative controls in histochemical methods.[25] Occasionally, it was reported that specific staining was obtained after tissue fixation.[26] This is opposed by the findings of other investigators who experienced that most fixatives, although reducing diffusion into the medium, reduced the amount of available ER.[9,27]

Several investigators have compared histochemical and biochemical methods for assessing human breast cancer. The reported results range from "a lack of correlation" to a 91% agreement between the methods.[20,28-34] The latter agreement, obtained by some investigators, might indicate a specificity for certain target tissues. Whether or not this is based on receptor-conjugate interaction, however, remains questionable.

III. THE STABILITY OF CONJUGATES

The controversy in the field prompted us to carry out experiments in order to study the properties of conjugates with respect to their suitability for receptor detection in tissue sections. The main criteria which must be met by any such conjugate are high receptor-binding affinity, low affinity for nonreceptor-binding proteins, and a reasonable stability. Therefore, we studied their stability with and without cytosol and the in vitro interaction with ER I by competitive binding analysis and by incubation of cytosol with a sepharose-bound conjugate.

Contamination of fluorescent ligands by noncovalently bound hormone has been implicated but not thoroughly investigated.[6] Free steroid in conjugates may occur due to inadequate purification after synthesis, instability in aqueous solutions, and enzyme actions. Conjugates are freed of nonreacted ligand by dialysis and/or gel chromatography. Purity checks based on spectrophotometry cannot be considered as sensitive enough for estradiol-BSA conjugates because of the relatively low extinction coefficient of E_2 (ξ_{280} = 2000/1 mol^{-1} cm^{-1}), on

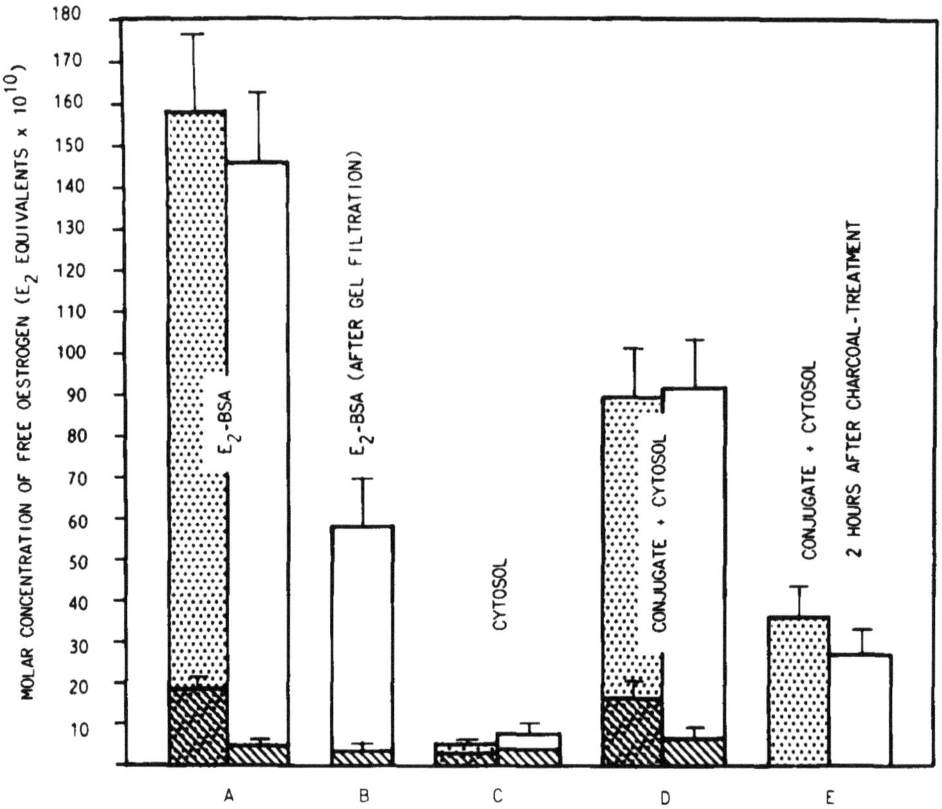

FIGURE 1. Measurement of free estrogen in conjugate solutions (A, B), cytosol (C), and conjugate-cytosol solutions (D, E) by means of RIA and RPBA. Stippled bars denote estrogen values obtained by RIA, open bars values obtained by RPBA, and hatched bars values after charcoal treatment. Freshly dissolved conjugate (E_2:BSA = 7.7:1; conjugate concentration in TE buffer = 0.2 mg/mℓ concentration of bound E_2 = 2.2×10^{-5} *M*) exhaustively dialysed previous to lyophilization (A). A conjugate, otherwise the same as specified under A, was additionally subjected to gel filtration (PD-10, Pharmacia) after dialysis (B). Calf uterine cytosol adjusted to 2 mg protein per milliliter with TE buffer (C). Conjugate-cytosol solution (E_2:BSA = 7.7:1; conjugate concentration = 0.12 mg/mℓ; concentration of bound E_2 = 1.3×10^{-5} *M*; concentration of cytosol = 0.8 mg protein per milliliter) obtained by mixing 3 mℓ A with 2 mℓ C (D). This conjugate-cytosol solution was charcoal treated, followed by incubation at room temperature for 2 hr (E). The bars represent means ± SEM of four separate experiments. (From Binder, M., *Histochem. J.*, 16, 1003, 1984. With the permission of the publisher, Chapman & Hall.)

the one hand, and the difference in affinity to the receptor between conjugate and free steroid, on the other. For the mixed anhydride method it has been reported that the synthesized steroid-BSA conjugates contained unbound or adsorbed steroid (2 to 4% of the weight of the conjugate).[35] Investigators who have used this, or other approaches for ligand synthesis, reported that their conjugate contained (after appropriate purification) either no uncoupled hormone or less than 1%.[12] In view of the minute amount sufficient for saturating ER binding sites and, furthermore, considering 1% of free hormone to be the detection level of the applied analytical methods, we have used receptor protein-binding assay (RPBA), RIA, and HPLC for the detection of free estrogen in aqueous conjugate solutions.[36,37]

With these methods (RIA, RPBA) we could detect considerable amounts of free estrogen in freshly prepared conjugates after intensive dialysis (Figure 1A). By subsequent gel filtration the estrogen level was lowered to approximately one third of that after dialysis (Figure 1B). Charcoal treatment of freshly dissolved conjugate resulted in a decrease of free estrogen to a value approximately equal to that of cytosol of immature calf uterus which is considered as estrogen free (Figure 1C). Again, charcoal completely removed estrogen. However, an increase in estrogen content could be observed when the charcoal-treated conjugate was left for 2 hr at room temperature (Figure 1E).

At first, we ascribed this increase solely to an enzymatic activity of cytosol. However, conjugates also released estrogen in solutions without cytosol (Figure 2). Storage of conjugate solutions for prolonged periods of time at room temperature resulted in an estrogen release (Figure 2A). The release rate was increased when conjugates were left in open vials (Figure 2B) or when cytosol was present (Figure 2C).

In an attempt to discriminate between the estrogen leakage by cleavage of covalent bonds and that of adsorptively bound estrogen, a mock conjugate was synthesized which was otherwise the same as that used for the experiments (E_2:BSA = 7.7:1) except that no isobutylchloroformate (coupling agent) was added to the reaction mixture. For this reason it can be assumed that there is actually no covalently bound estrogen in the mock conjugate. The similar quantity of free estrogen detected in freshly dissolved E_2-BSA and mock conjugate by successive ether extraction suggests that free estrogen initially present in E_2-BSA conjugates is the adsorptively bound ligand (estradiol-17β-hemisuccinate= E-HS) used for conjugate synthesis (Figure 3). After complete removal of adsorbed estrogen from the mock conjugate by repeated ether extraction it can be seen that — upon storage at ambient temperature — free estrogen remained only at the detection level. On the contrary, free estrogen in E_2-BSA solutions increased linearly upon storage at room temperature, thus indicating the cleavage of covalent bonds. Figure 3 further shows that ether extraction is more efficient compared to charcoal treatment. The above-made assumption, that free estrogen initially present in conjugate solution represents adsorptively bound ligand (E_2-HS, E_2-6CMO) from conjugate synthesis, was confirmed by HPLC (Figure 4). Apparently, after incubation with cytosol (presumably, due to enzymatic action) or after long-term storage at room temperature (probably formed by microbial contamination), E_2 could be detected in conjugate solutions. In the case of E_2-BSA conjugates this is readily explicable by cleavage of the ester bond at the 17 position, whereas the formation of E_2 from E_2-6CMO conjugates is not easily understood.

IV. IN VITRO INTERACTION OF CONJUGATES WITH ESTROGEN RECEPTOR TYPE I

The above-demonstrated imperfect stability gave rise to serious concerns which question whether an interaction between ER 1 and the conjugates will take place to any extent. For this purpose E_2-BSA, E_2-BSA-FITC, BSA, and BSA-FITC were covalently bound to CNBr-activated Sepharose-4B. Prior to use the gel was washed intensively either with 90% methanol or by extraction of the gel suspension with diethylether in order to remove any adsorbed free estrogen. The absence of estrogen was tested with RIA or RPBA. Immediately after purification the gel was incubated with calf uterine cytosol at 4°C for 2 hr. Receptor content of cytosol was determined with the DCC assay.

Incubation of calf uterine cytosol with sepharose-bound conjugates resulted in a significant decrease of cytosolic ER content (Figure 5). On the contrary, after treatment of cytosol with controls (sepharose, BSA-, and BSA-FITC-sepharose) ER content was only slightly diminished, probably due to nonspecific binding of ER to the control materials. This decrease in ER suggests an in vitro interaction of conjugate with cytosolic ER. A small percentage of this decrease (not exceeding 10% according to our release studies) is not due to conjugate-receptor interaction. Despite thorough purification of conjugates and incubation at low temperature (4°C), a small amount of estrogen is released from the conjugates during the incubation period, which masks some of the ER.

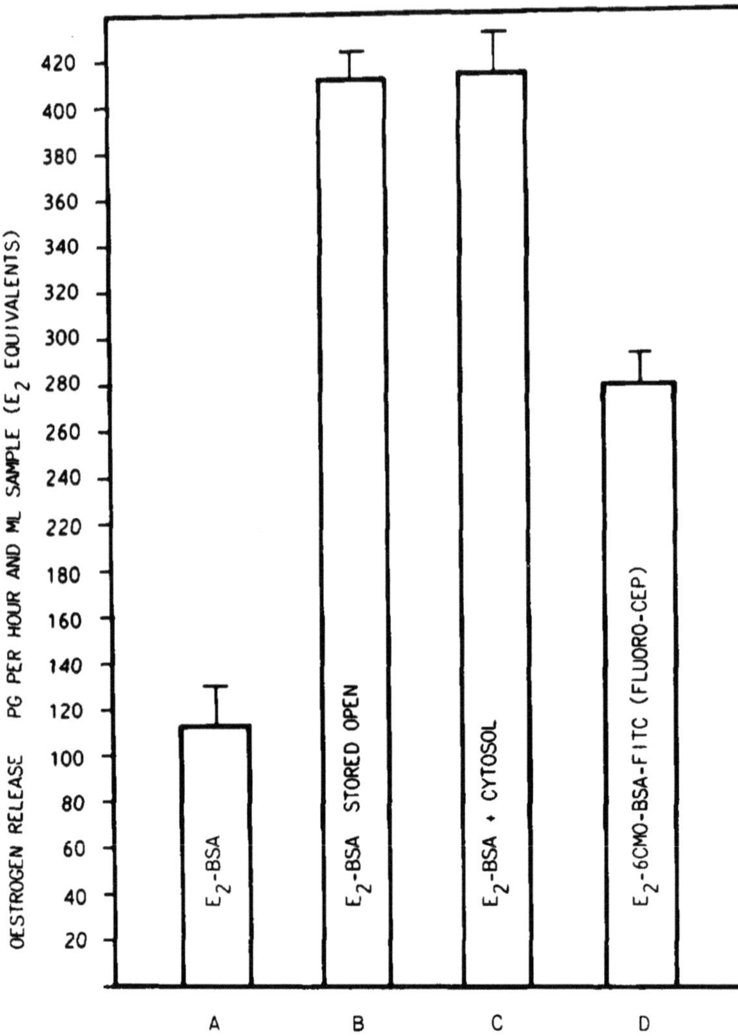

FIGURE 2. Amount of estrogen released per hour and milliliter from conjugates in solution with and without cytosol. E_2-BSA (E_2:BSA = 7.7:1; bound E_2 = 2.2×10^{-5} *M*) was dissolved in TE buffer (0.2 mg/mℓ), ether treated, and kept at ambient temperature (25°C) in screw-capped vials. At various times after dissolution up to 480 hr, 0.5-mℓ aliquots were withdrawn from the vials and processed for estrogen measurement by RPBA (A). The same conjugate was left at room temperature in open vials. Otherwise, as above (B). A conjugate-cytosol solution (specified in the legend of Figure 1D) was stored in screw-capped vials (25°C). Samples were taken and processed as described above (C). A commercial sample of a E_2-6CMO-FITC conjugate (Fluoro-Cep®, Zeus Scientific, Inc., lot CG526; storage recommendation, 25°C) was diluted with redistilled water (1:10) in order to adjust the concentration of bound E_2 to that of the E_2-BSA conjugate. Again, the sample was left stoppered at room temperature. At various times up to 210 hr, aliquots were removed and assayed for estrogen by RPBA (D). The values are means ± SEM of three to four tests. (From Binder, M., *Histochem. J.*, 16, 1003, 1984. With permission of the publisher, Chapman & Hall.)

The data from Figure 5 were used for the construction of Scatchard plots in order to determine the binding parameters of residual receptors in the cytosol after incubation with controls or sepharose-bound E_2-BSA and E_2-BSA-FITC, respectively (Figure 6). Compared with untreated or control-treated cytosols, conjugate-treated cytosols exhibited an increase

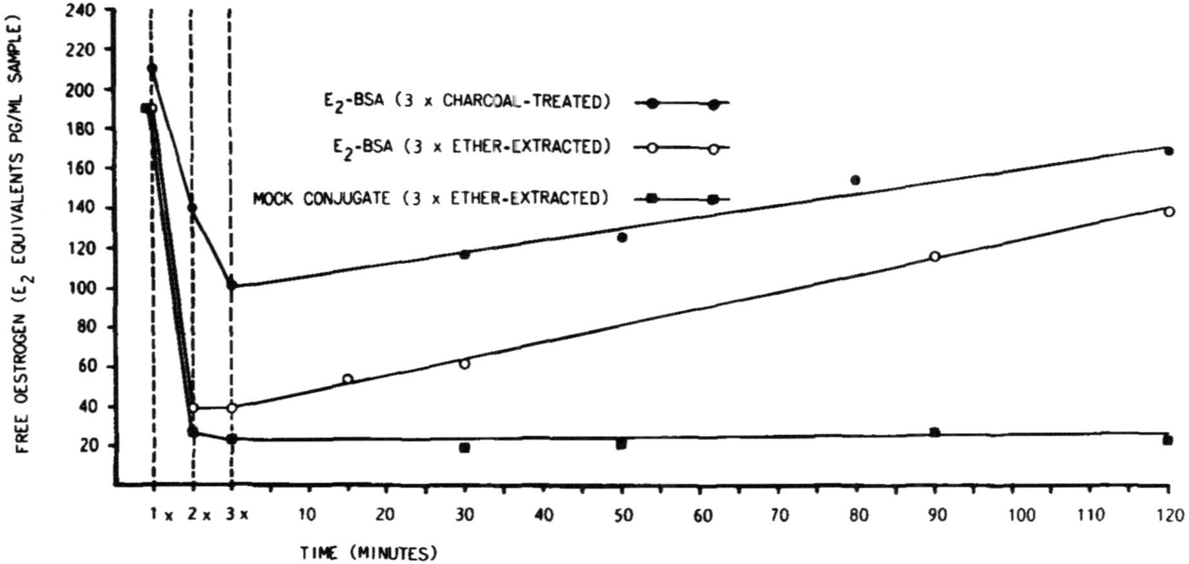

FIGURE 3. Time course of estrogen release from prepurified E_2-BSA in comparison with that from prepurified mock conjugate. E_2-BSA (E_2:BSA = 7.7:1; concentration of bound E_2 = 2.2 × 10^{-5} *M* and mock conjugate were freshly dissolved in TE buffer (0.2 mg/mℓ). Free estrogen was immediately measured by RPBA (E_2-BSA = 2900 pg/mℓ, mock conjugate = 4590 pg/mℓ). In a subsequent step, the same aliquots of E_2-BSA and mock conjugate were successively treated three times with charcoal instead of ether. Estrogen was, likewise, concomitantly measured. The treated samples were then stored in capped vials at room temperature. Estrogen was assayed by RPBA at the times indicated. (From Binder, M., *Histochem. J.*, 16, 1003, 1984. With permission of the publisher, Chapman & Hall.)

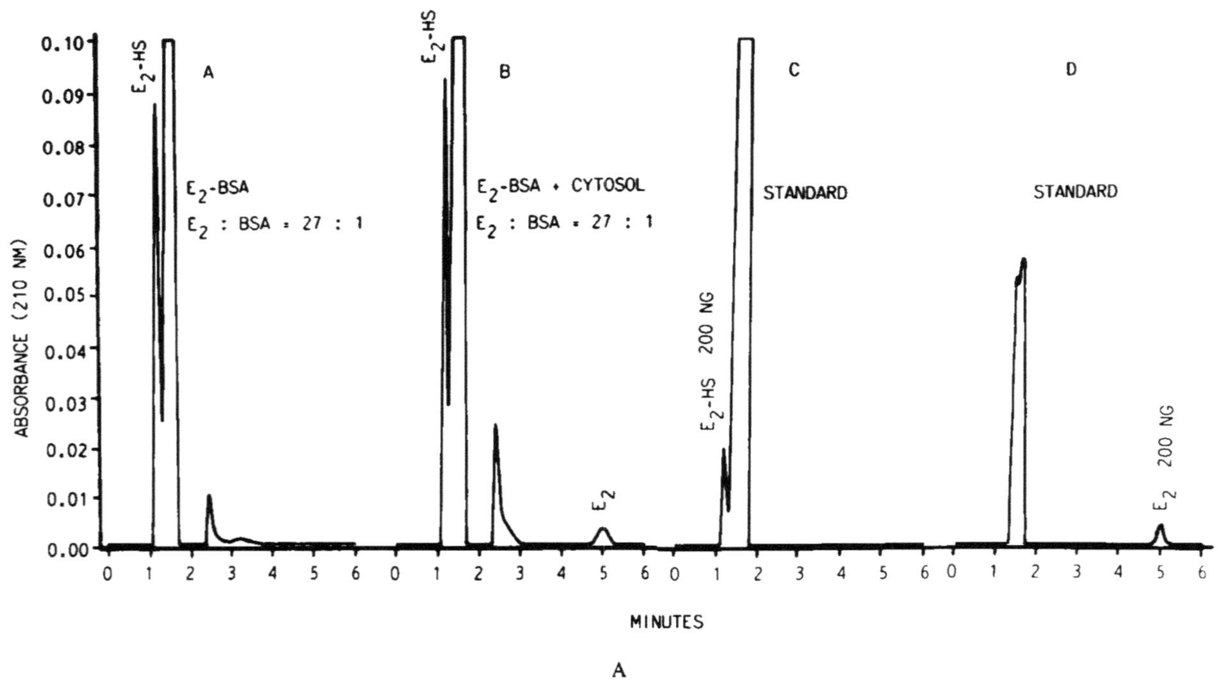

FIGURE 4. HPLC of ether extracts of conjugates. (a) 10 mℓ freshly dissolved E_2-BSA (E2:BSA = 27:1); conjugate (concentration = 2 mg/mℓ; concentration of bound E_2 = 7.4×10^{-4} *M*) was extracted with diethylether (5 vol of ether per sample volume, three times). The ether was evaporated to dryness and the residue redissolved in 1 mℓ ethanol. 10 $\mu\ell$ of the ethanolic solution were injected (A) . To the extracted conjugate solution (10 mℓ), 0.5 mℓ or calf uterine cytosol (2 mg protein per milliliter was added. Subsequently, this solution was incubated at 35°C for 2 days. Thereafter, the conjugate was reextracted with ether, which was evaporated. The residue was dissolved in 0.5 mℓ of ethanol of which 10 $\mu\ell$ were injected (B). Standards: E_2-HS (C), E_2 (D). (b) 0.65 mℓ of the same commerical sample (Fluoro-Cep®), which was used for the experiment described in Figure 2, was extracted with ether as above. After evaporation of ether and redissolution of the residue in 30 $\mu\ell$ ethanol, 20 $\mu\ell$ were injected (A). After ether extraction of 10 mℓ of a E_2-6CMO-BSA conjugate solution (E_2-6CMO:BSA = 11:1; conjugate concentration = 2 mg/mℓ; concentration of bound E_2 = 3×10^{-4} *M*, ether evaporation and redissolution of the residue in 100 $\mu\ell$ ethanol, 5 $\mu\ell$ were injected (B). Standards: E_2-6CMO (C), E_2 (D). The peak which eluted at 1.5 min represents ethanol (solvent peak). (From Binder, M., *Histochem. J.*, 16, 1003, 1984. With permission of the publisher, Chapman & Hall.)

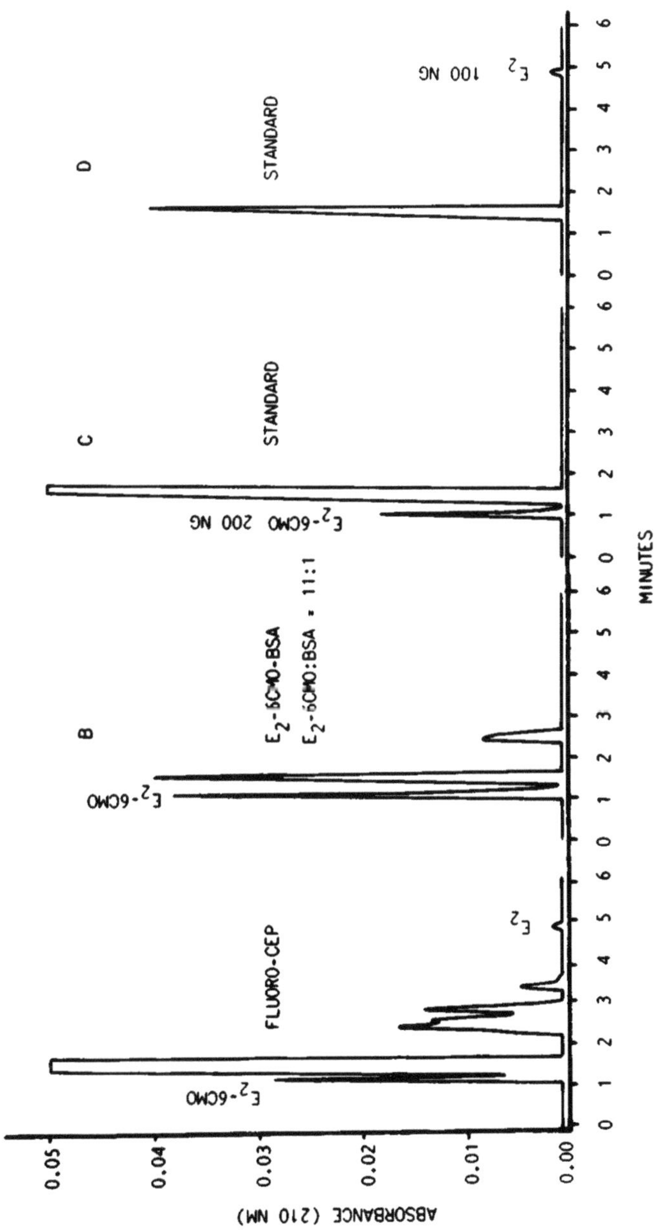
ABSORBANCE (210 NM)
0.05
0.04
0.03
0.02
0.01
0.00
MINUTES
0 1 2 3 4 5 6
FLUORO-CEP
E2-6CMO
E2
B
E2-6CMO
E2-6CMO-BSA
E2-6CMO:BSA = 11:1
C
STANDARD
E2-6CMO 200 NG
D
STANDARD
E2 100 NG

FIGURE 4B

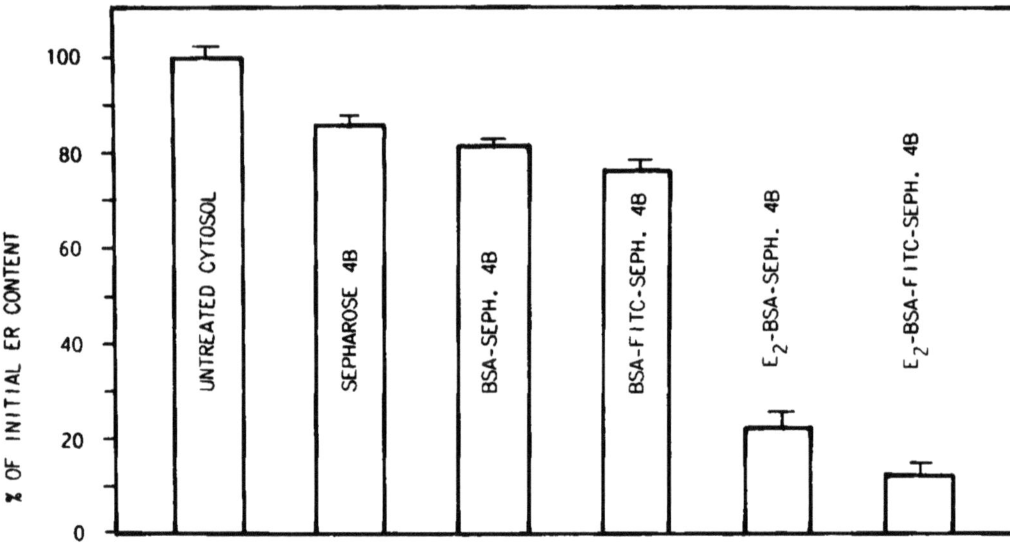

FIGURE 5. Interaction of ER (calf uterine cytosol) with sepharose-bound conjugate. Cytosol was incubated with sepharose, BSA-, BSA-FITC-, and E_2-BSA-FITC-sepharose. Receptor content of cytosol was measured with the DCC assay before and after incubation with the immobilized ligands or controls. The molar ratio of constituents of bound conjugates was E_2:BSA = 7.7:1 or E_2:BSA:FITC = 7.7:1:4.5. The concentration of sepharose-bound E_2 was approximately 4.2×10^{-5} *M* in the final incubation volume for the E_2-BSA and the E_2-BSA-FITC conjugate. Bars represent means ± SEM of four experiments. From Binder, M., *Histochem. J.*, 16, 1003, 1984. With permission.

in K_d value of approximately two orders of magnitude. It is noteworthy that this raised K_d value is approximately that of the so-called type II ER.[38]

The affinities for ER of a series of E_2-BSA and E_2-6CMO conjugates with and without FITC label have been determined by competitive binding analysis (Table 1). It can be seen that the affinities of both series of conjugates — besides being low — seem to be somewhat inflated by the presence of free estrogen. This is demonstrated by the difference in K_d values of charcoal-treated and noncharcoal-treated samples. Charcoal treatment decreases affinity by about one order of magnitude. Rao et al., using a single point competitive binding assay, found no effect of treatment on affinity (% inhibition of ^3H-E_2 binding).[39] The introduction of the FITC label further reduces affinity by roughly the same order. According to these K_d values, the criticized high ligand concentration used for incubation of tissue sections in histochemical methods seems to be justified. In constrast to Rao et al., we found that neither stoichiometry nor the position through which E_2 is linked to BSA has a significant influence on the affinity of both types of conjugates for ER.

V. CONJUGATE BINDING OF HUMAN BREAST CANCER CELL LINES

Human breast cancer cell lines as model systems are well suited for testing and validation of fluorescent ligands devised for histochemical receptor localization. They are appropriate for this purpose because there is a choice between receptor-positive and receptor-negative lines. The latter can serve as negative controls in receptor histochemistry. Furthermore, when kept under standardized conditions, receptor levels of ER-positive cell lines remain fairly constant and can be monitored with the DCC assay.

For our experiments we used MCF-7 cells, an ER-positive human mammary carcinoma cell line, which was originally provided by the American Type Culture Collection. These cells were routinely grown in RPMI 1640 medium supplemented with 10% fetal bovine

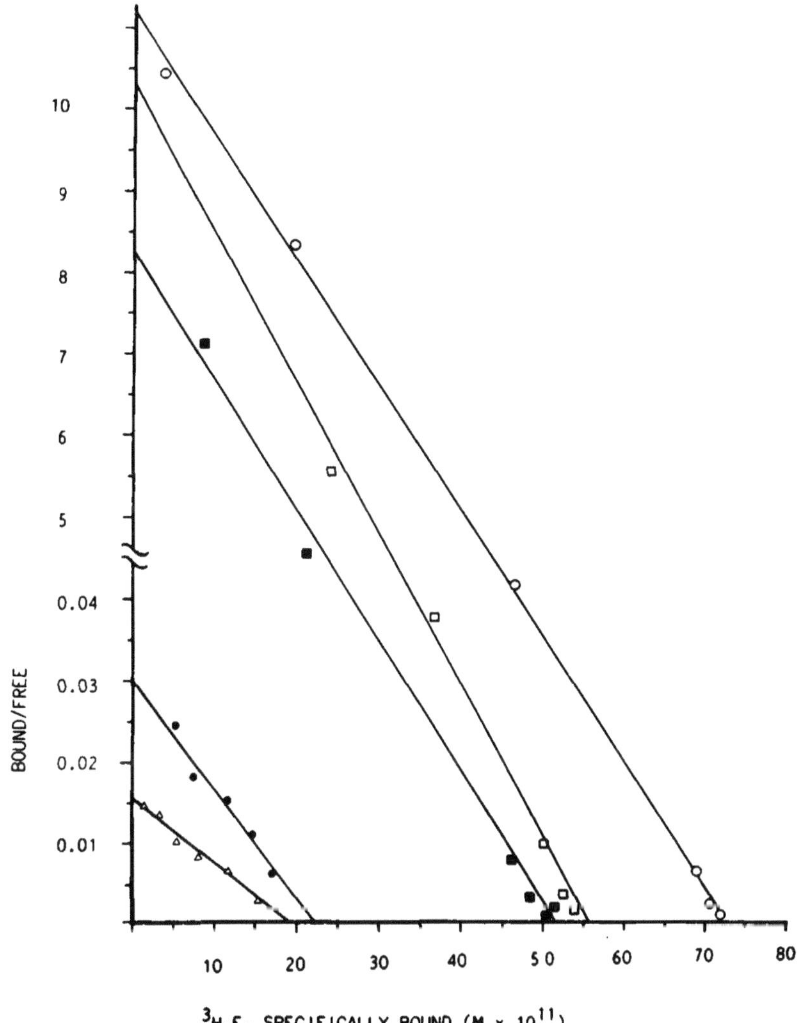

FIGURE 6. Decrease in K_d values of ER for ^3H-E_2 in cytosol after treatment with sepharose-bound conjugates. Scatchard plots from a representative experiment (described under legend of Figure 5). Untreated cytosol (-○-); $K_d = 6.4 \times 10^{-11}$ *M*, 450 fmol/mg protein. Cytosol treated with: BSA-sepharose (-□-); $K_d = 5.4 \times 10^{-11}$ *M*, 390 fmol/mg protein. BSA-FITC-sepharose (-■-); $K_d = 6.3 \times 10^{-11}$ *M*, 350 fmol/mg protein. E_2-BSA-sepharose (-●-); $K_d = 3.7 \times 10^{-9}$ *M*, 80 fmol/mg protein. E_2-BSA-FITC-sepharose (-△-); $K_d = 6.4 \times 10^{-9}$ *M*, 68 fmol/mg protein. From Binder, M., *Histochem. J.*, 16, 1003, 1984. With permission.

serum. For experimental studies this medium was replaced by medium containing charcoal-stripped serum 72 hr prior to experiments. For incubation with E_2 the culture medium was removed and replaced by a phosphate buffer containing CaCl, MgCl, and glucose (pH 7.3). E_2 was added in 1000 × concentrated solutions in absolute ethanol. In addition, two other receptor-negative human mammary carcinoma cell lines derived from the pleural effusions of patients with metastatic breast cancer were included for control experiments (MR-GA and LJ-LH). These cells were maintained as described above for the MCF-7 cells. The receptor content of each cell line was determined weekly with the DCC assay. ER content of MCF-7 cells proved to be 47 fmol/mg protein (x = 47 + 11 SD, n = 10) with a K_d of 5.2×10^{-10} *M*.

Table 1
EFFECT OF STOICHIOMETRY AND CHARCOAL PRETREATMENT OF E_2-BSA AND E_2-6CMO-BSA CONJUGATES ON AFFINITY FOR ER

A. E_2-BSA Conjugates

E_2-BSA			E_2-BSA-FITC		
	K_d value (*M*)			K_d value (*M*)	
	Charcoal treatment			Charcoal treatment	
Molar ratio (E_2:BSA)	Without	With	Molar ratio (E_2:BSA:FITC)	Without	With
3:1	1.9×10^{-7}	1.9×10^{-6}	3:1:4.7	1.1×10^{-5}	—
7:1	2.2×10^{-7}	2.2×10^{-6}	7:1:4.4	1.2×10^{-6}	9.6×10^{-6}
10:1	2.3×10^{-7}	2.4×10^{-6}	10:1:3.9	1.3×10^{-6}	8.4×10^{-6}
13:1	2.0×10^{-7}	3.9×10^{-6}	13:1:6.9	1.4×10^{-6}	7.9×10^{-6}
17:1	2.5×10^{-7}	6.9×10^{-6}	17:1:5.7	1.5×10^{-6}	9.2×10^{-6}
23:1	2.5×10^{-7}	5.0×10^{-6}	23:1:3.4	1.8×10^{-6}	1.1×10^{-5}

B. E_2-6CMO-BSA Conjugates

E_2-6CMO-BSA			E_2-6CMO-BSA-FITC		
	K_d value (*M*)			K_d value (*M*)	
Molar ratio (E_2-6CMO:BSA)	Charcoal treatment		Molar ratio (E_2-6CMO:BSA:FITC)	Charcoal treatment	
	Without	With		Without	With
4:1	1.0×10^{-8}	8.1×10^{-7}	4:1:5.2	4.6×10^{-7}	8.9×10^{-6}
5:1	2.2×10^{-8}	6.7×10^{-7}	5:1:5.1	5.9×10^{-7}	8.9×10^{-6}
7:1	2.5×10^{-8}	9.0×10^{-7}	7:1:6.9	7.3×10^{-7}	1.1×10^{-6}
10:1	2.8×10^{-8}	7.5×10^{-7}	10:1:7.4	9.7×10^{-7}	6.2×10^{-6}
14:1	2.5×10^{-8}	8.3×10^{-7}	14:1:8.3	1.0×10^{-6}	8.4×10^{-6}
19:1	2.3×10^{-8}	6.1×10^{-7}	19:1:7.7	8.1×10^{-7}	2.6×10^{-6}

Note: K_d values were determined by competition analysis using the DCC method and calf uterine cytosol. Each value represents the mean of at least four determinations. Intra- and interassay coefficient of variation were 19 and 27%, respectively. The K_d for the interaction of ER with E_2 itself is 2.9×10^{-10} *M*, for comparison (this value represents the mean of 30 calf uteri).

From Binder, M., *Histochem. J.*, 16, 1003, 1984. With permission of the publisher, Chapman & Hall.

Histochemistry was performed using frozen sections of cell pellets prepared from the cultured cells. The cells were harvested with EDTA and washed. Finally, the pellets were resuspended in 0.9% NaCl solution containing 2% ovalbumin. The suspended cells were frozen in liquid nitrogen and stored there until used. For the histochemical assays the staining protocols of Lee for E_2-6CMO-BSA and that of Pertschuk for E_2-BSA conjugates were followed with the exceptions outlined below.[11,12]

The staining intensity was found to be dependent on the duration of incubation, the conjugate concentration in the final working solution, and the molar ratio of constituents of the conjugates for the cell lines under investigation. With increasing incubation time staining intensity increases concomitantly up to 1 hr. Then the staining intensity remains unchanged, and no further increase can be observed. Staining intensity also increases with conjugate concentration and, thus, does not seem to be a saturable phenomenon. According to our experiments a concentration of bound E_2 in the working solution of 10^{-7} *M* and a molar ratio of FITC:BSA = 2 was found to represent the detection limit. In the present study a conjugate (E_2-BSA-FITC) with a molar ratio of constituents of E_2:BSA:FITC = 7 to 14:1:5 to 6 was used (concentration of bound E_2 in the working solution = 10^{-7} to 10^{-6} *M*. In addition, a E_2-6CMO-BSA-FITC conjugate (Fluoro-Cep® = Lee's conjugate) was also employed. For negative controls BSA-FITC was frequently used. Its usefulness as negative control depends on both the concentration of BSA-FITC and the ratio of FITC:BSA as is the case with the conjugates. Therefore, one has to consider that the BSA-FITC solution and the conjugate solution should have the same BSA concentration and the same number of FITC molecules per albumin molecule. Disregarding this can readily produce controls which are highly positive (with BSA-FITC with a higher FITC:BSA ratio than the conjugate). With MR-GA and LJ-LH cells as controls the same conjugate solution was used as for MCF-7 cells. With this design we experienced that although MCF-7 cells stained more brightly, the control cells were never devoid of staining. The same occurs with inhibition. Disregarding the fact that upon in vivo preincubation with E_2 or DES a nuclear translocation should have taken place with MCF-7 cells, inhibition was rather meager but, nevertheless, perceptible. Fixation of MCF-7 cells with acetone prior to incubation did neither abolish nor considerably reduce staining. Having in mind the imperfect stability of conjugates, we thought that staining might eventually benefit from shorter incubation times at 4°C. However, this modification did not induce a change in appearance.

VI. HISTOCHEMICAL RECEPTOR DETERMINATION IN HUMAN BREAST TUMORS: COMPARISON WITH THE DCC ASSAY AND HISTOPATHOLOGY

ER and PR levels have been measured in 104 human breast tumors by a biochemical and by a histochemical method. For the histochemical method steroid-BSA conjugates were used as ligands (for ER E_2-BSA-FITC and E_2-6CMO-BSA-FITC and for PR 11-Pg-BSA-FITC or 11 Pg BSA TMRITC). Binding specificity was assessed with CI-628 (ER) and R 5020 (PR) as competitors. Nonspecific binding was assessed with BSA-FITC or BSA-TRMITC. The test procedures that followed were essentially that of Lee (for E_2-6CMO-BSA-FITC) and that of Pertschuk (for E_2-BSA-FITC). The specimens were classified as negative, + − = borderline, +, + + according to percentage of brightly fluorescing cells (10 to 30% = borderline). For the biochemical assay a DCC assay was used.

The tumor samples examined by us for the presence of ER and PR consisted of 67 primary breast carcinomas (1 carcinoma-sarcoma included), 10 metastases, and 5 recurrences. Breast specimens from 22 patients with benign breast disease were also examined. These were made up of 16 mastopathies, 2 fibroadenomas, and 1 case of papilloma, mastitis, gynecomastia, and 1 axillary lymph node each. Of the 67 postmenopausal women with primary breast carcinomas, more than half had tumors of the common invasive duct cell variety with the subtypes of solid to scirrhous (not otherwise specified type) (Table 2). The majority of the benign cases were mastopathies with a borderline receptor content as measured by both methods. Although over 50% of the malignant cases were ER and PR positive by both techniques (Table 3), only three received antiestrogen therapy. All others were given chemotherapy (Cooper-Schema), probably because of their advanced stages. Of the three patients with endocrine therapy, all are alive, but one developed a recurrence.

The overall agreement between the histochemical and the biochemical test was 87.5% for ER and 75% for PR.

Table 2
HISTOPATHOLOGICAL TYPES

Post-Histopathological (UICC) Classification (Stages)

ILC	CLIS/ILC	ILC / IDC	IDC	ICD / DCIS
I: 1		I: 2	I: 9	
II: 5	II: 1	II: 5	II: 18	II: 1
		IIIa: 1	IIIa: 8	
IIIb: 1	IIIb: 1		IIIb: 12	IIIb: 1
Cases: 7	2	8	47	2

Carcinoma-sarcoma	Recurrences	Metastases
I: 1		
Cases: 1	5	8

Note: ILC, invasive lobular carcinoma; CLIS, carcinoma lobular *in situ;* DCIS, ductal carcinoma *in situ;* IDC, invasive ductal carcinoma; NOS, non-otherwise specified; pTNM, post-surgical histopathological classification; T, primary tumor; N, regional lymph node; M, distant metastases; and I—IIIa,b, stages.

Table 3
82 MALIGNANT TUMORS (fmol/mg PROTEIN CYTOSOL)

Biochemical results			Histochemical results		
	ER	PgR		ER	PgR
3	27	31	Neg.	26	28
3—10	6	5	±	27	25
10—30	9	12	+	20	16
30—100	24	19	+ +	9	13
—100	6	15			

Total Comparison of the 104 Malignant and Benign Cases in the Histochemical and Biochemical Methods

	ER	PgR
Good Correlation	91/87.5%	78/75%
Poor Correlation	13/12.5%	26/25%

VII. EDITORS' COMMENT

In spite of the generally negative tone set in this chapter, it is of interest that these workers obtained very high correlations between their histochemical and biochemical ER and PgR breast cancer assays.

REFERENCES

1. **DeSombre, E. R., Carbone, P. P., McGuire, W. L., Wells, S. A., Wittliff, J. L., and Lipsett, M. B.,** Special report. Steroid receptors in breast cancer, *N. Engl. J. Med.*, 301, 1011, 1979.
2. **Glascock, R. F. and Hoekstra, W. G.,** Selective accumulation of tritium-labelled hexestrol by the reproductive organs of immature female goats and sheep, *Biochem. J.*, 72, 673, 1959.
3. **Jensen, E. V. and Jacobson, H. I.,** Fate of steroid estrogens in target tissues, in *Biological Activities of Steroids in Relation to Cancer*, Pincus, G. and Vollmer, E. P., Eds., Academic Press, New York, 1960, 161.
4. **Folca, P. J., Glascock, R. F., and Irvine, W. T.,** Studies with tritium labelled hexoestrol in advanced breast cancer, *Lancet*, 2, 796, 1961.
5. **Seibert, K. and Lippman, M.,** Hormone receptors in breast cancer, *Clin. Oncol.*, 1, 735, 1982.
6. **Chamness, G. C., Mercer, W. D., and McGuire, W. L.,** Are histochemical methods for estrogen receptor valid?, *J. Histochem. Cytochem.*, 28, 792, 1980.
7. **McCarty, K. S., Reintgen, D. S., Seigler, H. F., and McCarty, K. S., Sr.,** Cytochemistry of sex steroid receptors: a critique, *Breast Cancer Res. Treat.*, 1, 315, 1982.
8. **Stumpf, W. E. and Sar, M.,** Histochemical approaches for the localization of steroid hormone receptors, *Acta Histochem. Cytochem.*, 15, 560, 1982.
9. **Underwood, J. C. E., Sher, E., Reed, M., Eisman, J. A., and Martin, T. J.,** Biochemical assessment of histochemical methods for oestrogen receptor localisation, *J. Clin. Pathol.*, 35, 401, 1982.
10. **Binder, M., Boehm, M., and Czerwenka, K.,** Is estrogen receptor — conjugate interaction relevant for the histochemical detection of intracellular estrogen-binding?, in *Proc. 11th Int. Congr. Clin. Chem.*, Kaiser, E., Gabl, F., Mueller, M. M., and Bayer, P. M., Eds., Walter de Gruyter, Berlin, 1982, 519.
11. **Lee, S. H.,** Cytochemical study of estrogen receptor in human mammary cancer, *Am. J. Clin. Pathol.*, 70, 197, 1978.
12. **Gaetjens, E. and Pertschuk, L. P.,** Synthesis of fluorescein labelled steroid hormone-albumin conjugates for the fluorescent histochemical detection of hormone receptors, *J. Steroid Biochem.*, 13, 1001, 1980.
13. **Hawkins, R. A. and Penney, G. C.,** Histochemical detection of oestrogen receptors: the Edinburgh experience, in *Proc. 11th Int. Congr. Clin. Chem.*, Kaiser, E., Gabl, F., Mueller, M. M., and Bayer, P. M., Eds., Walter de Gruyter, Berlin, 1982, 513.
14. **Lee, S. H.,** The histochemistry of estrogen receptors, *Histochemistry*, 71, 491, 1981.
15. **Sierralta, W. D. and Szendro, P. I.,** Origin and quantification of cytoplasmic estradiol receptor in resting target cells, *Hoppe-Seyler's Z. Physiol. Chem.*, 364, 1497, 1983.
16. **Lee, S. H.,** Uterine epithelial and eosinophil estrogen receptors in rats during the estrous cycle, *Histochemistry*, 74, 443, 1982.
17. **Daxenbichler, G., Weiss, P., and Piegger, E.,** Critical evaluation of histochemical "receptor" assays, in *Proc. 11th Int. Congr. Clin. Chem.*, Kaiser, E., Gabl, F., Mueller, M. M., and Bayer, P. M., Eds., Walter de Gruyter, Berlin, 1982, 473.
18. **Jensen, E. V., Suzuki, T., Kawashima, T., Stumpf, W. E., Jungblut, P. W., and DeSombre, E. R.,** A two-step mechanism for the interaction of estradiol with rat uterus, *Biochemistry*, 59, 632, 1968.
19. **Pertschuk, L. P., Tobin, E. H., Carter, A. C., Eisenberg, K. B.,** Immunohistologic and histochemical methods for detection of steroid binding in breast cancer: a reappraisal, *Breast Cancer Res. Treat.*, 1, 297, 1982.
20. **McCarty, K. S., Woodward, B. H., Nichols, D. E., Wilkinson, W., and McCarty, K. S., Sr.,** Comparison of biochemical and histochemical techniques for estrogen receptor analyses in mammary carcinoma, *Cancer*, 46, 2842, 1980.
21. **Eriksson, H. A.,** Estrogen-binding sites of mammalian liver: endocrine regulation of estrogen receptor synthesis in the regenerating rat liver, *J. Steroid Biochem.*, 17, 471, 1982.
22. **West, N. B., Anderson, D. J., Stunz, L. L., and Hoch, E. J.,** A specific estrogen receptor in the mouse spleen. Characterization and evidence of physiological regulation, *J. Steroid Biochem.*, 16, 557, 1982.
23. **Dahlberg, E.,** Characterization of the cytosolic estrogen receptor in rat skeletal muscle, *Biochim. Biophys. Acta*, 717, 65, 1982.
24. **Rosen, T. S., Maciorowski, Z., Wittlin, F., Epstein, A. L., Gordon, L. I., Kies, M. S., Kucuk, O., Kwaan, H. C., Vriesendrop, H., Winter, J. N., Fors, E., and Molteni, A.,** Estrogen receptor analysis in chronic lymphocytic leukemia, *Blood*, 62, 996, 1983.
25. **Walker, R. A.,** The use of peroxidase-labelled hormones in the study of steroid binding in breast carcinomas, in *Proc. 11th Int. Congr. Clin. Chem.*, Kaiser, E., Gabl, F., Mueller, M. M., and Bayer, P. M., Eds., Walter de Gruyter, Berlin, 1982, 507.
26. **VanMarle, J., Lindeman, J., Ariens, A. T., Labruyere, W., and van Weeren-Kramer, J.,** Estrogen receptors in human breast cancer. I. Specificity of the histochemical localization of estrogen receptors using an estrogen-albumin FITC complex, *Virchows Arch. (Cell Pathol.)*, 40, 17, 1982.

27. **Penney, G. C. and Hawkins, R. A.**, Histochemical assay of oestrogen receptor, *Lancet*, 1, 930, 1980.
28. **Lee, S. H.**, Cancer cell estrogen receptor of human mammary carcinoma, *Cancer*, 44, 1, 1979.
29. **Berger, G., Frapport, L., Berger, N., Bremond, A., Ferold, J., and Rochet, Y.**, Localisation cytoplasmique des recepteur steroidiens des carcinomes mammaires par histofluorescence, *Arch. Anat. Cytol. Pathol.*, 28, 341, 1980.
30. **Tominaga, T., Kitamura, M., Saito, T., Itoh, I., and Takikawa, H.**, Comparative histochemical and biochemical assays of estrogen receptors in breast cancer patients, *Gann*, 72, 60, 1981.
31. **Boehm, M., Binder, M., and Czerwenka, K.**, Estradiol-albumin conjugates and coumestrol as ligands of binding proteins for the histochemistry of estrogen receptor (ER) in breast cancer and calf uterus, in *Proc. Symp. on the Analysis of Steroids*, Eger, Goeroeg, S., Ed., Publishing House of the Hungarian Academy of Sciences, Budapest, 1981, 121.
32. **Hanna, W., Ryder, D. E., and Mobbs, B. G.**, Cellular localization of estrogen binding sites in human breast cancer, *Am. J. Clin. Pathol.*, 77, 391, 1982.
33. **Raam, S., Nemeth, E., Tamura, H., O'Briain, D. S., and Cohen, J. L.**, Immunohistochemical localization of estrogen receptors in human mammary carcinoma using antibodies to the receptor protein, *Eur. J. Cancer Clin. Oncol.*, 18, 1, 1982.
34. **Pertschuk, L. P., Eisenberg, K. P., Leo, V. C., Rainford, E. A., Carter, A. C., and Macchia, R. J.**, Immunofluorescence detection of estrogen receptors with monoclonal antibodies. Clinical correlations of steroid binding by histochemistry in breast and prostate carcinoma, in *Proc. 11th Int. Congr. Clin. Chem.*, Kaiser, E., Gabl, F., Mueller, M. M., and Bayer, P. M., Eds., Walter de Gruyter, Berlin, 1982, 493.
35. **Erlanger, B. F., Borek, F., Beiser, S. M., and Lieberman, S.**, Steroid-protein conjugates. I. Preparation and characterization of conjugates of bovine serum albumin with testosterone and with cortisone, *J. Biol. Chem.*, 228, 713, 1959.
36. **Binder, M.**, Oestradiol-BSA conjugates for the receptor histochemistry: problems of stability and interactions with cytosol, *Histochem. J.*, 16, 1003, 1984.
37. **Binder, M., Boehm, M., and Halbritter, H.**, Type I estrogen receptors cannot be histochemically localized in frozen tissue sections using estradiol-albumin-FITC conjugates, *Acta Histochem. Suppl.*, 29, 71, 1984.
38. **Clark, J. H., Eriksson, H. A., and Hardin, J. W.**, Uterine receptor-estradiol complexes and their interaction with nuclear binding sites, *J. Steroid Biochem.*, 7, 1039, 1976.
39. **Rao, B. R., Patrick, T. B., and Sweet, F.**, Steroid-albumin conjugate interaction with steroid-binding proteins, *Endocrinology*, 106, 356, 1980.

INDEX

A

Acidified alcohol, 127—131
Actinomycin D, 20—21
Adrenals, antibody studies of, 135—136
Affinity, see also Binding capacity; Saturability; Specificity, 12
of estradiol, 80
with radiohistochemical techniques, 16—18
of estrogen receptors in cryostat frozen sections, 63—64
measuring, 11, 17
radioautographic techniques and, 12, 15—18
saturability, see also Saturability and, 12
over spectrum, 12
Affinity cytochemistry, see Cytochemical techniques
Affinity/potency relationships, 19—20
Agarose beads, 114—115
Alcohol, acidified, 127—131
Alkylating agents (CY), 48—50
Androgen receptors
in prostate diseases, 120, 121
as predictor of clinical response to endocrine therapies, 146—147
Antibodies, monoclonal, 113
Antibody studies, see also Immune response; Immunohistochemical techniques
advantages and disadvantages of, 126
biochemical techniques vs., 139
of progesterone receptors
advantages and disadvantages of, 137, 139—140
artifactual translocation and, 131, 134
in chicken adrenals, 136
in chicken brains, 136—137
in chicken Bursa of Fabricius, 134
in chicken ovaries, 135—136
in chicken oviducts, 126—132
in chicken pituitary glands, 136—137
effects of fixative and, 127—131
frozen section techniques, 131—132
intranuclear translocation and, 133—134
limitations on, 137
nuclear reactions and, 135—136, 139
paraffin section techniques, 127—131
saturation criteria and, 131, 134
specificity criteria and, 131
variation of cell types and, 134
quantification problems with, 144
techniques of, see also specific techniques, 168
as tool for localization, 3, 113
usefulness of, 126, 135
Autoradiography, 168
problems with, 168—169
of progesterone receptors in chicken oviducts, 131

B

Benign hyperplasia (BPH), see Prostate diseases
17 Beta ligands, see Fluorescent conjugate techniques
BF4, see Antibody studies
Binding, see Receptors; specific receptors
Binding capacity, see Competition experiments; Saturability
Biochemical techniques, 2—3, 168
advantages and disadvantages of, 6, 8, 40, 60—62, 133—134
antibody studies vs., 139
in conjunction with immunocytochemical techniques, 162—164
definition of, 57
histochemical techniques vs., 45, 47—51, 60—61, 80, 145—147
morphological techniques vs., 3
problems with, 52—56
in human tumors, 96
specificity criteria and, 169
Biological response, see Response, biological
Blood eosinopenia, 22, 23
Bovine serum albumin (BSA) ligands, see also Fluorescent BSA conjugate techniques, 62
BPH, see Prostate diseases
Brain, antibody studies of, 136—137
Breast cancer, see Mammary tumors
B subunit, see Progesterone receptors, subunits of
BSA, see Bovine serum albumin ligands
BSA conjugates, see Fluorescent conjugate techniques
Bursa of Fabricius, antibody studies of progesterone receptors in, 134

C

Capacity, see Competition experiments; Saturability
Clinical response, see specific therapies and treatments
Colchicine, effects of on estrogen binding, 21—22
Competition experiments, see also Saturability; Specificity, 9
in antibody studies, 131, 134
in fluorescent BSA conjugate studies, 44—46, 102, 114, 115, 156
in histochemical studies, 77, 79
for hormone specificity, 64, 114
in radioautographic studies, 12
Competitive binding, see Competition experiments
CY, see Alkylating agents
Cyclic AMP, estrogen binding and, 7, 23
Cytochemical techniques, see also specific techniques, 112
definition of, 57
evolution of, 112—113

in mammary tumors, 115—117
advantages of, 121
usefulness of, 118—120
problems with, 54—56
in prostate diseases, 120
validating, 57
Cytosol-nuclear receptors, 6
for estrogen, 7, 13—14
factors affecting, 18, 22
genomic response to estrogen and, 19
radioautographic localization of, see also Radioautography, 9, 10, 13, 14
radiohistochemical studies of, see also Radiohistochemical techniques, 9, 11, 16—18

D

DCC assays, see Dextran-coated charcoal assays
Denaturants, 18
Deoxyribonucleic acid, see DNA synthesis
DES, see Diethylstilbestrol binding
Dextran-coated charcoal (DCC) assays, 40
fluorescent conjugate techniques vs., 45, 47—51
in mammary tumors, 52—54, 147—148, 179—180
in immuno-galactosidase techniques, 162
semiquantified ERICA techniques vs., 148
Diethylstilbestrol (DES) binding
estrogen response and, 19
as measure of specific E_2 binding, 102
by membrane receptors, 19—20
Dimethylbenzanthracene (DMBA), 86
DNA synthesis, relationship of to estrogen response, 48—51
Double immunoperoxidase technique, 113, 114, 116

E

E-BSA-FITC, see Fluorescent conjugate techniques
Endocrine response
estrogen receptors as measure of, 96
in human tumor cell lines, 99
Endocrine therapies
clinical response to, semiquantitative ER assessments as predictor of, 145—146
clinical response to in mammary tumors, 144
quantified fluorescent conjugate techniques, as predictor of, 155—156
clinical response to in prostate diseases, semiquantitative androgen binding techniques as predictor of, 146—147
usefulness of receptor studies and, 158
Eosinopenia, blood, 22, 23
Eosinophilia, uterine, 22
Eosinophil receptors, 6, 7
for estrogen, 7, 13—16, 18, 23—27
migration and, 27, 28
in uterine blood vessels, 23, 24
fluorescent conjugate studies of, see also Fluorescent conjugate techniques, 65
genomic response to estrogen and, 19
for glucocorticoids, 27, 28
immune response and, 26, 27
migration and, 27, 28
radioautographic localization of, see also Radioautography, 9—11, 13—16
radiohistochemical studies of, see also Radiohistochemical techniques, 9, 11, 16
affinity criteria and, 16—18
ER, see Estrogen receptors
ERI, see Estrogen receptor index
ERICA, see Estrogen Receptor Immunocytochemical Assay
ER proteins, see Estrogen receptor proteins
Estradiol binding
affinity/potency relationships and, 19—20
to cell surfaces
factors affecting, 114
specific membrane sites and, 19—20, 114—115
demonstration of, 112
fluorescent BSA ligand studies of, see also Fluorescent conjugate techniques, 113—115
levels of interaction and, 115
radiohistochemical studies of, see Radiohistochemical techniques
17β-Estradiol 6-O-carboxymethyloxime-bovine serum albumin-fluorescein isothiocyanate, see Fluorescent conjugate techniques
17β-Estradiol 17-hemisuccinate-bovine serum albumin-fluorescein isothiocyanate, see Fluorescent conjugate techniques
Estriol binding, 15
affinity/potency relationships and, 19—20
by membrane receptors, 19—20
radiohistochemical studies of, see Radiohistochemical techniques
Estrogen binding
classic test for, 79
in rat uterus, 65—74
augmentation and, 70, 74
coexistence of with progesterone binding, 65
effects of estrous cycle on, 13—14, 65—69
eosinophils and, 65
replenishment hypothesis of, 13—14, 70—74
two-step mechanism of, 6
Estrogen Receptor Immunocytochemical Assay (ERICA), 148
problems with interpretation of, 157—159
Estrogen receptor index (ERI), 40
Estrogen receptors
affinity spectrum for in membranes, 19—20
agents affecting, 19
biochemical studies of, see Biochemical techniques
clinical response to endocrine therapy and, see Endocrine therapies
correlation of measures of, 52—53
cytosol-nuclear type, see Cytosol-nuclear receptors

DES measures and, 19, 102
detection of, see also specific techniques, 20, 168
in DMBA-induced mammary tumors
loss of with serial transplantation, 86, 91—93
measurement of, 86—87
effects of nonsteroidal antiestrogens on rat uterus and, 74—79
eosinophil type, see Eosinophil receptors
fine needle aspiration measurement of, see Needle aspiration
flow cytometry measurements of treatment and, 104
fluorescent conjugate studies of, see Fluorescent conjugate techniques
in human mammary tumors
correlation of measures of and, 52—56
in human tumor cell lines, 96—98
in identification of endocrine responsiveness, 96
immunocytochemical studies of, see also Immunocytochemical techniques, 112—113
multiple mechanisms of, 27
evidence for, 7, 20
mediation of, 7
multiple types of, 2
evidence for, 6—8, 13, 18—20
in uteri, 6
progesterone and, 13—14
in prostate diseases, 120—121
quantification of, 144
radioautographic studies of, see also Radioautography, 8
radiohistochemical studies of, see Radiohistochemical techniques
relation of to mammary tumors, 162
soluble vs. tissue-bound types of, 61—62
tumor growth states and, 104, 107
usefulness of cryostat sections for studies of, 61
validating measures of, 57
Estrone receptors, 15
DCC assays of, see Dextran-coated charcoal assays
fine needle aspiration measurement of, see Needle aspiration
fluorescent conjugate studies of, see Fluorescent conjugate techniques
in mammary tumors, 42
effects of treatment on, 48—51
relationship of to cell proliferation, 48—50
types of by fluorescence in animals, 43
types of by fluorescence in humans, 52
variation of ERI and, 41—42, 51—52
relationship of to cell proliferation in intact livers, 50—51
Ethanol extraction, 16

F

17-FE, see Fluorescent conjugate techniques
Flow cytometry (FCM)
advantages of, 109
applications of, 104, 107—109
binding intensities and, factors affecting, 102
early development of, 96
enhanced fluorescence and, 108—109
light scatter histograms in
for channel gating checks of staining heterogeneity, 100—102
of fresh vs. frozen-thawed cells, 100
procedures for, 96—100
resonance energy transfer (RET) and, 107—108
Fluorescent conjugate techniques
in competitive binding studies, see Competition experiments
DCC assays vs., 45, 47—54
effects of cell freezing on, 100
effects of incubation time and temperature on, 42, 44, 102
effects of tissue type on, 63
flow cytometry in, see Flow cytometry
fluorescence determination and, 42
fluorescence enhancement and, 108—109
in human tumor cell lines, 97—99
ligand affinity for ER proteins and, 80
ligand concentration and, 42, 44, 80, 102
ligand diffusion and, 62
saturability criteria and, 64, 77, 79, 102
specificity criteria and, 64, 102, 104
specificity of tissue criteria and, 64—65
type I interactions and, 171—176
usefulness of, 40, 79
DCC assays in mammary tumors vs., 52—54, 147—148, 179—180
for estradiol receptors, 113—115
saturability criteria and, 114, 115
specificity criteria and, 114
for estrogen receptors, 62—63, 112
affinity criteria and, 63—64
for estrone receptors
cells utilized in, 40—41
effects of incubation variation on, 42, 44, 45
ligand characteristics and, 40
ligand concentration and, 42, 44
preparation of cells for, 41
saturability criteria and, 44—46
specificity criteria and, 42—43
usefulness of, 45, 47—51
in immuno-galactosidase method, 162, 163
interaction with ER type I and, 171—176
as measure of clinical response, see specific therapies and treatments
problems with, 156—159, 168—169
purification and, 169—171
quantification of
equipment for, 148—149
problems with, 144, 178—179
procedures for, 149
semiquantified results vs., 149—155
semiquantification of, DCC assays in mammary tumors vs., 147—148
stability and, 169—172
usefulness of, 40

validation of in human mammary tumors
effects of incubation time and, 178—179
procedures for, 176—178
saturability criteria and, 178—179
specificity criteria and, 178
1-(N)-Fluoresceinyl-estrone-thiosemicarbazone, see Fluorescent conjugate techniques

G

Glucocorticoids
effects of on estrogen response, 21—23
eosinophils and, 27, 28
Glutaraldehyde, 127—131

H

Heat shock, 118
^{3}H-estrogen binding, see Estrogen binding
Histamine
estrogen response and, 22
role of, 26
Histochemical techniques, see also Radiohistochemical techniques; specific techniques, 168
advantages and disadvantages of, 61—62, 79—80
in human tumors, 96
biochemical techniques vs., 45, 47—54, 60—61, 80, 145—147
for determining estrogen activity, 79—80
interpretation of, 148
specificity criteria and, 169
usefulness of, 2, 3
Histologic techniques, usefulness of, 158—159
Hormone receptors, see Receptors; specific receptors
Hormone response, see Response, biological; specific hormones
Hormones, see Steroids; specific hormones
Hormone therapies, see also specific therapies and treatments
predictors of clinical response to, 119
in mammary tumors, 146
in prostate diseases, 121

I

βI-G, see Immuno-galactosidase techniques
Immune response and eosinophils, 26, 27
Immunochemical techniques, basis of, 112
Immunocytochemical techniques
basis of, 112—113
in conjunction with biochemical techniques, 162—164
Immuno-galactosidase techniques
advantages and disadvantages of, 163—164
DCC procedure and, 162
fluorescent conjugate method in, 163
Immunohistochemical techniques, see Antibody studies; Histochemical techniques
Immunoperoxidase techniques, for progesterone receptors, 127, 131
Indomethacin, effects of on estrogen binding, 20—21
Inhibition, see Saturability
Insulin
effects of on estrogen binding, 19
effects of on estrogen response, 21, 23
Iodination, effects of on estrogen binding, 18

K

90K protein, see Progesterone receptors, subunits of

L

LH, see Luteinizing hormone
Ligand-binding techniques, see specific techniques
Luteinizing hormone (LH), association of with steroids, 136—137

M

Malignant neoplasms, see Tumors; specific tumors
Mammary tumors
clinical response to endocrine therapies for quantified fluorescent conjugate techniques as predictor of, 155—156
semiquantitative assessments as predictor of, 145—146
DCC vs. semiquantified fluorescent conjugate assessments of, 147—148
effects of cell variation on quantification of, 144
effects of specimen-removal method, 117—118, 120—121
ERICA measures for, 157—159
estrone binding in
cell proliferation and, 48—50
types of by fluorescence, 43
variation of ERI in, 41—42
fine needle aspiration of in estrone binding measurement, 54—57
fluorescent conjugate studies of, see Fluorescent conjugate techniques
intracellular interactions and, 115—117
plasma membrane interaction and, 113—115
relation of to estrogen receptors, 162
Mammary tumors, DMBA-induced
decline of estrogen receptors in, 86—93
fluorescent conjugate studies of, 87—91
serial transplantation of, 86
transition from epithelial to spindle-shaped cells, 88—93
transition from hormone dependency to autonomy in, 86, 91—93
Mammary tumors, human

abnormal responsiveness in, 121
correlation of ERI with biochemical measures of estrogen binding in, 52—56
estrone binding in
types of by fluorescence, 52
variation in ERI and, 51—52
fluorescent conjugate studies of
DCC assays vs., 52—54, 179—180
validation of, 176—178
receptors in
effects of heterogeneity of on clinical management, 119—120
impaired nuclear translocation and, 119
intratumor diversity of, 118—120
Monoclonal antibodies, see also Antibody studies, 113
Morphological techniques, see also specific techniques, 112
biochemical techniques vs., 3
classical, 2

N

Needle aspiration, 54—57
advantages of, 56—57
effects of on steroid localization, 117—118
Neoplasms, see Tumors; specific tumors

O

Ovaries, antibody studies of hormone receptors in, 135—136

P

Passive diffusion, 3, 113
PCA, see Prostate diseases
PEP, see Polyestradiol phosphate
Pituitary glands, antibody studies of, 136—137
Plasma membrane, interaction of steroids with, 113—115
Polyestradiol phosphate (PEP), as antigen, 112
Potency, see Affinity/potency relationships
PR, see Progesterone receptors
Progesterone, effects of on estrogen binding, 13—14, 20—21, 70, 74
Progesterone binding
coexistence of with estrogen binding in rat uterus, 65
two-state theory of, 139—140
Progesterone receptors
antibody studies of, see Antibody studies
in chicken oviduct
effects of primary vs. secondary stimulation in, 132—133
usefulness of, 126
forms of, 126
subunits of
antibody reactions with, 126
effects of fixation procedures on demonstration of, 127—131
effects of fixative on, 131
evidence for in nuclei of chicken oviducts, 132—133
variation in distribution of, 140
Propranolol, effects of on estrogen binding, 20—21
Prostate diseases
androgen binding in, 120, 121
clinical response to endocrine therapies in semiquantitative androgen binding techniques as predictor of, 146—147
differential distribution of estrogen binding sites in, 120
effects of cell variation in on quantification of, 144
effects of specimen-removal method, 117—118, 120—121
estrogen binding in, 120—121
fluorescent conjugate studies of quantified vs. semiquantified results from, 149—155
hormone response in
prediction of, 121
relation of to tumor grade, 120

Q

Quantitative analyses, see Fluorescent conjugate techniques

R

Radiation, effects of on estrone binding in mammary tumors, 48—50
Radioautography
advantages and disadvantages of, 60
affinity criteria and, 12
binding capacity criteria and, 12
of cytosol-nuclear receptors, 13—14
dry technique of, 9, 10
for estrogen binding
affinity criteria and, 15
early techniques for, 13
eosinophils in, 13, 15—16
specificity criteria and, 15—16
interpretation of, 11—12
principles of, 9—10
quantitative evaluation of radioautograms and, 11, 17
usefulness of, 8
wet technique of, 9—11, 13
Radiohistochemical techniques, see also Histochemical techniques, 9, 11
definition of, 9
for estradiol binding, 16—18
for estriol binding, 16—18
for estrogen binding in eosinophils and affinity criteria, 16—18

Radiolabeled ligand-binding techniques, see Autoradiography
Receptor binding, translocation process and, 116—118
Receptors
affinity of, see Affinity
antibody studies of, see Antibody studies
association of with particulate cytoplasmic components, 116
binding capacity of, see Saturability
characteristics of, 6
cytoplasm-to-nucleus system of, 113—117
definition of, 3
effects of heterogeneity of on quantification, 144
in human mammary tumors
diversity of, 118—120
impaired nuclear translocation and, 119
nuclear localization of, 116—117
qualitative vs. quantitative evaluations of, 145
radioautography studies of, see Radioautography
response and, see Response, biological
semiquantification procedures for, 144—145
specificity of, see Specificity
two-state theory of, 139—140
visualizing problems with, 126
Receptosomes, definition of, 116—117
Replenishment regulation, 13—14, 70—74
Response, biological
as evidence for multiple estrogen receptors, 18—20
measuring, 12—13

S

Saturability, see also Affinity; Binding capacity; Competition experiments, 9, 12
affinity and, 12
in antibody studies of progesterone receptors, 131
in Bursa of Fabricius, 134
BSA binding and, 64
competition experiments and, see Competition experiments
fluorescent conjugates and, 44—46, 102, 114, 115, 178—179
radioautographic technique and, 12
of target cells in rat uterus, 77, 79
Specificity, see also Affinity; Binding capacity; Competition experiments; Saturability
in antibody studies of progesterone receptors, 131
biochemical techniques and, 169
fluorescent conjugates and, 42—43, 102, 104, 114, 178
histochemical techniques and, 169
of hormone, 12
with fluorescent conjugates, 64
with radioautographic techniques, 12
radioautographic studies and, 15—16
of tissue, 12
with fluorescent conjugates, 64—65
with radioautographic techniques, 12
8S-PR, see Progesterone receptors, subunits of
Steroid binding
artifactual induction of, 117—118
estrogenic, see Estrogen binding
mechanisms of, 3
nuclear membrane as regulator of, 119
quantification of, 2
Steroid receptors, see Receptors
Steroids, see also specific hormones
modulation of peculiar membrane function by, 115
Sulfhydryl groups, effects of on estrogen binding, 18

T

Tamoxifen, effects of on estrone binding, 49—51
Theophylline, effects of on estrogen binding, 19, 21, 23
Therapy, see also specific therapies and treatments
effects of on estrogen receptors, 104
multimodality of, 57
Thyroid hormones, effects of on estrogen binding, 19, 21, 23
Translocation process, 116—117
effects of specimen-removal method on, 117—118
Transurethral electrocutting, effects of on steroid localization, 117—118
Treatments, see Therapies; specific therapies and treatments
Tryptophan blockers, effects of on estrogen binding, 18
Tumor removal, effects of on estrone binding, 48—50
Tumors
breast, see Mammary tumors
effects of tissue heterogeneity on quantitative analyses of, 145
human
endocrine sensitivity in, 99
estrogen receptors in cell lines of, 96—98
measuring hormone responsiveness in, 96
progression to autonomy in hormone dependent types of, 121
prostate, see Prostate diseases
relation of estrogen receptors to growth states of, 104, 107

U

Uterine eosinophil leukocytes, see Eosinophil receptors

For Product Safety Concerns and Information please contact our EU
representative GPSR@taylorandfrancis.com
Taylor & Francis Verlag GmbH, Kaufingerstraße 24, 80331 München, Germany

www.ingramcontent.com/pod-product-compliance
Ingram Content Group UK Ltd.
Pitfield, Milton Keynes, MK11 3LW, UK
UKHW052221190926
475702UK00006B/1142
9781315895048